Continental Rifted Margins 1

SCIENCES

Geoscience, Field Director – Yves Lagabrielle

Dynamics of the Continental Lithosphere,
Subject Head – Sylvie Leroy

Continental Rifted Margins 1

Definition and Methodology

Coordinated by
Gwenn Peron-Pinvidic

iSTE

WILEY

First published 2022 in Great Britain and the United States by ISTE Ltd and John Wiley & Sons, Inc.

ISTE Ltd
27-37 St George's Road
London SW19 4EU
UK

www.iste.co.uk

John Wiley & Sons, Inc.
111 River Street
Hoboken, NJ 07030
USA

www.wiley.com

Library of Congress Control Number: 2021950855

British Library Cataloguing-in-Publication Data
A CIP record for this book is available from the British Library
ISBN 978-1-78945-061-3

ERC code:
PE10 Earth System Science
PE10_5 Geology, tectonics, volcanology
PE10_7 Physics of earth's interior, seismology, volcanology
PE10_11 Geochemistry, crystal chemistry, isotope geochemistry, thermodynamics
PE10_12 Sedimentology, soil science, palaeontology, earth evolution

Contents

Preface

Gwenn PERON-PINVIDIC[1,2]
[1] *Geological Survey of Norway (NGU), Trondheim, Norway*
[2] *Department of Geoscience and Petroleum, Norwegian University of Science and Technology, Trondheim, Norway*

Continental margins mark the transition between continents and oceans, which are the two first-order types of land masses on Earth. Active continental margins are associated with convergent lithospheric plates where collisional processes occur – they are the sites of intense tectonic, seismic and volcanic activity; while passive continental margins correspond to the regions that mark the rupture of a continent and the opening of an ocean basin, ultimately leading to the establishment of an oceanic ridge. Passive margins archive two of the most fundamental geological processes operating on Earth: "rifting" and "breakup". Passive continental margins, also often referred to as "rifted margins", register the extension and ultimate separation of the continental lithosphere and the beginning of oceanic crust accretion. After continental breakup, they are tectonically quieter than the active margins.

Rifted margins contribute to our understanding of extensional processes in the continental lithosphere in several ways. First, they represent the most widespread manifestation of Mesozoic and Cenozoic extensional deformation, archiving the structures created during the dismembering of the supercontinent Pangea, a major stage in the tectonic history of our planet. In addition, they border almost all of the land masses that surround the Atlantic, Indian, Antarctic and Arctic Oceans and represent significant surfaces of today's Earth. Moreover, since they represent the

---

For a color version of all the figures in this chapter, see www.iste.co.uk/peronpinvidic/margins1.zip.

Continental Rifted Margins 1,

final stage of evolution of an intracontinental rift zone, they can provide a complete record of extensional history and are therefore unique geological features.

Rifted margins are studied by various disciplines of Earth Science (geology, geophysics, geochemistry). Thanks to an increasing amount and better public access to high-quality data, there has been a great improvement in our understanding over the last two decades. However, there are still important debates happening in the scientific community on several questions. Their structural architecture, as well as their spatial and temporal evolution, are still being discussed. No strict consensus has been reached on the composition of the basement, tectonic structures, sedimentary geometries or magmatic content. Correspondingly, their relationship with the adjacent oceanic areas, specifically the first kilometers of oceanic crust, is still largely unconstrained.

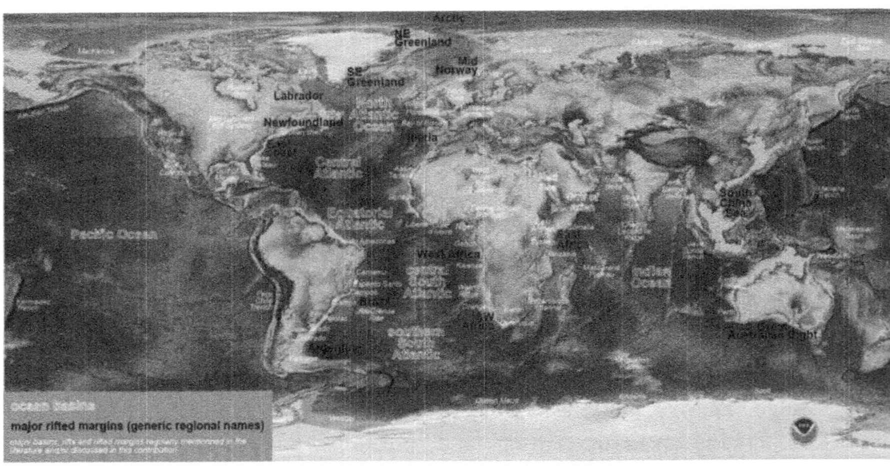

Figure P.1. *Topographic map of the world with the major ocean basins, rifted margins and main geological structures regularly mentioned in the literature and/or in this book*

How to use this book

This book focuses on continental rifted margins, with the aim of explaining the status of knowledge and the way in which understanding has evolved over the last few decades, thanks to the scientific community. Concise text descriptions are used throughout to ensure the content is practical and accessible. We have also incorporated myriad schemas and illustrative figures in all chapters in order to reach a wide audience, particularly students. For further detailed information, the reader is referred to key publications listed under the **Further reading** sections.

The book is designed to be a useful tool for students and those who are new to the rifted margin community – a sort of "cookbook" to facilitate the reading of scientific publications and provide basic definitions and explanations on a series of key scientific questions. No detailed information is given on the methods: only succinct reminders are provided. If interested, readers are referred to books and contributions specifically dedicated to these topics. Here, the focus is directed towards the rifted margin itself by providing fundamental definitions and explaining how we study these features, how the current theories were developed, where our understanding of rifted margins currently stands and where we are headed as a research community.

Overview

The book comprises two volumes and is structured into three main parts:

Volume 1 includes Parts 1 and 2. **Part 1** summarizes all definitions and descriptions necessary for understanding rifted margins. Chapter 1 introduces the basic definitions of the different rift types, followed by a list of the main fault and basin types encountered in rift settings. This list is non-exhaustive but provides basic definitions. The aim is to summarize fundamental terminology so as to avoid any misunderstanding on the use of some terms in the rest of the book, not to propose full definitions. Chapter 2 describes in further detail how rifted margins are classified – spanning the different categories usually found in the literature (e.g. divergent vs. oblique, volcanic vs. non-volcanic, sediment rich vs. sediment starved). The chapter summarizes the evolution of knowledge and understanding achieved by the scientific community over the last few decades, from the initial models proposed in the 1980s to the modern views discussed today. The final part of the chapter then lists and discusses a series of scientific questions selected to illustrate the future scientific challenges our community is facing.

Part 2 explains how we study rifted margins concretely, presenting the four main methods of investigation commonly used to study rifted margins: onshore (Chapter 3), offshore (Chapter 4), numerical modeling (Chapter 5) and analogue modeling (Chapter 6). Chapter 3 explains how rifted margin settings can be studied onshore based on remnants outcropping in orogenic structural contexts. Chapter 4 lists and defines, with illustrative data examples, the main geophysical methods used on a regular basis to image and study modern offshore margins. Chapters 5 and 6 focus on the modeling strategy, including both analogue and numerical approaches.

Volume 2 includes **Part 3**, which is made up of nine chapters. These chapters illustrate the topics by reviewing a series of case examples considered to be representative of the most discussed study sites at present. The aim of this part of the book is to cover a wide range of rifted margin categories, with both onshore archives and modern offshore cases.

Color code and geological timescale

The following figures provide details of the color code used in most figures in Chapters 1 and 2, if not specified otherwise (Figure P.2), and the geological timescale used throughout this book (Figure P.3).

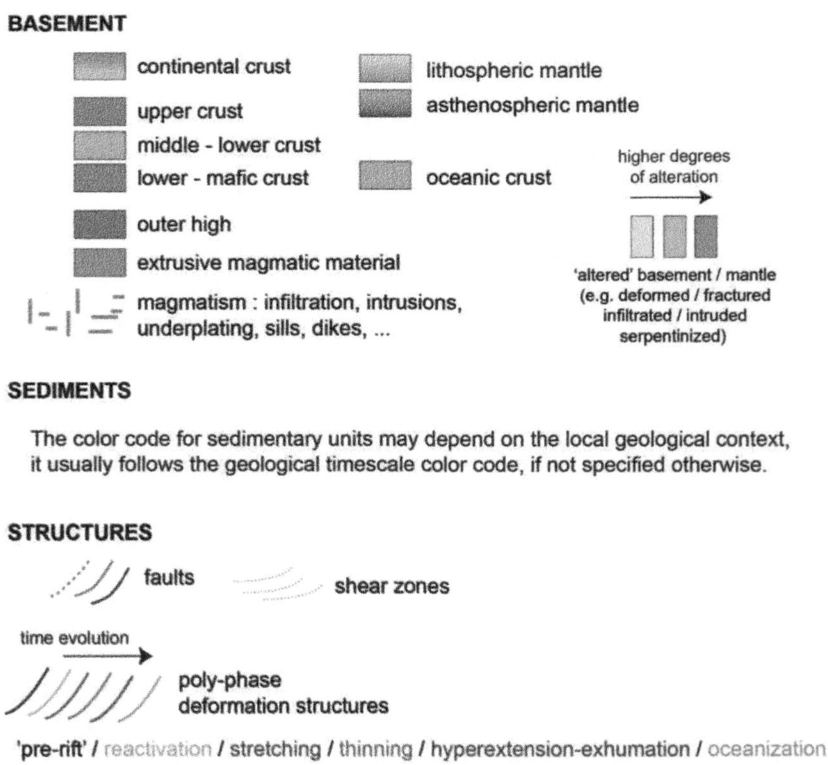

Figure P.2. *Color code used in most figures of this book if not specified otherwise*

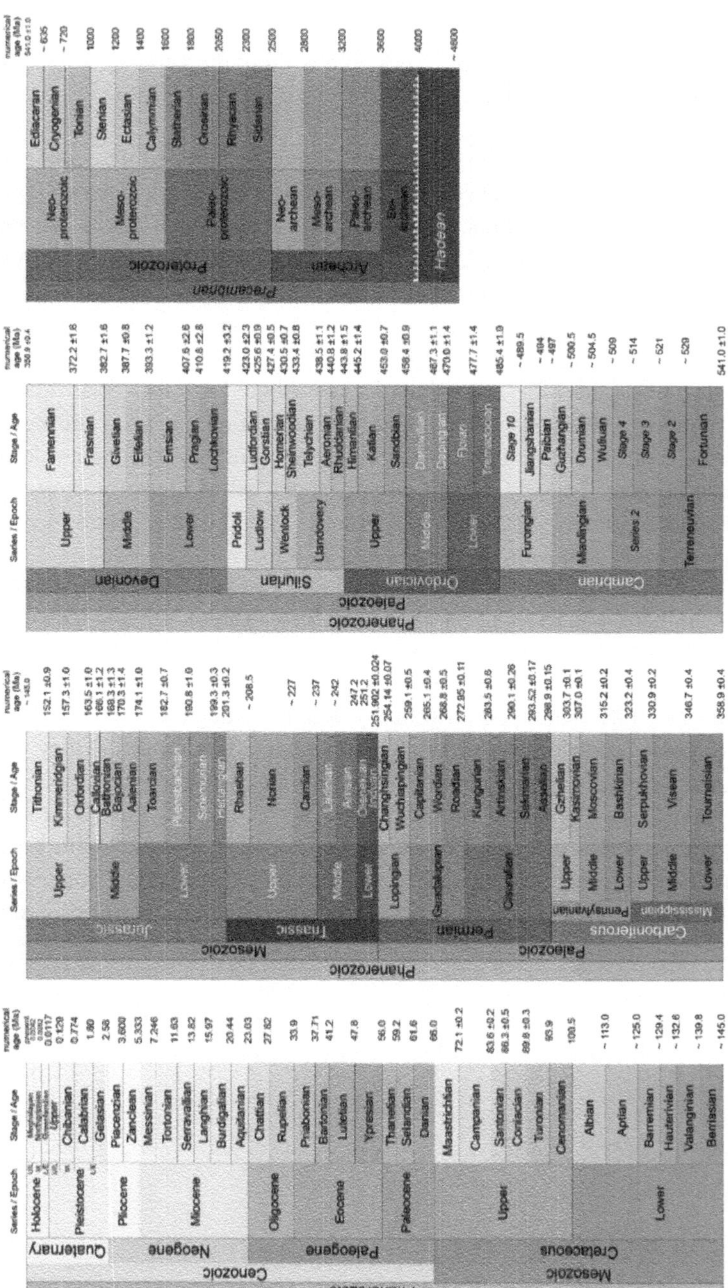

Figure P.3. *Geological timescale (source: International Commission on Stratigraphy)*

Acknowledgments

This book would never have been possible without the valuable help and support of many colleagues and friends.

I would first like to thank the 25 authors who contributed to the various chapters, for their enthusiasm, confidence, professionalism and patience (in alphabetical order): T. Andersen, D. Aslanian, J. Autin, S.P. Chang, P. Chenin, M. Dlescluse, C. Gaina, J. Jakob, H.J. Kjøll, Y. Lagabrielle, S. Leroy, G. Lymer, A. McCarthy, G. Mohn, M. Moulin, C. Nonn, P.T. Osmundsen, M. Perez-Gussinye, B. Petri, M. Pubellier, T. Reston, G. Schreurs, J. Tugend, L. Watremez and F. Zwaan.

Various colleagues also significantly contributed to the finalization of the chapters: reading, reviewing, commenting and correcting our texts and descriptions. Reviewers are rarely mentioned in scientific contributions, yet their role is key to shaping the content. I would like to warmly thank all of them for their significant help (alphabetical order): M. Bonini, D. Franke, M. Gouiza, J. Jakob, A. McCarthy, M. Nirrengarten, A. Rotevatn, K. Senger and Z. Sun.

A special thank you to Joya Tetreault, Per Terje Osmundsen and a dear friend for their corrections, comments and suggestions for Part 1.

I would also like to thank Sylvie Leroy and Yves Lagabrielle for inviting me to write this book, contributing to the Geosciences Encyclopaedia they supervise. Many thanks for your support and confidence during the whole process.

Finally, personally, I would like to thank specific colleagues who have helped me to grow as a researcher through collaborations, discussions, supervisions and debates: Per Terje Osmundsen, Gianreto Manatschal, John Naliboff, Loic Fourel, Tim Redfield, Emmanuel Masini, Espen Torgersen, Katia Svendby and Suzanne McEnroe.

Thank you.

December 2021

PART 1

Rifting and Rifted Margins: Definitions

1

What is Rifting? Introduction and Basic Definitions

Gwenn PERON-PINVIDIC[1,2]
[1] *Geological Survey of Norway (NGU), Trondheim, Norway*
[2] *Department of Geoscience and Petroleum, Norwegian University of Science and Technology, Trondheim, Norway*

1.1. Introduction

The Earth's interior is composed of multiple layers of differing composition and thickness. The exterior layers are the **oceanic** and **continental crust**, followed by the **upper mantle** – together constituting the **lithosphere** (Figure 1.1). The Earth's lithosphere is divided into tectonic plates that are in constant motion relative to each other, following a repeated evolution of opening, widening and closing oceanic basins. This plate tectonic cycle became known as the "**Wilson Cycle**" in honor of J. Tuzo Wilson (1908–1993), who actively participated in the important evolution of knowledge and understanding achieved in geoscience in the 1960s (Wilson 1966). By this point, breakthroughs in geophysical techniques had produced observations of magnetic anomalies, seafloor spreading ridges and hot spots that had established plate tectonics and plate motions as fact (e.g. Atwater 1970; Molnar Atwater 1973). This, in turn, allowed J.T. Wilson to identify the systematic separation and reassembly of continental masses.

For a color version of all the figures in this chapter, see www.iste.co.uk/peronpinvidic/margins1.zip.

Continental Rifted Margins 1,
coordinated by Gwenn PERON-PINVIDIC.

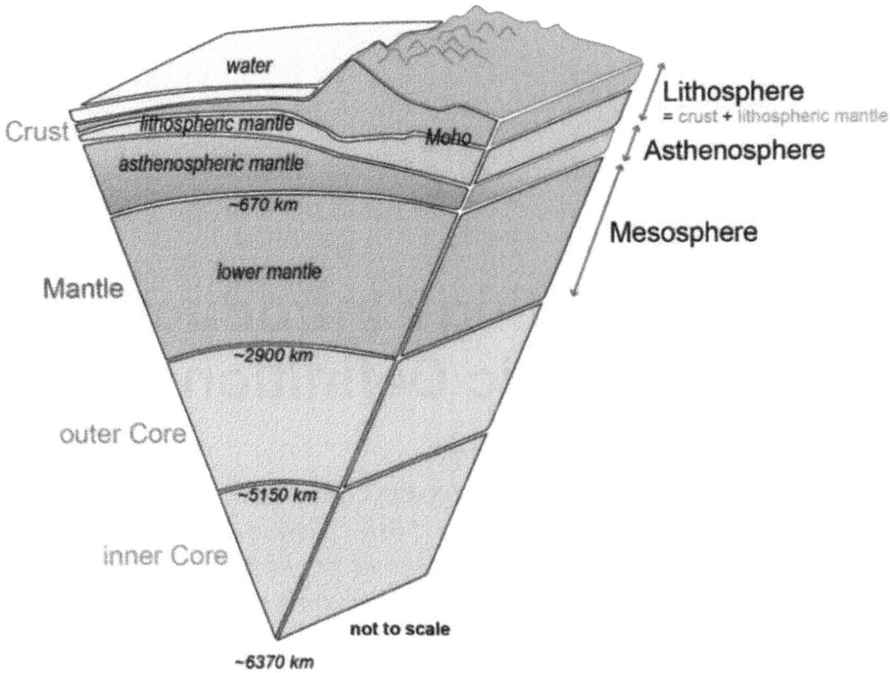

Figure 1.1. *Schematic representation of the main layers constituting the Earth*

Within the Wilson Cycle, **continental rifting** corresponds to the extensional stage that leads to thinning and rupture of the continental lithosphere (Figure 1.2). The primary product of rifting is the **rift**, which may subsequently become a **rifted margin** if the extensional processes lead to ridge emplacement, genesis of oceanic crust and the opening of an ocean basin.

Rifts were identified as unique geological occurrences before plate tectonics were established (e.g. Suess 1891; Gregory 1896; de Lapparent 1898; Willis 1928; Bullard 1936). Initially, they were described as topographic depressions flanked by arrays of subvertical faults (e.g. Gregory 1923). It was not until the last century that rifts were recognized as extensional features rather than as results of compressional deformation (e.g. Vening Meinesz 1950).

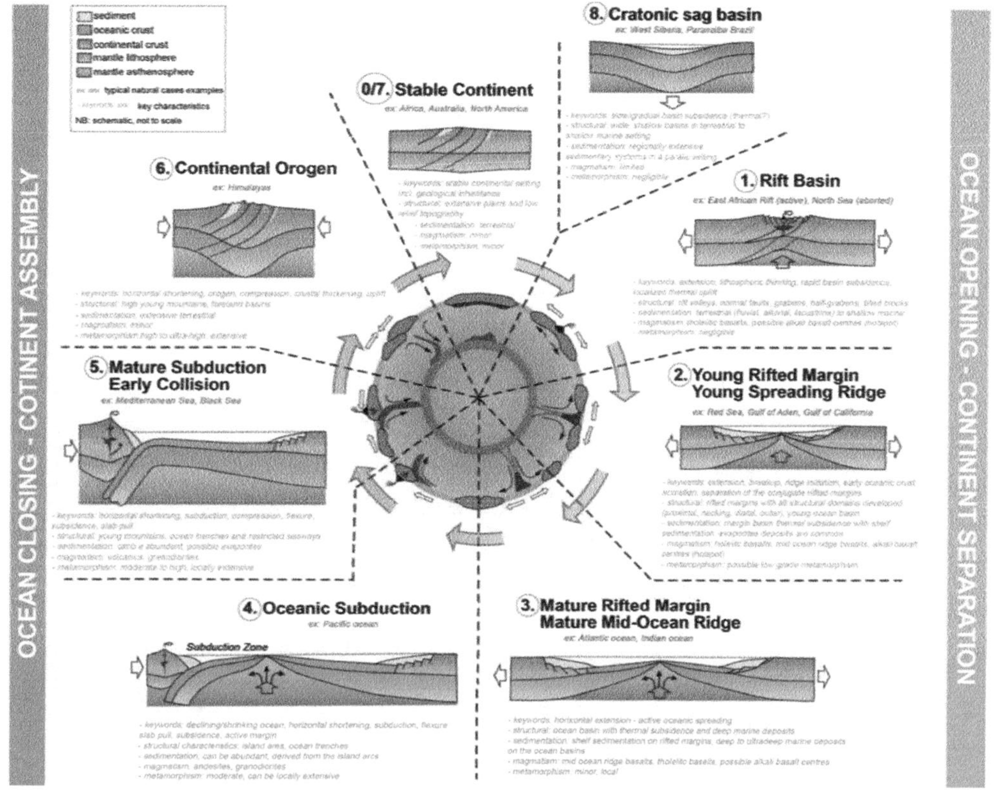

Figure 1.2. *Schematic representation of the Wilson Cycle in eight steps illustrating the various stages of the successive opening and closing of ocean basins (source: modified after Wilson* et al. *2019)*

Rifts can be summarized as geographical regions consisting of extensional sedimentary basins of various sizes, with various tectonic and sedimentary geometries that are linked in various structural contexts. The repeated use of the adjective "various" in this basic definition leads to a description that may sound rather vague and unprecise. This is deliberate, since the extension applied to a continental domain can lead to the formation of multiple structures with different geometries. If rifts are defined as topographic depressions that are created by normal faulting and associated subsidence, then the dimensions, physical characteristics and linkage characteristics between the different extended subdomains can span a wide variety of options. Because the range of observed architectural possibilities in rifts is broad, numerous classification systems for rifts have been proposed in the literature: passive versus active, narrow versus wide, intracontinental, diffuse, synorogenic and so on. However, even though each rift type may appear unique, a series of geologic features are commonly observed in most rift settings. These include specific basin types (grabens, half-grabens, pull-apart/strike-slip basins), extensional structures (normal faults, detachment faults, core complexes), sedimentary geometries (growth structures, unconformities) and salt- and magma-related features (e.g. Neumann and Ramberg 1978; Keen 1985; Ziegler 1992; Ruppel 1995; Corti *et al.* 2003; Merle 2011).

This chapter provides a brief summary of these features, a list of the main types of rifts and case examples. Our aim is to clarify the basic notions required to study rifted margins – this is done to avoid any misunderstanding of the use of some terms in the following chapters. Regularly, a **Further reading** section will be provided as an illustration of the many contributions published on the various topics. The reader interested in finding more in-depth explanations is referred to these contributions for further detailed information.

1.2. Rift classifications

This section begins with the definition of the active and passive rifting categories, as these are regularly mentioned in the literature and are the primary designations of extensional rift settings. However, because these two categories are very loosely defined and present-day data coverage is limited, it is difficult to fit all rifts in this classification scheme. For this reason, we prefer a classification that is based on the general tectonic setting of the rifts, as proposed by Ruppel (1995), which will be described in section 2.2.

1.2.1. *Rift mechanism classification: active versus passive rifting*

Originally, rifts were classified by the process of lithospheric thinning: in **active rifts**, deformation was thought of as driven by active asthenospheric upwelling, whereas in **passive rifts**, the asthenospheric upwelling was itself viewed as a passive response to lithospheric extension (Neumann and Ramberg 1978; Şengör and Burke 1978) (Figure 1.3).

Active rifting processes are governed by the upwelling of warm asthenospheric mantle beneath the base of the continental lithosphere, causing uplift and thinning of the continental plate by heating, basal shear and melting (Burke and Dewey 1973; Turcotte and Emerman 1983; Buck 1991). The idea is that in the earliest stages of this process, the lithospheric mantle is thinned by thermal erosion, while the continental crust is uplifted and affected by volcanic products, but preserves its initial thickness. The crust is thinned by extension in secondary stages (Merle 2011). This model was developed to explain various observations in rifts such as uplifted plateaus and abundant volcanism (e.g. flood basalts). Depending on the setting, the cause of asthenospheric rise is often related to a mantle plume, although the observations and constraints remain debated. For instance, the Rio Grande Rift (US–Mexico) is often listed as a type example of an active rift, with a narrow geometrical structural setting that is formed by laterally distributed pure shear extension (Wilson *et al.* 2005), with no observations suggesting the presence of a plume.

On the other end of the spectrum, passive rifting is driven by distant stresses that cause crustal and lithospheric mantle extension and thinning, resulting in a passive asthenospheric mantle rise (Turcotte Oxburgh 1973; Keen 1985; Ruppel 1995). The driving forces are generated by plate boundary motions (e.g. subduction pull and ridge push) or by convective motions at the base of the lithosphere. The main contrast between passive and active rifting is that in passive rifting, the rise of the asthenosphere occurs only in response to the thinning of the overlying lithosphere – as opposed to active upwelling, driving the extension for active rifting. This passive upwelling can then generate various secondary processes affecting the rift-like decompression melting and consequent mantle-derived magmas which are added to the crust (as underplating, intrusions and extrusions) and specific lateral thermal gradients (Ruppel 1995; Huismans *et al.* 2001).

Based on these definitions, it must theoretically be possible to distinguish the two rift types based on volcanism and the timing of the extensional phases or crustal uplift parameters. However, modern observations show that these models, although useful, are too simplistic. Various rifts show tectonic evolution and structural settings that combine "passive" and "active" characteristics (Ruppel 1995). These passive or active rifting classifications rather appear as end-member types in a continuous spectrum of continental extension processes.

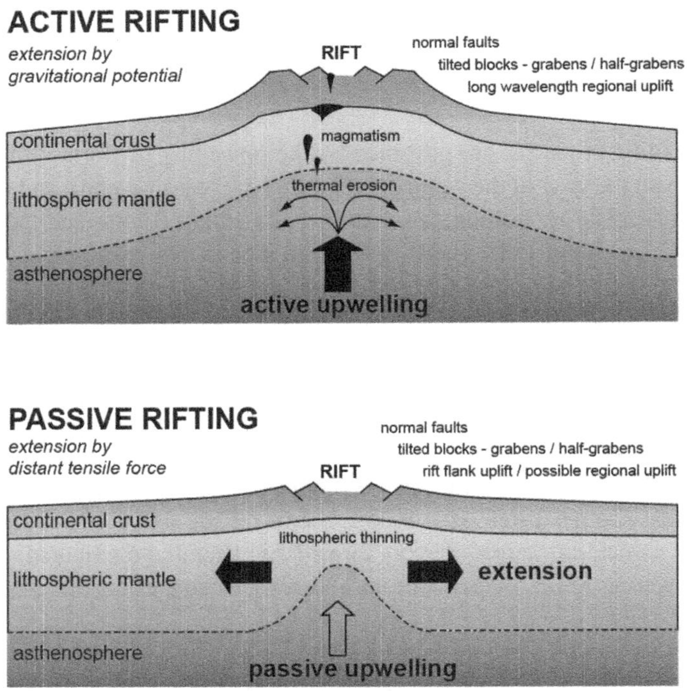

Figure 1.3. *Schematic illustration of the two main theoretical categories of rifting: active and passive. Active rifting is controlled by the warm asthenospheric mantle that rises beneath the base of the continental lithosphere, causing thinning of the plate and uplift at the rift zone. Passive rifting is controlled by distant extensional stresses, causing stretching and lithospheric thinning and passive rise of the asthenosphere below the rift zone*

1.2.2. *Plate tectonic setting classification*

Rifts develop under extensional stress regimes. Even though the structural context sounds straightforward, the regional tectonic setting can span a wide range of situations between the end-member convergent (with fore-arc, back-arc, retro-arc and intra-arc basins) and divergent settings (with rift and rifted margin basins). Hybrid structural contexts are also encountered. Within each setting, several types of basins can be categorized, depending on the tectonic stress regime of the basin itself (divergent, strike-slip, transform, transtensional, transpressional), the composition of the basement (continental, cratonic, oceanic), the sediment supply (sediment-rich, sediment-starved), the magma supply (magma-rich, magma-poor) and the presence of salt. Numerous other parameters are also recognized as strongly influencing the development and final geometries of the rift, such as the thermal structure, lithospheric rheology, inherited structures, crustal composition, opening geometries and extension rate. The interplay between these parameters is still poorly understood.

We review the main rift classifications and their characteristics. In order to construct a clear list of the various types of rifts, we follow Ruppel's (1995) early approach and propose that we first classify rifts based on their regional tectonic setting: intracontinental, convergent and divergent. This results in an overarching basic classification that has the advantage of embracing the majority of rift basin types, although some hybrid and alternative cases are always possible. The influencing parameters will be summarized in the next chapter.

1.2.2.1. *Intracontinental rifts*

Intracontinental rifts develop in the interior of a continental plate, far from any plate boundary, thus with no apparent direct tectonic connection to the regional stress fields produced by plate tectonics. However, continental plates can deform internally in response to far-off changes in stress conditions, such as changes in the plate boundary configuration (e.g. changes in the dip/velocity/direction of the subduction slab). Intracontinental rifts are classified according to their geometry (narrow vs. wide) or their structural setting (aulacogens and intracratonic rifts).

Narrow intracontinental rifts

Narrow rifts (Buck 1991) are characterized by a thinning in the crust and lithospheric mantle over a relatively narrow zone of about 100–150 km wide (Figure 1.4). The formation of narrow rifts is attributed to local weakening factors such as thermal thinning, strain weakening and local magmatism, where deformation is focused at the weakest or thinnest region of the lithosphere (Kusznir 1987; Buck 1991, 2004). The resulting rifts developed under this extensional mode are characterized by distinct lateral gradients in crustal thickness, topography and heat flow (Bonatti 1985; Corti *et al.* 2003). Archetypal examples of narrow rifts are the East African Rift system (Chorowicz 2005), the Rhine Graben (Brun *et al.* 1992; Dèzes *et al.* 2004), the Baikal Rift (ten Brink Taylor 2002), the Rio Grande Rift (van Wijk *et al.* 2018) and the Gulf of Suez-Red Sea Rift (Ragab and El-Kaliouby 1992; Bosworth *et al.* 2021).

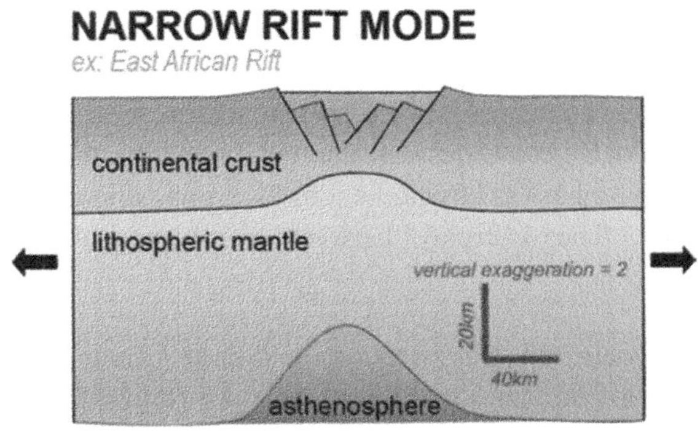

Figure 1.4. *Illustration of the "narrow rift mode" introduced by Buck (1991). In this extensional case, the lithospheric thinning affects a narrow zone of up to 100–150 km wide*

Case example: The East African Rift

The East African Rift (EAR) covers a broad region of East Africa and encompasses a series of discrete fault-bounded valleys organized into distinct rifts, branches and plateaus (Figure 1.5). The EAR is considered to be the archetype of active intracontinental rifts. The EAR contains several distinct sectors, from the Red Sea, Gulf of Aden and Afar region in the North to the eastern and western branches southward on the African continent. The eastern branch crosses the Ethiopian Plateau and includes two main segments: the Main Ethiopian Rift and the Kenyan rift. The western branch contours along the East African Plateau and Tanzania Craton and includes the East African Great Lakes and particularly the Lake Albert and Albertine Rifts. The western branch continues south-eastward with the Malawi Rift towards the Mozambique Channel.

The regional architecture can be described as a series of grabens of various dimensions, arranged in an en-échelon geometry within the different branches. Typically, the basins are flanked by chains of normal faults which shift polarity along-strike through transfer zones (Chorowicz 2005; Ebinger 2005; Corti 2009). The basins are usually asymmetric, up to 7 km deep, and bordered by 70–120 km long normal faults that accommodate dip-slip movements. The Kenyan and Ethiopian branches are magmatically dominated and the successive stages of rift evolution are preserved (Keranen *et al.* 2004; Rooney *et al.* 2005). The extensional geometries are so evident at the tectonic plate-scale that researchers have already named the new "plates-to-be": the Nubian plate, which corresponds to the western lands and encompassing most of Africa, and the Somalian plate, which is pulling away eastward (Bird 2003). The plates are separating at a rate of 6–7 mm yr^{-1}

(Corti 2009) and should reach a lithospheric rupture in about 10 Ma. These two plates meet the Arabian plate in the north at a triple junction situated in the Ethiopian Afar region (Figure 1.5).

The geological reason and exact mechanisms responsible for the formation of the EAR are debated. Dating volcanic rocks from the Main Ethiopian Rift indicates that the rifting process probably began ~30 Ma ago with the eruption of voluminous flood basalts, which resulted in the formation of the Ethiopian and Somalian plateaus (Corti 2009; Rooney 2017). The basins' initiations are diachronous along the EAR with distinct episodes of extension followed by periods of relative tectonic quiescence. The fault systems responsible for the formation of the separate rift basins are understood to date from Oligocene and Mio-Pliocene times (Chorowicz 2005).

The observation and mapping of thick volcanic sequences apparently emplaced before the rift-faulting events favored the development of plume theories. Since the observations are not straightforward to interpret and the models multiple, several plume hypotheses have been developed. These invoke the presence of several distinct localized plumes (Kenya, Ethiopia, Arabia) (Montelli *et al.* 2006), a super-plume (Ebinger and Sleep 1998; Bastow *et al.* 2008) or a composite scenario including a deep plume originating at the core–mantle boundary and feeding multiple plume-stems in the upper mantle beneath the EAR (Hansen *et al.* 2012). Tectonic theories focus instead on the regional tectonic setting; the development of extensional stresses is due to plate reorganization from a ridge jump in the Indian Ocean (Burke 1996). Mantle convection cells activated due to lithospheric thinning (producing thermal gradients) and/or similar, cratonic-edge-driven cells can then explain melt genesis and volcanic intrusions and extrusions (King and Ritsema 2000). Ongoing discussions and debates may today favor a composite origin with both plate tectonics and plume magmatism contributing to the extensional geometries, with feedback effects on one another (Rooney 2017, 2019, 2020a, 2020b, 2020c).

The aulacogen case

Sometimes, when a region is submitted to extensional strain, three main rift axes can develop in response to the far-field plate movements. As extension continues, usually approaching breakup stage, one of the three rift axes typically fails or stops extending – this one is called the "aulacogen" (Burke 1977; Şengör and Burke 1978) (Figure 1.6). Aulacogens are often reactivated in the post-rift stages under renewed extension or compression. A well-studied example is the Benue Trough, that extends from the equatorial Atlantic Ocean into the continental hinterland of Nigeria (Figure 1.7). It formed in the Early Cretaceous as a major rift arm accommodating the separation of South America from Africa (Olade 1975; Robert 2008).

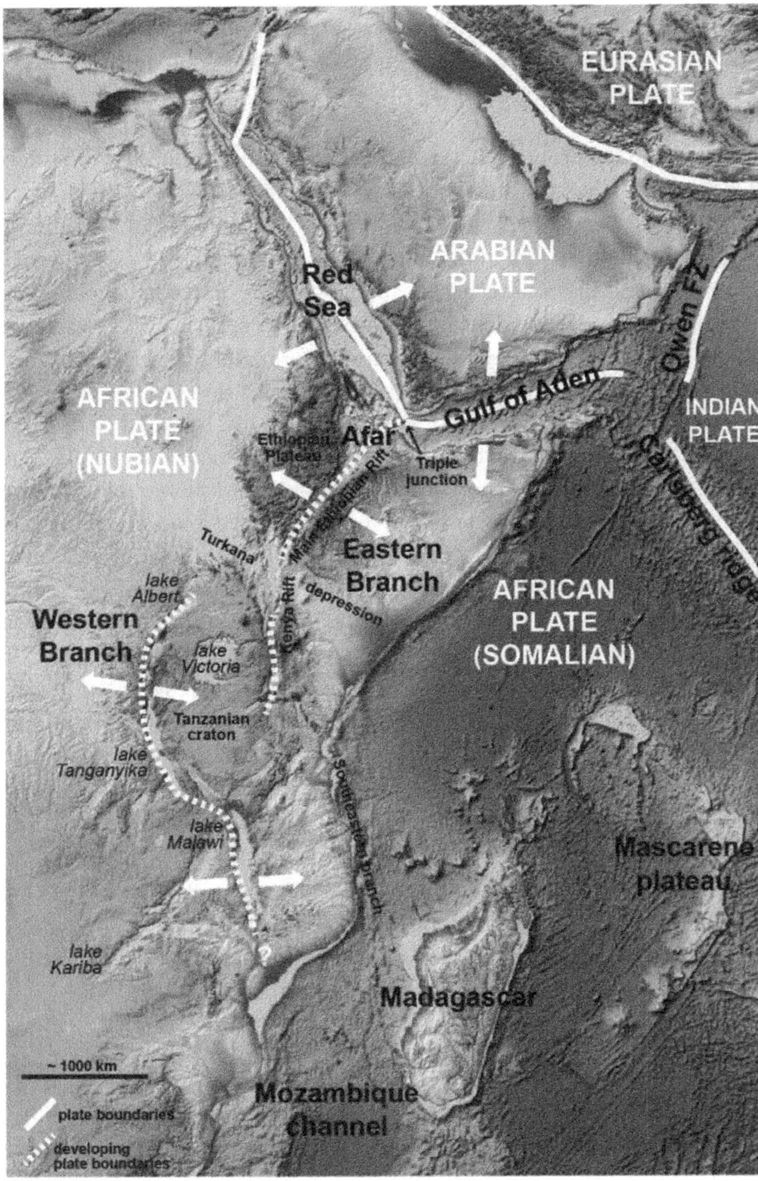

Figure 1.5. *Topographic map of the East African Rift system with indication of some of the major segments, basins and structures regularly mentioned in the literature. Elevation and bathymetric map is based on the NOAA Etopo1 (source: Amante and Eakins 2009)*

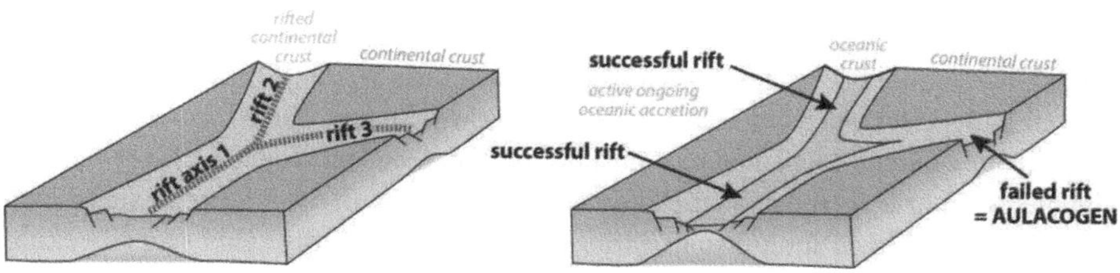

Figure 1.6. *Schematic representation of an Aulacogen: a typically narrow intracontinental rift in a triple junction that ceased activity*

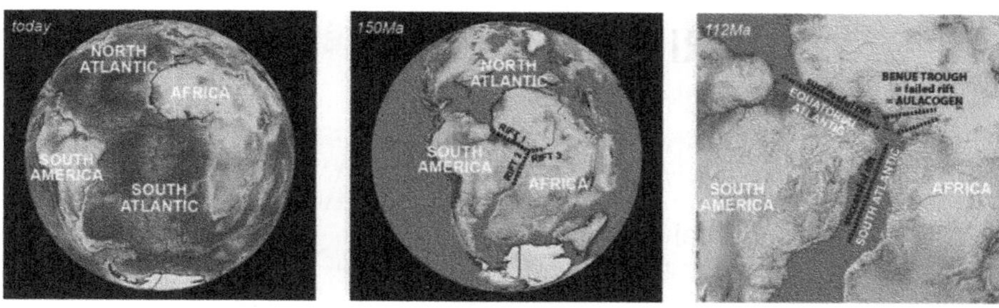

Figure 1.7. *Maps locating the Benue Trough aulacogen in the South Atlantic plate tectonic evolution. The maps have been created with the GPlates software (Müller* et al. *2018), using the Matthews* et al. *(2011) sample data sets. Elevation and bathymetric map is based on the NOAA Etopo1 (source: Amante and Eakins 2009)*

Wide intracontinental rifts

Wide/diffuse rifts are characterized by uniform thinning of the crust and lithospheric mantle over a width greater than the thickness of the lithosphere (Figure 1.8). The resulting geometry typically corresponds to isolated basins that are distributed over broad regions (Brun Choukroune 1983), as in the Basin and Range Province (e.g. Dickinson 2002; Parsons 2006). Unlike narrow rifts, wide/diffuse rifts are characterized by relatively small lateral gradients in crustal thickness and topography variations, with a relatively flat Moho.

Case example: The Basin and Range Province

The Basin and Range Province is a region covering a vast area in the western United States and northwestern Mexico (Figure 1.9). The physiography is defined by changes in elevation, alternating between north–south oriented, narrow-faulted mountain chains (15–20 km wide) and flat arid valleys (~30 km wide). The subparallel mountain ranges correspond to the crests of fault-blocks bounded by normal faults developed from east–west-directed extension. The blocks are exposed to erosion, leading to sediment transport to the adjacent valleys. However, the climate is so arid in most of the province that the runoff water is not significant enough to proceed to long-distance transport of the sediments. This special setting results in the building of thick alluvial fans, which led to the definition of the famous "Gilbert deltas" (Gilbert 1928).

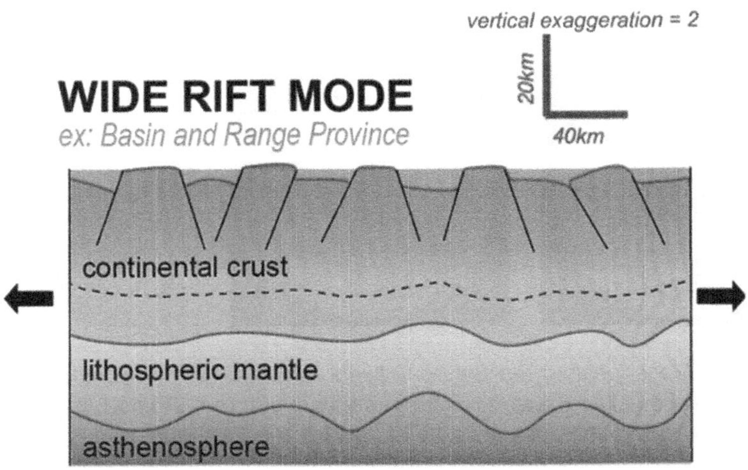

Figure 1.8. *Illustration of the "wide rift mode" (source: Buck 1991)*

The tectonic reason for the formation of the province is interpreted to be due to crustal buoyancy forces and stress changes at plate boundaries (Sonder and Jones 1999). The entire region is characterized by high heat flow, slightly thinned continental crust (27–35 km) and a relatively flat Moho with reduced upper mantle P wave velocities (7.5–8.0 km s^{-1}) (Holbrook 1990; Gilbert 2012). Structurally, the region displays distinct modes of deformation (high-angle faulting, low-angle detachment and metamorphic core complexes) and the rocks have experienced syn-deformational high-grade metamorphism (amphibolite, granulite, eclogite facies) (Axen *et al.* 1993). The observations led to the development of the simple shear model (Wernicke 1981, 1985) (see Chapter 2) where a detachment surface separates an upper plate setting (hanging wall) from a lower plate setting (footwall), which are mostly made of upper crust dissected into blocks by high-angle normal faults and highly-deformed lower crust, respectively (Davis and Lister 1988; Lister and Davis 1989).

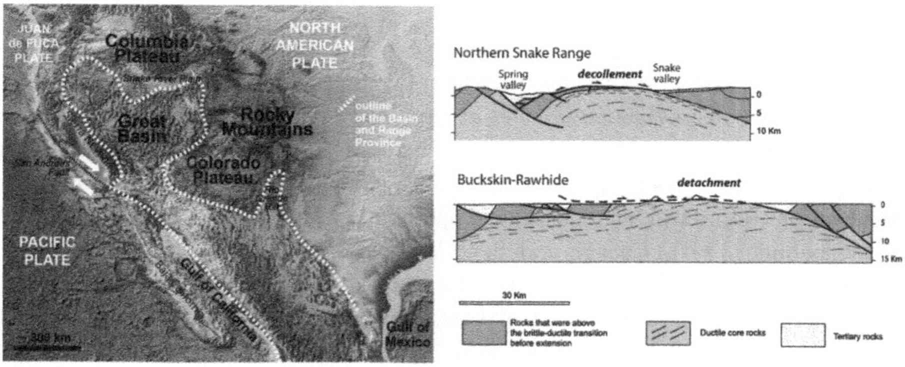

Figure 1.9. *Topographic map of the Basin and Range Province. Elevation and bathymetric map is based on the NOAA Etopo1 (Amante and Eakins 2009). Cross-sections modified from Brun* et al. *(2018), based on Miller* et al. *(1999) and Spencer and Reynolds (1990) to illustrate the combined effects of detachment, decollement and normal faults into the formation of the Basin and Range physiography*

Intracratonic rifts

Cratons are regions of the Earth where the continental lithosphere is much thicker than the average thickness of 35–40 km and are considered to be tectonically stable. The lithospheric mantle roots of cratons have unusually high thicknesses, reaching 200–400 km – more than double the standard values observed at the non-cratonic continental lithosphere (Artemieva Mooney 2002). Cratons are usually divided into two categories: the shields, where crystalline basement rocks outcrop and platforms, where the basement is overlaid by younger sediments. An intracratonic rift is thus a rift that develops on a specific lithospheric context away from active tectonic boundaries (Figure 1.10).

The driving forces creating intracratonic rifts remain extremely poorly understood. Their characteristics are not explained by standard rift models, as the extensive basement faulting and graben and/or half-graben basins appear to be absent (Klein and Hsui 1987; Hartley and Allen 1994). Uniform stretching models could explain the basic configuration of intracratonic basins, but these models fail at explaining the subsidence characteristics. The subsidence is extremely slow, and subsidence curves typically lack the characteristic two-step structure usually observed for rifted margins with the distinction between the first phase of tectonic-related subsidence (the effect of the faulting events) and the thermal subsidence (see Chapter 2). Various hypotheses have been posited to explain these peculiar subsidence curves of intracratonic basins. The main theories include subsidence caused by density changes within the lower crust and lithospheric mantle due to mineral phase changes (Gac *et al.* 2012), thermal relaxation and subsidence due to (negative) dynamic topography (Heine *et al.* 2008), thermal relaxation of a thick lithosphere undergoing low strain rates (Armitage and Allen 2010), tectonic reactivation (Braun and Shaw 2001) and uplift followed by subaerial erosion (Burov and Cloetingh 1997). Cacace and Scheck-Wenderoth (2016) reviewed all these hypotheses and determined that the long life span and slow subsidence of intracratonic basins are the result of feedback effects between sedimentation and thermal re-equilibrium at deeper crustal and mantle depths.

The basins of intracratonic rifts are often described as bowl-shaped depressions hundreds of kilometers across. The sedimentary successions correspond to terrestrial and shallow-water sediments (carbonates, shales, sandstones), typically exhibiting a layer cake infill architecture, supposedly thicker and more complete than in adjacent areas of the craton but still relatively thin. The accumulation rates are extremely slow (2–20 m/Ma), and sedimentation usually keeps pace with basin subsidence throughout the rift history. A key characteristic of intracratonic basins is that they preserve their stratigraphy over very long time periods.

Other than the Congo Basin, other examples of intracratonic basins include the Michigan basin, the East Barents Sea and the Chad basin.

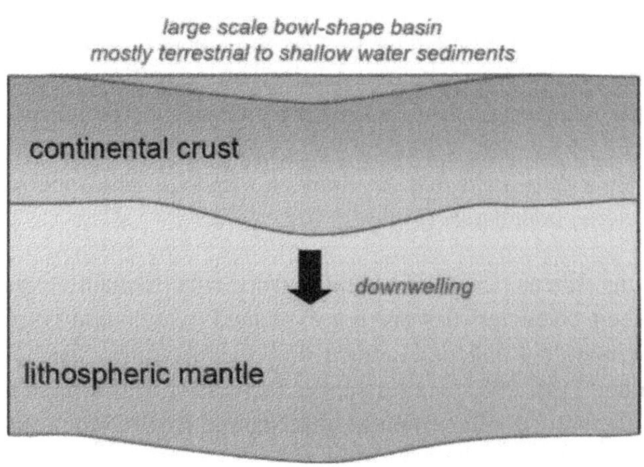

Figure 1.10. *Illustration of the intracratonic rift basin case*

Case example: The Congo Basin

The Congo Basin is a large, circular intracratonic sedimentary basin located in Central Africa (Figure 1.11) (Giresse 2005; Kadima *et al.* 2011; Watts *et al.* 2018). The depth of the lithosphere–asthenosphere boundary is modeled at 220 ± 30 km (Crosby *et al.* 2010). The basin is composed of at least three sub-basins separated by ~100 km wide basement ridges. The dimensions of the tilted basement blocks are unusually wide, theorized to be a result of the cold thermal gradients and thus higher strength of the brittle lithosphere (Foster and Nimmo 1996). The average sedimentary thickness is about 4 km, but up to 9 km of sedimentary rocks have been observed locally. The stratigraphic record spans 0.55 Ga, starting from the late Pre-Cambrian, and the basin is still subsiding today (Daly *et al.* 1992). The drilled stratigraphic sequences show consistently lagoonal and shallow-marine environments, with some truncation events that are correlated to far-field tectonic events (Daly *et al.* 1992). No magmatism has been reported, except at Early Cenozoic times with the emplacement of kimberlites at the borders of the basin.

1.2.2.2. *Rifts at convergent plate boundaries*

Sedimentary basins can also develop in close spatial and temporal relation to an active orogen in a convergent tectonic setting. These include basins in trenches, fore-arcs and foreland, intra-arcs, back-arcs/hinterland, retro-arcs/-foreland, pro-foreland and wedge-top regions. The list of basins in convergent settings is very long, often confusing and sometimes misleading. The various descriptors can identify the different structural settings in which the basins form (e.g. if the basin develops over the subducting plate, on the edge of the orogen, far from the orogen, within the orogen, etc.), and/or the different types of substratum (e.g. continental or oceanic crust). However, given the regular structural complexity of orogenic settings, the description of the basin into one category or the other is often not straightforward.

In general, basins in convergent plate boundaries can be produced in areas under local extension or in areas under compression where accommodation space is generated. For instance, extension can occur in a convergent tectonic boundary in the back-arc region (Figure 1.12). Typical examples of basins related to orogenesis include the Upper Rhine Graben, the North Alpine foreland basin, the Ganges basin and the Andean basins.

In a post-orogenic context, the overthickened lithosphere tends to a re-equilibration of its gravitational and thermal status by lithosphere thinning and, hence, extension (e.g. Dewey 1988; Rey *et al.* 2001). This is the so-called orogenic collapse, a specific tectonic mechanism that can lead to the formation of extensional basins inside the orogen (see Chapter 2).

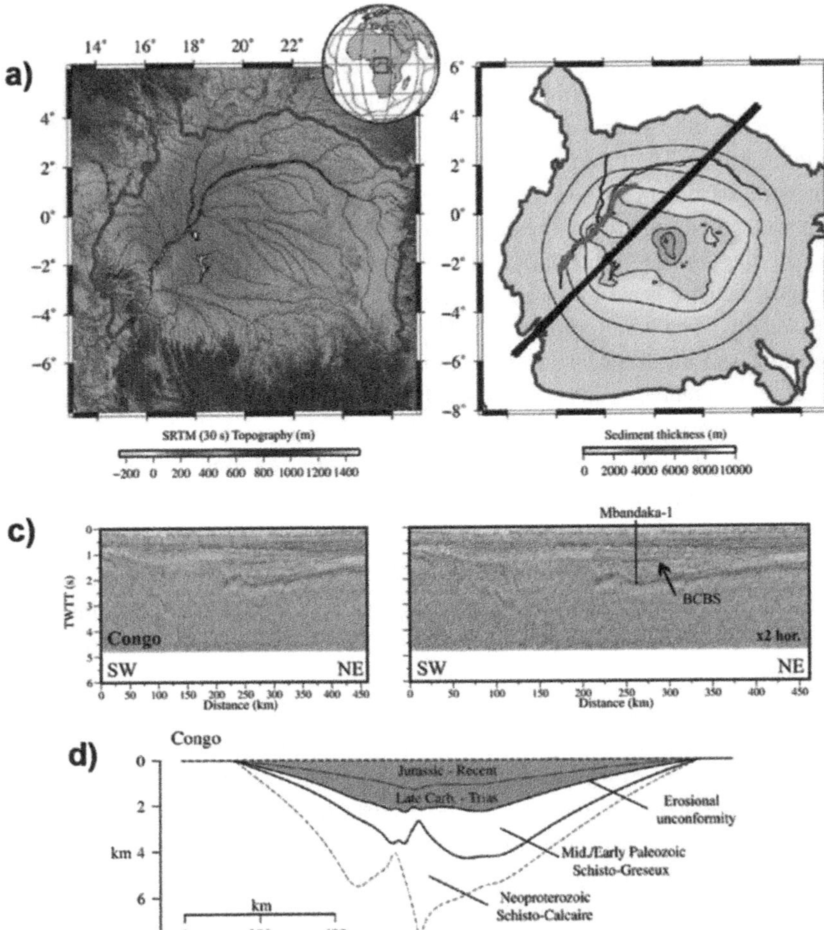

Figure 1.11. *a) Topographic map of central Africa showing the outline of the main cuvette of the Congo Basin (blue line). The inset globe shows worldwide location; b) sediment thickness map of the Congo Basin. The red line shows the location of the c) seismic profile, and the thick black line shows the d) section; c) seismic profile, without (left) and with (right) vertical exaggeration. BCBS: base of the cratonic basin sequence and d) geological cross-section showing the stratigraphic architecture of the Congo Basin (source: Watts* et al. *2018)*

Back-arc basins can be further highlighted, as these can evolve into actual rifts and rifted margins. Back-arc basins form above subduction zones in response to tensional forces generated by the down-going slab and related dynamics, and/or the collapse of the adjacent edge of the continent (Figure 1.12) (Stern 2002). Back-arc basins have peculiar geometries; they are usually very long (several hundreds to thousands of kilometers) and relatively narrow (a few hundred kilometers) and are often associated with magmatism and significant hydrothermal activity. Basalts in these basins have geochemical signatures similar to mid-ocean ridge basalts (MORBs), although they are significantly richer in magmatic water. In extreme cases, the back-arc rifting can be very pronounced, with a significant rise of the asthenosphere leading to a lithospheric breakup and the genesis of new oceanic crust. Typical examples of active back-arc basins include the Tyrrhenian Sea, the Mariana Trough, the Black Sea, Tonga-Kermadec, the south Fiji basin, with most being found along the western margin of the Pacific Ocean.

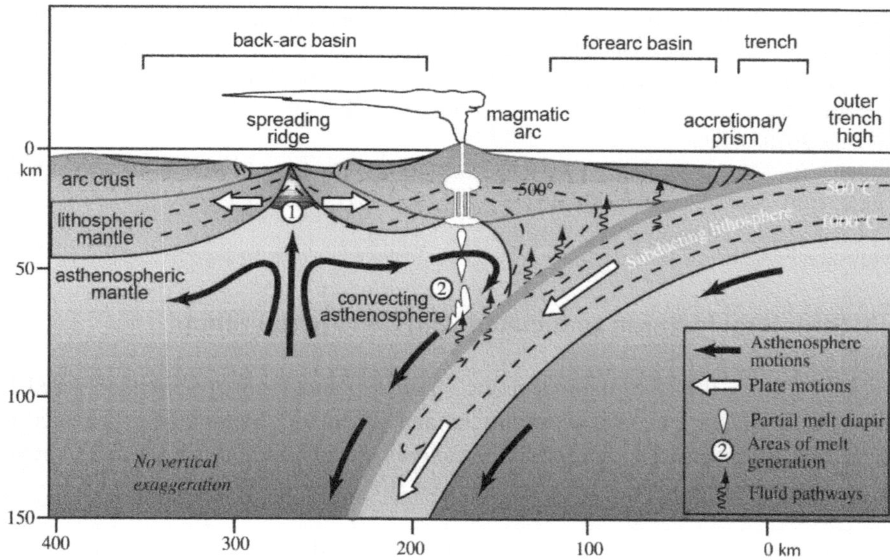

Figure 1.12. *Illustration of the back arc, forearc and trench basins related to a subduction zone. The figure is a schematic section through the upper 140 km of a subduction zone, showing the principal structural components (source: modified from Stern 2002)*

1.2.2.3. *Rifts at divergent-plate boundaries*

Rifted margins – also called "continental margins", "passive margins" or "Atlantic-type margins" – were defined by Ruppel (1995) as regions *"where continental lithosphere meets oceanic lithosphere without an intervening plate* *boundary*". This rift category can be considered as the successful version of the intracontinental rifts, in the sense that the extensional processes continued on to the ultimate separation of the continental lithosphere and the emplacement of an oceanic spreading system. Rifted margins are then an archive of the entire spectrum of deformation processes leading to complete continental lithosphere separation. They bound all continental masses in the Atlantic, Indian and Arctic Oceans and thus record the separation of the supercontinent Gondwana. Chapter 2 is dedicated to the full description of their architecture and evolution.

Further reading.– The above descriptions are abbreviated and often simplified. If interested in reading and learning further, the reader is referred to the following list of publications and references.
– General: (Wilson 1966; McKenzie 1978; Brun and Choukroune 1983; Keen 1985; Wernicke 1985; Lister *et al.* 1986; Buck 1988, 1991; Dewey 1988; Ruppel 1995; Rey *et al.* 2001; Foulger 2002; Corti *et al.* 2003; Axen 2004; Cacace and Scheck-Wenderoth 2016).
– East African Rift: (Chorowicz 2005; Ebinger 2005; Corti 2009; Rooney 2017, 2019, 2020a, 2020b, 2020c).
– Basin and Range Province: (Wernicke 1981, 1985; Davis and Lister 1988, 1989; Axen *et al.* 1993; Sonder and Jones 1999).
– Congo Basin: (Crosby *et al.* 2010).

1.3. Structural features associated with continental rifting

This section lists the major tectonic structures and basin types encountered in rifts and rifted margins. Basic definitions and schematic diagrams are provided to clarify the main terminology used in the following chapters of this book and to avoid any misunderstanding.

Note that this section does not aim to describe the details of all structural features that can be encountered in extensional settings. The descriptions below cover a biased and non-exhaustive selection of structures that are considered to be representative of rifts and rifted margins. For a proper primer in structural geology, the reader is referred to contributions specifically dedicated to this theme (see the **Further reading** sections, and, for example, Fossen 2010).

1.3.1. *Extensional mechanisms*

The lithospheric mechanisms that accommodate extension can be roughly divided into three categories: **pure shear**, **simple shear** and **crustal flow** (Figure 1.13). Pure shear refers to a homogeneous flattening of a rock body that is elongated in one direction and shortened perpendicularly. First applied by McKenzie

(1978) to regional rift settings, this mechanism assumes that the lithosphere deforms uniformly. Simple shear is a deformation mode involving a rotational component, in which parallel markers in a rock body remain parallel while translating relative to each other. (see Chapter 2 for further description of the models applied to rift settings). Numerical, analogue and conceptual models have been developed and tested based on these core mechanisms. Even though each deformational mechanism is based on an idealized version of the lithospheric structure and rheology, they are considered to represent most of the structural settings encountered on Earth.

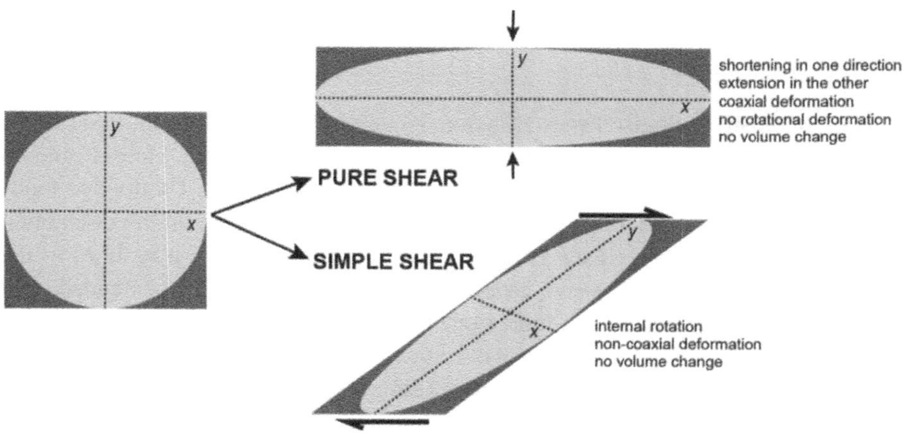

Figure 1.13. *Illustration of the pure shear and simple shear deformation mechanism (source: modified after Fossen 2010)*

1.3.2. *Main structural geometries*

The following list contains most of the deformation structures that characterize extensional settings. Schematic figures are provided to relate each structural geometry with field and seismic examples in order to illustrate their typical identification and interpretation in rifted margin studies.

1.3.2.1. *Normal faults*

In geology, faults correspond to discontinuities in rocks along which displacement occurs: when a rock mass is submitted to forces, discontinuities may develop. Depending on various parameters of the rock (e.g. rigidity, rheology), these discontinuities can accumulate a certain amount of stress. When the stress exceeds the strength limit of the rock, the discontinuity can rupture, creating a fracture, and strain energy is released. Depending on various parameters, such as the rheology of the rock (brittle, ductile), pore pressure, temperature and strain rate (low, high), the strain release can be gradual or instantaneous.

Fractures can be of various sizes (**shear fractures, joints, faults, detachment faults**) and geometries (**high angle, low angle, listric or spoon-shaped, concave upward/downward**), with large or little to no movement, and accommodate various types of displacements (**normal, reverse, strike-slip and oblique-slip**). The two blocks separated by the fault plane are called the **hanging wall** (above) and **footwall** (below).

The Scottish geologist Ernest Masson Anderson (1877–1960) introduced the basic definitions of fault mechanics (Anderson 1905), using the Mohr–Coulomb theory to explain the different dips of the various types of faults. He divided faults into three principal types (normal, reverse, strike-slip) depending on the orientation of the compressive stress axes (Figure 1.14): in an idealized earth, if σ_1 – the maximum principal compressive stress – is vertical, the crust is extended, and normal faults are generated at a high angle (dip around 60°). When σ_1 is horizontal and σ_3 – the lowest principal compressive stress – is vertical, shortening occurs and reverse faults are generated at a low angle (dip around 30°). If both maximum and minimum principal stresses are horizontal, or tangential to the Earth's surface, and σ_2 – the intermediate principal stress – is vertical, then near-vertical strike-slip faults are generated. The angle between the σ_1 axis and the shear plane is called the angle of internal friction (α). This relationship between faulting and stress is known as **Anderson's theory**.

Depending on the overall geometry of the fault, additional terms can be used such as: 1) **thrust faults** are reverse faults with a dip under 45°; 2) **listric faults** are normal faults with a fault plane that curves with depth and flattens into a sub-horizontal layer often called **decollement**; 3) **synthetic** and **antithetic** faults describe minor or secondary faults associated with a major fault: synthetic faults dip in the same direction as the primary fault, whereas the antithetic fault dips in the opposite direction.

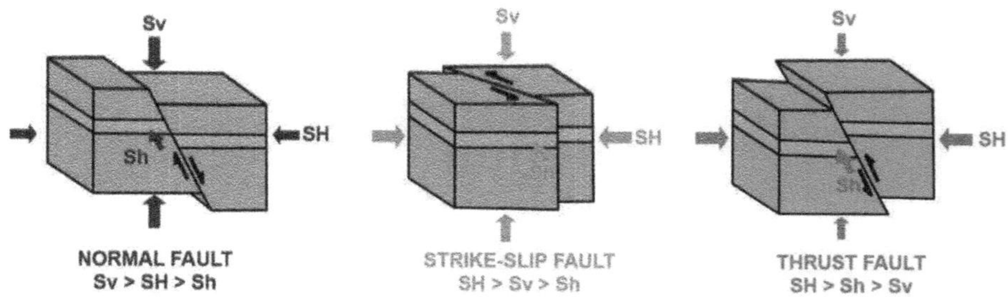

Figure 1.14. *Illustration of the three ideal Andersonian fault types. S refers to the principal stress axes*

Faults are actually complex structures, and it is often impossible to define a single fault surface for the fault plane, so interpreters regularly use the term fault zone to refer to the complex deformation associated with multiple fault surfaces. The fault core is where most of the displacement has been accommodated, a damage zone, where the rock is highly deformed and a drag zone, with kinematic indicators that show the amount and/or direction of displacement (Figure 1.15) (e.g. Fossen 2010).

Faults are described by their **length** (L) and their **displacement** (D) (Figure 1.16) (Elliott *et al.* 1976; Watterson 1986; Scholz 2019), and various questions remain on the displacement–length relationships and fault growth mechanisms: faults rarely originate with their total length extent. This is the reason why we talk about "fault nucleation". The rock block separation is supposed to initiate on a nucleus and extend in a certain direction. However, the mechanism and time–space evolution are not fully understood. In terms of normal faults in rifts, fault growth models can be divided into two categories: (1) the "propagating fault" or "isolated fault" model and (2) the "constant-length" fault model (Rotevatn *et al.* 2019). In the propagating fault model, the displacement synchronously increases and accordingly the fault grows in length (Cartwright *et al.* 1995; Walsh *et al.* 2003). On the other hand, with the constant-length model, the fault is supposed to reach almost its full length extent early in its evolution, and displacement occurs at later stages by successive increments (Jackson and Rotevatn 2013; Childs *et al.* 2017). Based on case examples, Jackson *et al.* (2017) show that normal fault growth may incorporate both models, with a predominance of the constant-length model and periods of minor fault-tip propagation and coeval displacement accumulation.

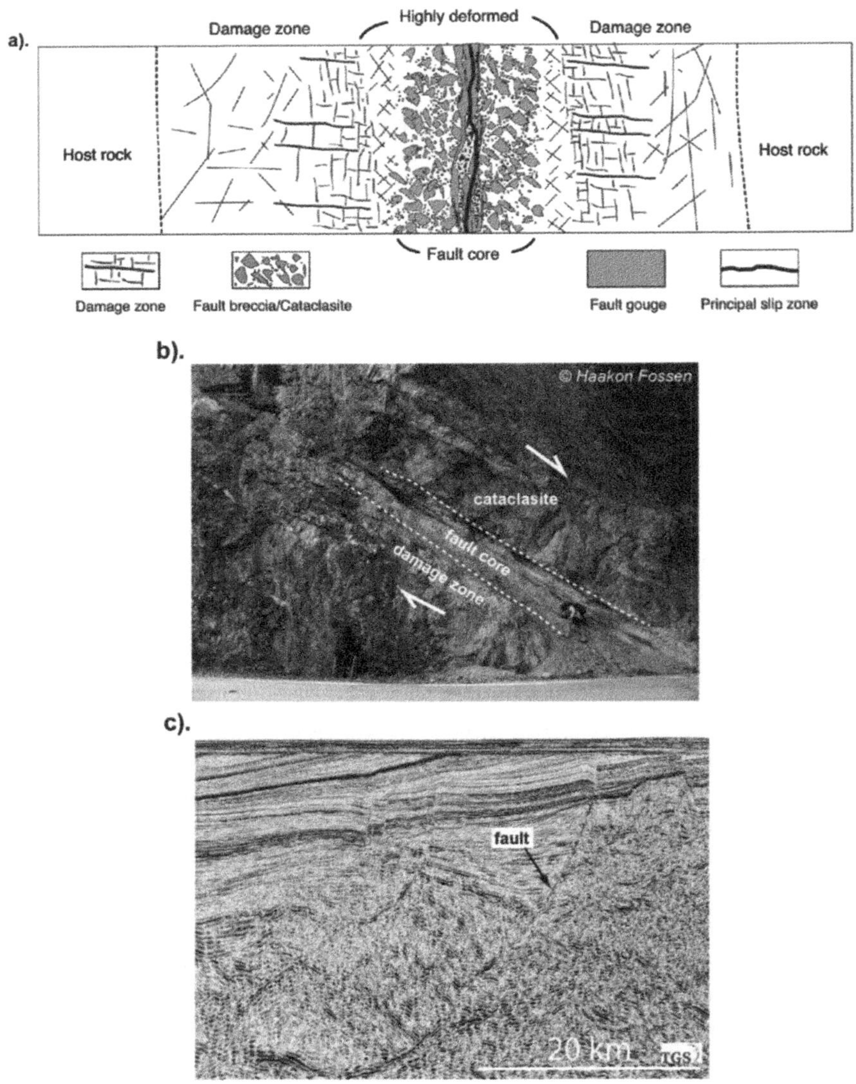

Figure 1.15. *Illustration of normal faults: a) cartoon illustrating the fault core and damage zone (source: Yang* et al. *2020); b) field observation photo by @Haakon Fossen from the Lærdal-Gjende normal fault, Norway and c) offshore seismic line, showing a series of normal faults with up to several km offset, North Sea (source: Haakon Fossen, available at: https://folk.uib.no/nglhe/)*

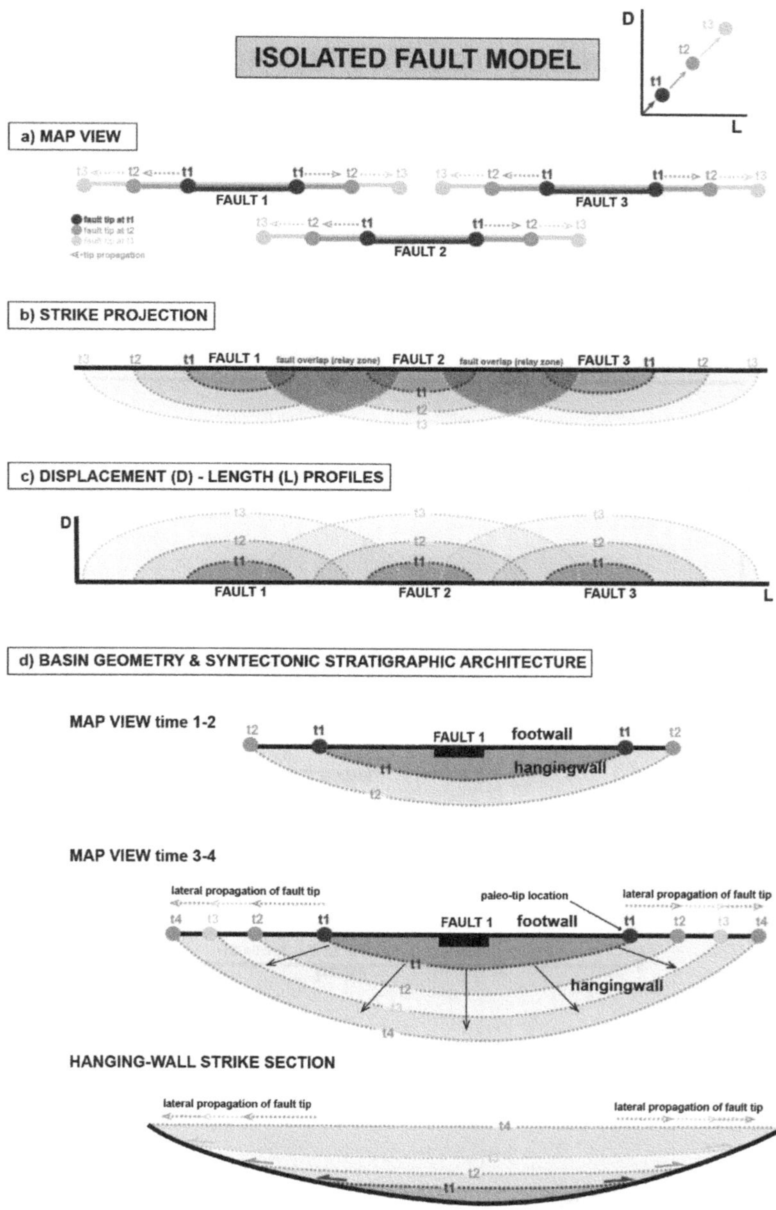

Figure 1.16a. *Conceptual model of the development of normal fault systems. The isolated fault model shown is one of the main types: a) map view; b) strike projection; c) displacement–length (D–L) plots and d) basin geometry and syntectonic stratigraphic architecture (source: modified from Jackson* et al. *2017)*

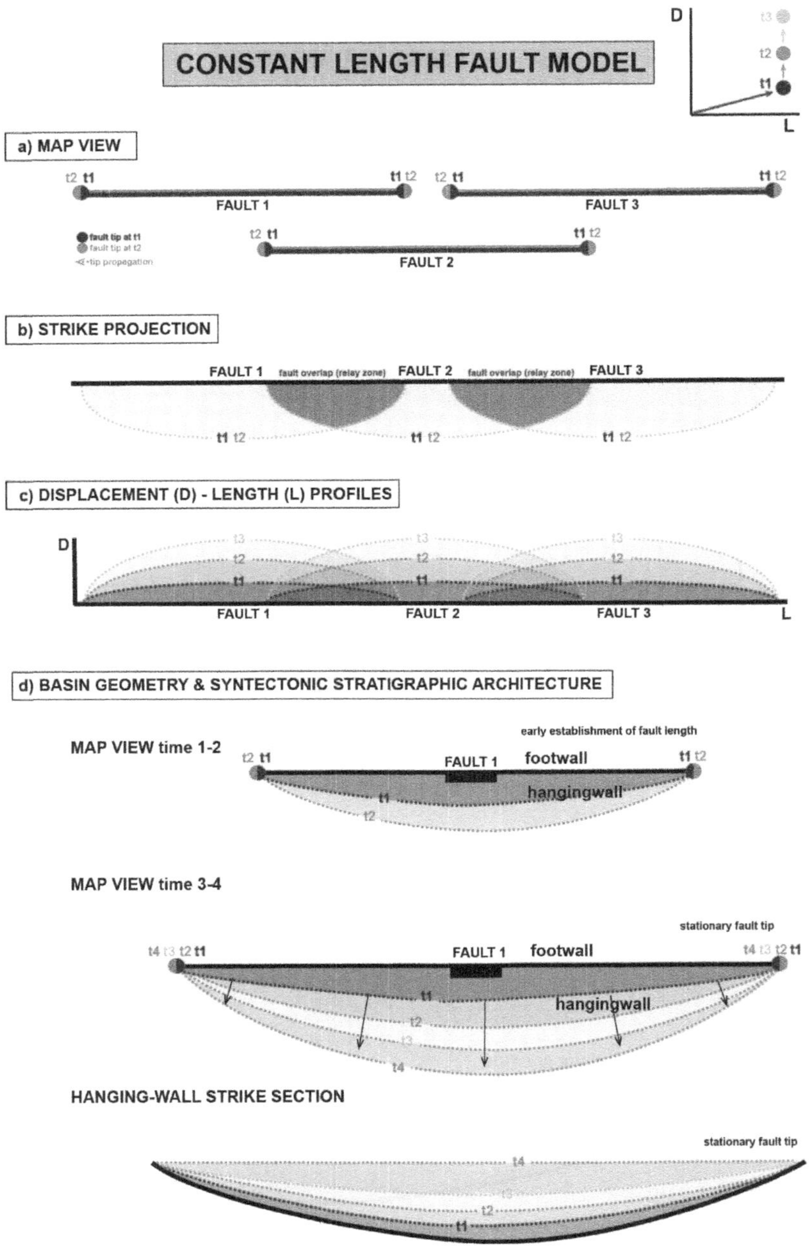

Figure 1.16b. *The constant length fault model, another commonly used modeling type: a) map view; b) strike projection; c) displacement–length (D–L) plots and d) basin geometry and syntectonic stratigraphic architecture (source: modified from Jackson* et al. *2017)*

Anderson's law remains extremely elegant for studying fault dips. However, this theory is best applied to ideal cases with isotropic rocks and coaxial deformation (the stress axes do not rotate). However, in reality, fault blocks are often composite, with different rheology and lithology, and thus different mechanical strengths. Therefore, faults have various dips that cannot be explained by Andersonian mechanics and debates remain on various aspects of the theory, notably on the question of the activity of low-angle faults.

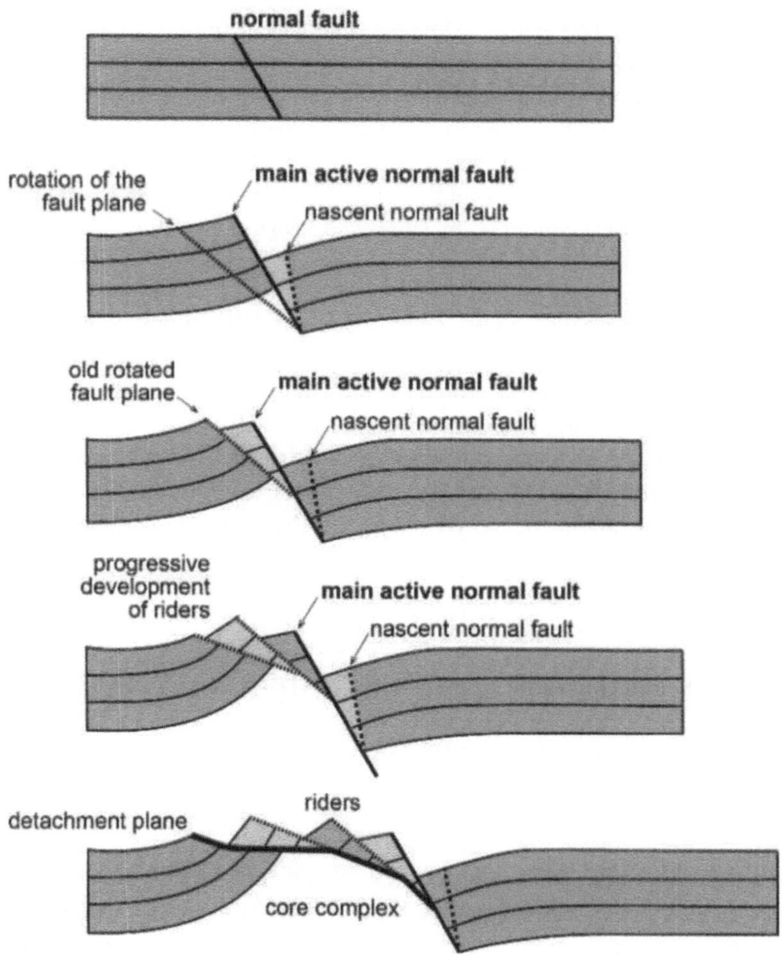

Figure 1.17. *Schematic illustration of the formation of a low-angle fault (LANF) plane/detachment plane by rotation of higher-angle normal faults in the context of the development of a metamorphic core complex (source: based on a figure from J.P. Burg, after Buck 1988)*

While the simple shear model of extension developed in the 1980s (e.g. Wernicke 1985), numerous observations of faults with very low-angle surfaces, but with apparent large-scale extension and significant amounts of displacement accumulated. This led to fundamental discussions on the feasibility of tectonic displacement along such low-angle surfaces. Faults were categorized as Andersonian and non-Andersonian, depending on whether they were "correctly" oriented with respect to the regional stress-field or not. Normal faults with an apparent dip-angle below 30° contradict Anderson's fault theory and are often termed "non-Andersonian faults". Doubts remained on the actual activity of such low-angle faults and it was regularly proposed that the low-angle normal faults (LANFs) are flexurally rotated segments of larger-scale complex fault systems and that the regional extension is actually operated on the higher-angle segments (>60°) (Figure 1.17), thereby respecting Anderson's theory (Buck 1988). However, others argue that displacements on low-angle surfaces are possibly due to elevated pore fluid pressures and weakening factors (Wernicke 1981; Axen 1992; Scott and Lister 1992).

In rift and rifted margin studies, extension across the rift can be defined by the parameters **beta factor** (β), **thinning factor** (ε), **elongation** (*e*) or **stretching factor** (*S*) (McKenzie 1978; Hellinger and Sclater 1983; Davis Kusznir 2004; Reston 2007; Reston 2009; Reston and McDermott 2014). These terms are used to constrain the amount of extension accommodated over a certain area defined by section lengths before (L_0) and after (L_f) the deformation episode (Figure 1.18). The initial and final crustal thicknesses (t_0 and t_f) can also be used to define the area that has extended.

The elongation is equivalent to: $e = (L_f\text{-}L_0)/ L_0$.

The beta or stretching factor: $\beta = S = 1+e = 1+(L_f\text{-}L_0)/L_0 = L_f /L_0$.

Thus, $L_f = \beta L_0$, wherein β goes from 1 ($L_f = L_0$) to infinity ($L_f >> L_0$).

And the thinning factor: $\varepsilon = (t_0\text{-}t_f) / t_0$ with $t_f = t_0 / \beta$, thus $\varepsilon = 1\text{-}1/\beta$.

Note that alternative definitions of these parameters have been proposed in the literature (and/or some relationships between L and t can be assumed). Therefore, to avoid any misunderstanding, special attention should be paid to the definition listed by the authors.

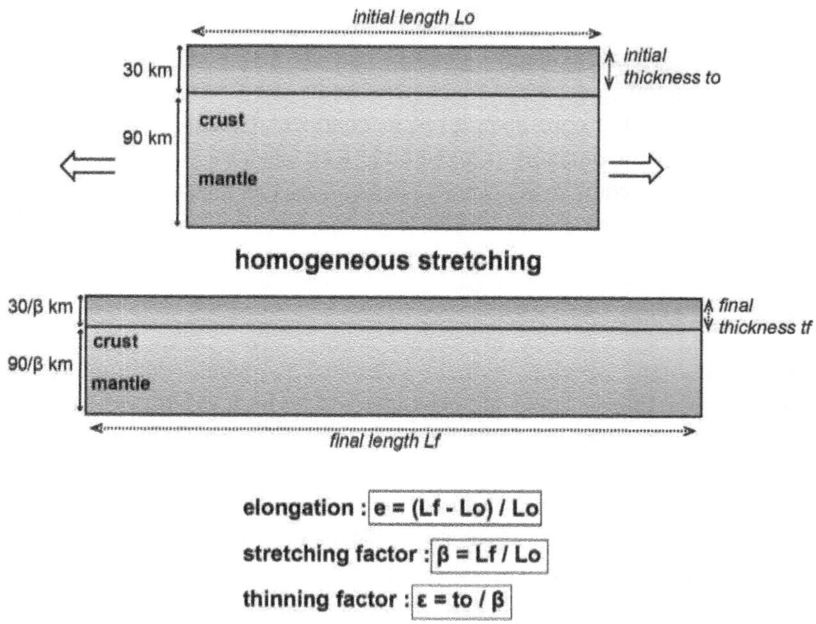

Figure 1.18. *Illustration of the elongation, stretching and thinning factors*

1.3.2.2. *Detachment faults*

"Detachment fault" is a term often used in rifted margin studies to refer to a specific fault plane that is distinguished from standard normal faults by its sub-regional to regional-scale geometry and low-angle apparent slip (Figure 1.19). First identified in the Basin and Range Province (Longwell 1945; Wernicke 1981), a series of publications in the 1980s and early 1990s proposed the occurrence of similar structures in other continental rifted margins (Wernicke 1985; Lister *et al.* 1986, 1991).

A detachment fault is a normal fault of very large-scale extent, accommodating significant displacements (tens of km), with a dip usually varying from higher angles at its breakaway to lower angles at its decollement level (Figure 1.20). The geometry, either concave-upward or concave-downward, typically includes multiple segments – the most active being the ones at higher angles, whereas the lower-angle segments are instead interpreted as segments that have been flexurally rotated during unroofing after their main phase of activity. Whether or not the low-angle segments can actually be considered as active remains debated, as these are not explained by Andersonian fault mechanics (see the section above) (Axen 2004). Detachment faults are often understood to begin at middle-lower crustal depths in zones with weak rheologies, where mylonitic structures form. Ductile shear is the main deformation mechanism at depths greater than 20 km, whereas brittle behavior dominates at shallower depths. Detachment faults are often associated with the genesis of metamorphic core complexes in continental settings and oceanic core complexes in oceanic domains (see below). Key onshore examples include the Nordfjord-Sogn Detachment in south-west Norway, the Snake Range detachment system of the Basin and Range Province in the western USA and the Whipple Mountains detachment in California.

1.3.2.3. *Shear zones*

Shear zones may develop in rocks submitted to intense deformation, where some planar to sub-planar zones are characterized by high strain rates compared to the surrounding rocks that have undergone lower finite strain (Figure 1.20). Shear zones are often considered to be the deeper crustal counterpart of the shallow brittle faults. The motion of the more rigid surrounding blocks may imply a rotational, non-coaxial component in the shear zone. Depending on the rheological context, shear zones can be brittle, semi-brittle, ductile or a combination. Ideally, the deformation is concentrated either in a narrow fracture (brittle) or distributed over a wider zone (ductile). The change from brittle to ductile behavior is classically associated with depth, with the development of intermediate deformation styles where brittle fracturing coexists with plastic flow. However, other parameters can strongly influence that transition, such as the local lithospheric thermal state, the presence of fluids, compositional changes (e.g. serpentinization), strain rates, stress field orientation and compositional anisotropies.

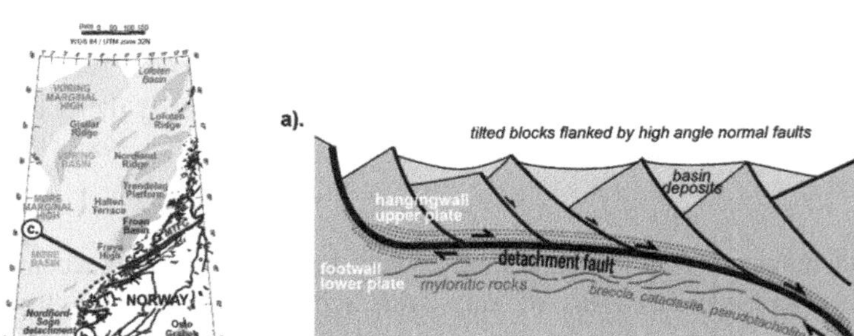

Figure 1.19a. *Diagram illustrating a detachment fault accommodating displacement at low angles, with slightly to highly deformed rocks in its footwall (i.e. mylonitic rocks, breccias, cataclasites), and high-angle normal faults bounding tilted blocks on its hanging wall*

b).

Figure 1.19b. *Photo showing a field example of the Nordfjord-Sogn detachment zone (NSDZ), which is a major detachment fault in Norway. WGR: Western Gneiss region (source: photo by Per Terje Osmundsen)*

c).

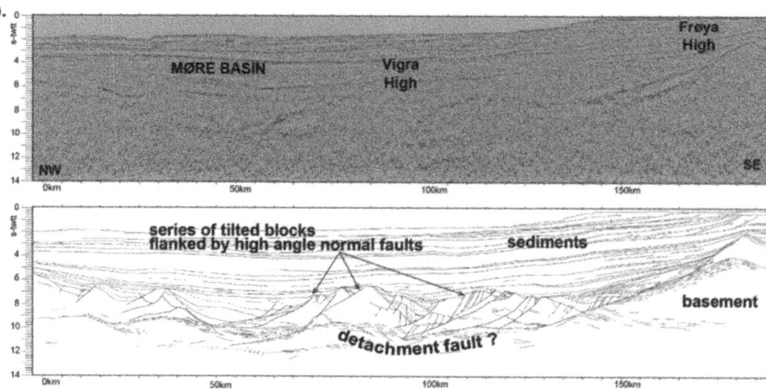

Figure 1.19c. *Extract from a seismic reflection profile in the mid-Norwegian rifted margin (top left) with the interpreted structures (lower left) including a major regional detachment fault and tilted fault blocks. Locations on the map inset, upper left (source: from Peron-Pinvidic and Osmundsen 2020)*

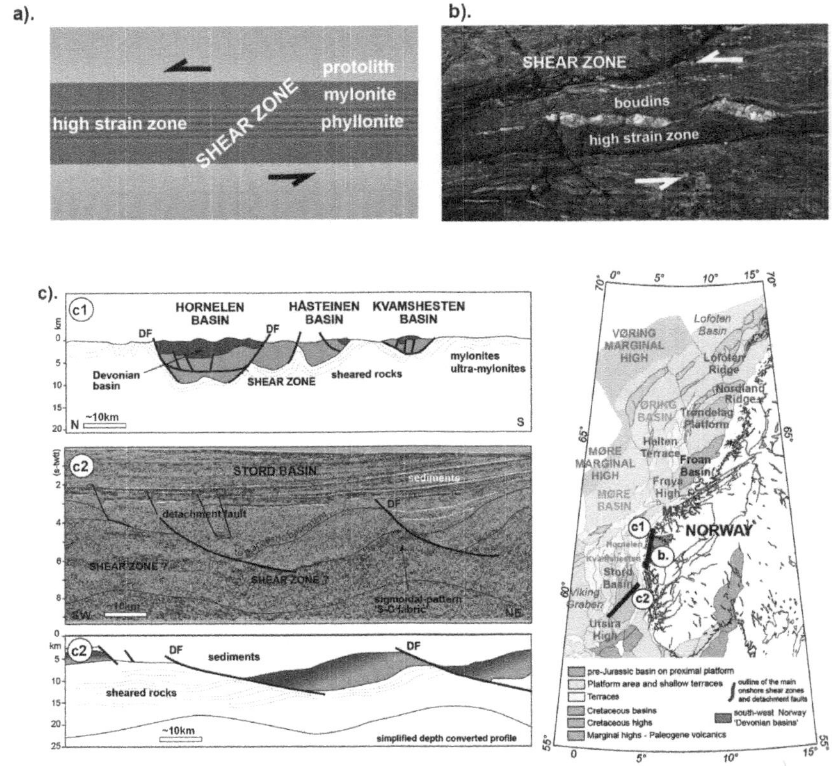

Figure 1.20. *Illustration of a shear zone: a) diagram illustrating the concentration of the deformation in a preferred zone that generates the formation of highly deformed rocks; b) field photo (photo credits © Haakon Fossen) of the Nordfjord-Sogn detachment (Rutledal, Norway) and c) c1: schematic cross-section illustrating the structural setting of the south-west Norway Devonian basins (source: Osmundsen and Andersen 2001); c2: extract from a seismic reflection profile in the Stord Basin in the North Sea showing the possible presence of shear zones at basement depths (source: Peron-Pinvidic and Osmundsen 2020). Locations on the map inset, lower right*

1.3.2.4. *Metamorphic core complexes*

When a region is submitted to large amounts of extension, mid- to lower-crustal levels can be dragged out from beneath the brittle fractured upper crust. The resulting dome-shaped structure is called a **metamorphic core complex** (Whitney *et al.* 2014; Brun *et al.* 2018). Typically, the exhumation process includes a concave-downward detachment fault, exhibiting ductile deformation in the footwall with amphibolite- to greenschist-facies metamorphism, and brittle deformation in the hanging wall with high-angle normal faults (HANFs) and tilted unmetamorphosed crustal blocks. These HANFs often merge with the detachment fault at depth (Figure 1.21). Depending on the lithospheric parameters (e.g. thermal state, extensional rate), the core complex may be capped by a shear zone rather than a simple fault plane, with brittle to ductile deformation features. Deformation products associated with such structures are multiple and can include breccias, mylonites and phyllonites. Metamorphic core complexes are encountered in various extensional settings and are often interpreted in seismic data of rifted continental margins.

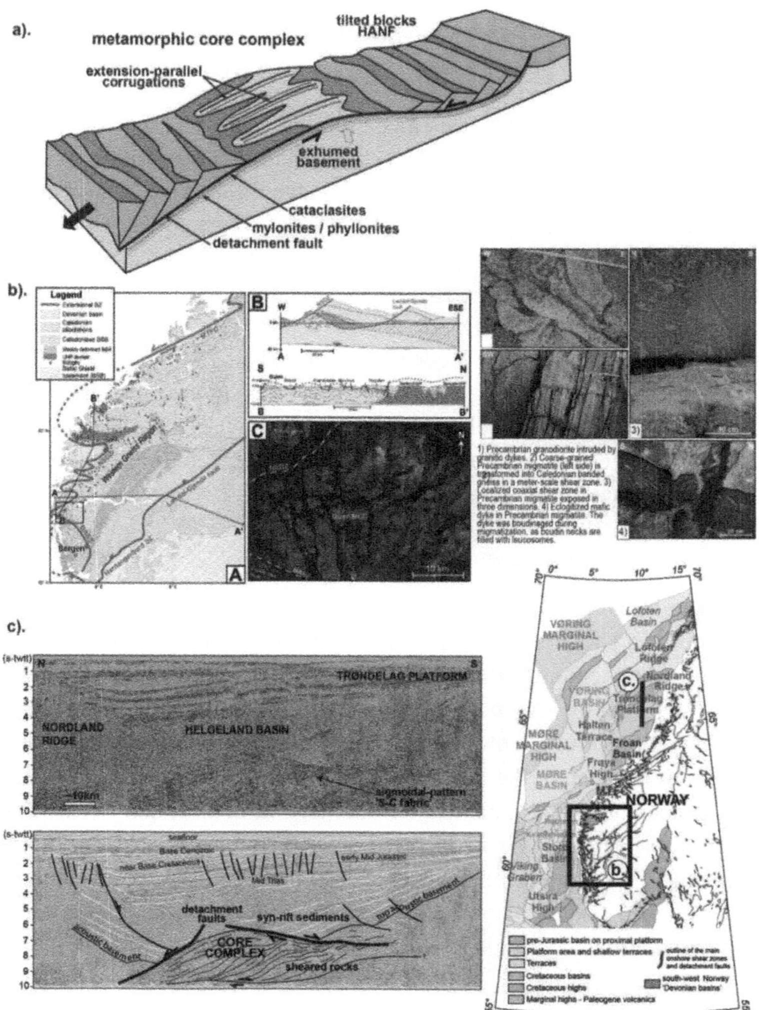

Figure 1.21. *Illustration of a metamorphic core complex*

CONTINUATION OF CAPTION FOR FIGURE 1.21.– *a) Schematic representation of a metamorphic core complex with a window of exhumed basement capped by a detachment fault and overriding tilted blocks flanked by high-angle normal faults (HANF) (source: modified from Osmundsen* et al. *2003 and Fossen 2010); b) field example with map, sections and photos from the Western Gneiss Region in southwestern Norway (A, B) and focus on the Gulen Metamorphic core complex (MCC) with outcrop photos (right-hand inset photos). NSDZ: Nordfjord-Sogn detachment zone. BASZ: Bergen Arc Shear Zone (source: from Wiest* et al. *2019) and c) seismic reflection profile (TGS) from the Trøndelag Platform; in time (s-twtt), without (top) and with interpretation (bottom) (location on the map inset on the right) (source: from Peron-Pinvidic and Osmundsen 2020).*

The **rolling-hinge model** for metamorphic core complexes (Buck 1988; Wernicke and Axen 1988; Lavier *et al.* 1999) includes the definition of a large-scale concave-downward detachment fault which evolves from high to low angles as the footwall flexes upward during unroofing from below the hanging wall (Figure 1.22). This model is often used to explain basement exhumation in continental rifted margin and oceanic ridge studies. Reston and McDermott (2011) further refined the rolling-hinge concept by having successive detachment faults at the rift switch polarity and thus cut into their predecessor's footwall (the so-called "flipping" or "flip-flop" model). The resulting alternating dip orientation of the main detachment then provides an elegant model to explain how extensive surfaces of symmetric exhumed basement can be generated by an asymmetric tectonic structure (for further information, see Chapter 2).

1.3.2.5. *Boudinage structures*

In structural geology, **boudinage** refers to a deformed structure produced during the extension of a rock that is characterized by the presence of sheared rigid parts enclosed in a less competent surrounding matrix (Figure 1.23). The "boudins" correspond to pieces of the more rigid rocks that were once part of a continuous layer and have been broken into individual units, isolated laterally by anastomosing shear zones deformed in the ductile or semi-ductile stress-field. Some boudins exhibit a "pinch-and-swell" geometry. Boudinage structures can occur at all scales from microscopic to lithospheric. In rifted margin studies, it has been proposed that the continental lithosphere may deform at the regional scale under such a deformation mode. It is usually associated with specific geological settings including high thermal conditions.

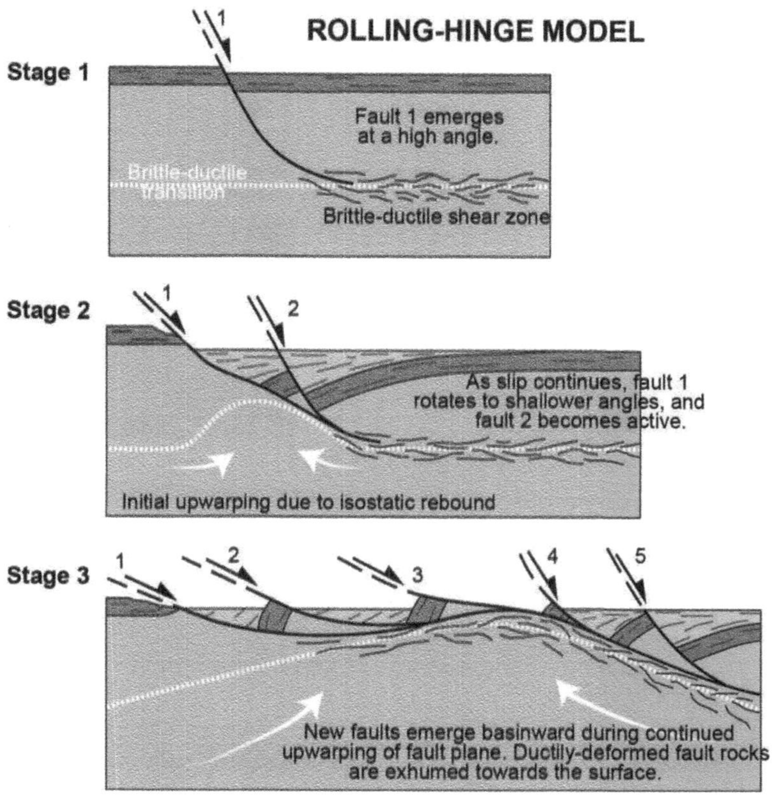

Figure 1.22. *Schematic illustration of the evolution of low-angle normal faults (LANFs) with the domino-block rotation model and the rolling-hinge model (source: modified from Ricketts* et al. *2015)*

1.3.2.6. *Folds*

Folds can occur at various geologic scales and in various tectonic settings. Intuitively, folds are often associated with shortening rather than extension. However, they may also develop in extensional settings in specific structural contexts, such as in fault propagation folds (folding above a fault tip) or related to the intrusion of igneous material (e.g. above a laccolith). They can also be associated with differential compaction of sedimentary layers. Constrictional strain can also produce folds and compressional geometries in basins under regional extension (e.g. SW Norway Devonian basins, Osmundsen Andersen 2001). Folds are analyzed based on their overall geometry. The three main categories of folds are **monocline**, **anticline** and **syncline**. Geometrically, they are described by how the limbs dip to the hinge axis (Figure 1.24).

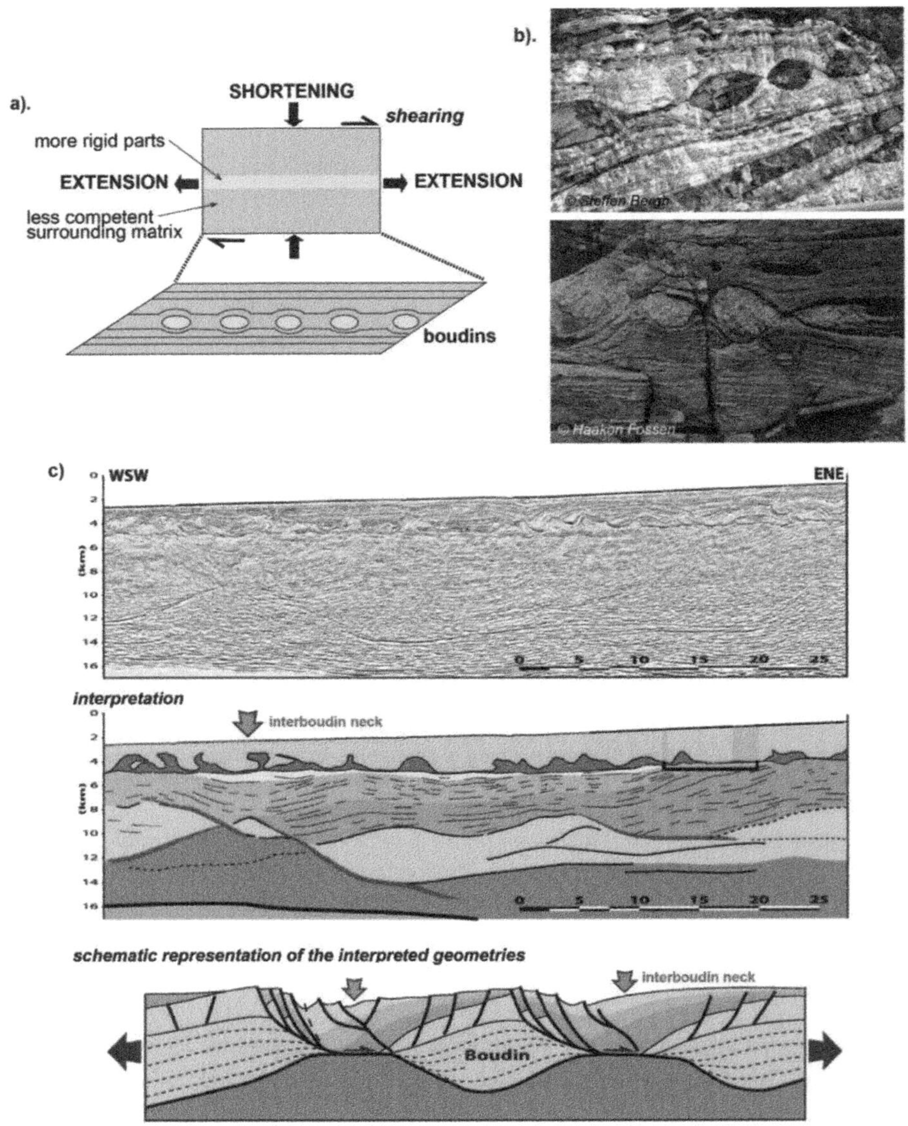

Figure 1.23. *a) Illustration of a schematic boudinage structure; b) field examples of "pinch-and-swell" boudins from Norway with amphibolite boudins within quartz schists (top: Finnmark, Norway) and a deformed pegmatite dike (bottom: Roan, Norway) (photo credits © Steffen Bergh and © Haakon Fossen) and c) an offshore seismic reflection example from the Gabon rifted margin (source: modified from Clerc* et al. *2018)*

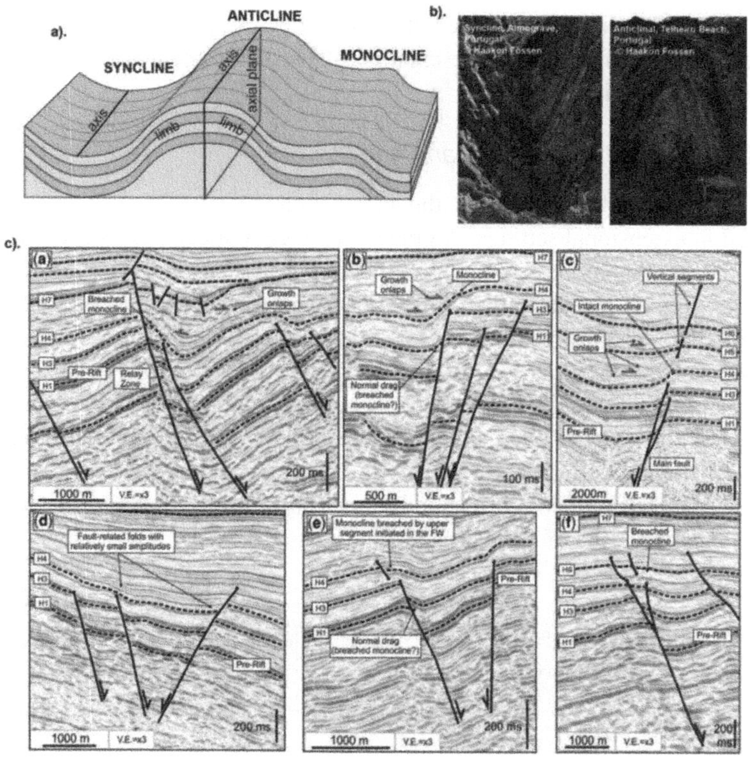

Figure 1.24. *a) Illustration of the main fold geometries; b) field example photos from Portugal (photo credits © Haakon Fossen) and c) seismic refection profile examples of fault-propagation folds associated with different sets of "simple" and "conjugate" normal faults, from the Inner Moray Firth, UK (source: from Lapadat* et al. *2017)*

Further reading.– The above descriptions are abbreviated and often simplified. If interested in reading and learning further, the reader is referred to the following list of publications and references:

– General: (Wernicke 1985; Lister *et al.* 1986; Buck 1988, 1991; Lavier *et al.* 1999; Axen 2004; Davis and Kusznir 2004; Reston 2007, 2009; Reston and McDermott 2011; Whitney *et al.* 2014; Jackson *et al.* 2017; Brun *et al.* 2018; Rotevatn *et al.* 2019).

– For structural geology definitions, we recommend the excellent book (Fossen 2010).

1.3.3. *Main basin types*

Rifts and rifted margins are usually composed of several basins of various types and linked in various ways. Below is a list of the major types of basins that can be encountered in rift and rifted margin studies. Again, this list does not aim at being entirely complete, but representative of the main geometries encountered in rifts and rifted margins.

1.3.3.1. *Grabens, half-grabens and horsts*

Grabens and half-grabens are the most common type of sedimentary basins generated in extensional settings. A graben is typically identified as a valley flanked by distinct topographic escarpments. Structurally, grabens are formed by arrays of oppositely dipping normal faults. In the ideal case, the flanking faults are parallel and antithetic and grow at the same rate so that there is no rotation of the hanging wall. The overall downward movement of the common hanging wall leads to the formation of a symmetric depression: the graben (Figure 1.25). However, faults rarely grow at the same rate, so grabens are usually asymmetric. If several faults form, the hanging wall platform can be further dissected and can generate a horst (Figure 1.25). Half-grabens correspond to the topographic depression developed on the tilted hanging walls of normal faults.

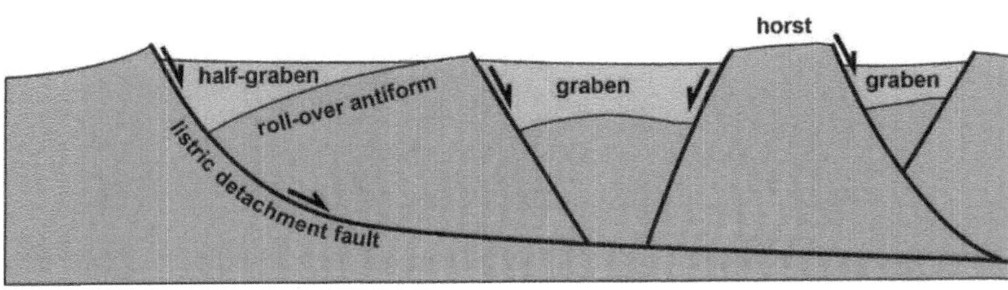

Figure 1.25. *Diagram illustrating the structural geometries of a graben, horst and half-graben*

1.3.3.2. *Pull-apart basins*

In some cases, extension can occur related to strike-slip faults. In regions with multiple segments of strike-slip faults, the lateral displacement between the faults will lead to stepovers and, eventually, the development of pull-apart basins – also named strike-slip basins (Figure 1.26). Typically, pull-apart basins are strongly compartmentalized, with each individual basin sharing similarities with discrete intracontinental rifts. The standard basin geometry corresponds to a rhomb-shaped graben or half-graben. In theory, these basins tend to be fairly deep with a fast subsidence. Type examples of strike-slip faults with pull-apart basins include the San Andreas Fault, the Dead Sea Transform and the North Anatolian Fault (Şengör *et al.* 2005). The most cited examples of strike-slip rifts include the Salton Sea, the Death Valley, the Jordan Valley-Dead Sea Rift – Gulf of Eilat and the onshore northernmost part of the Red Sea Rift.

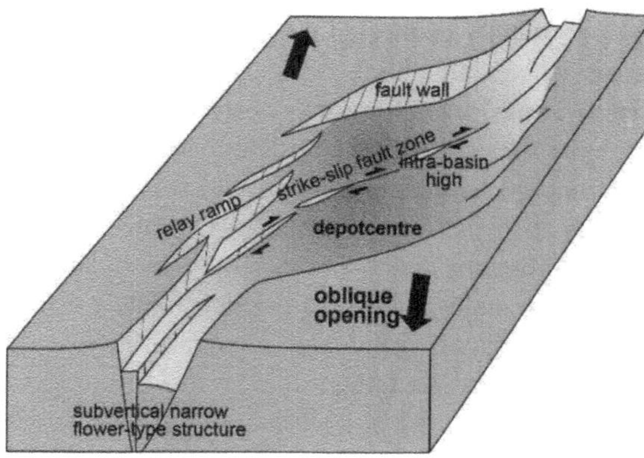

Figure 1.26. *Diagram illustrating the main structural geometries associated with a pull-apart basin, at its early stage of development (source: modified from Wu* et al. *2009, based on analogue modeling results)*

1.3.3.3. *Sag basins*

In geology, the term "sag basin" is used to describe deep, large-scale basins with an overall circular/elliptical shape (Kingston *et al.* 1983; Middleton 1989). Interior cratonic basins are often referred to as sag-type basins, displaying accumulations of sediments in an overall oval, long-lasting depression. The formation of sag basins is hypothesized to be driven by long-lasting thermal subsidence (Sleep 1971) and not involve major faulting. In the recent literature, the term "sag basin" has been widely used to describe basins encountered in distal rifted margins, usually referring to large-scale (>100 km wide) basins displaying conformable sedimentary sequences that are draped over underlying faulted topography (see the expanded description in Chapter 2).

1.3.3.4. *Linkage of rift faults and basins*

Faults rarely form as isolated features and a rift system is built through the activity of several families, or populations, of faults. With ongoing extension and crustal thinning, faults and basins may interact and link in various geometries. Depending on the scale of the study, various terms are used in the literature to refer to these distinct areas. The series of faults that interact and/or accommodate extension under similar structural settings are often known as "**fault complexes**", "**fault systems**" or "**fault zones**". At one stage, as faults grow during ongoing extension, they may come close to each other and specific interactions are then observed. Distinct vocabulary is then used to characterize the faults: unlinked faults that do not mechanically interact, soft-linked faults that form an overlapping structure but are not physically connected and hard-linked faults that are physically connected. The areas where interaction occur are called **overlapping structures**. These are called **relay zones**, **transfer zones** or **accommodation zones**, depending on the geometrical characteristics of the structural connection between the faults (Figure 1.27). Each zone is characterized by special structural features: relay ramps form in relay zones between synthetic faults, transfer faults are oriented perpendicular to the overlapping faults and usually display strike-slip displacement (not to be confused with transfer zones), en-échelon faults are series of faults lying oblique to the overall structural trend and organized in a series of parallel or subparallel, overlapping or step-like faults.

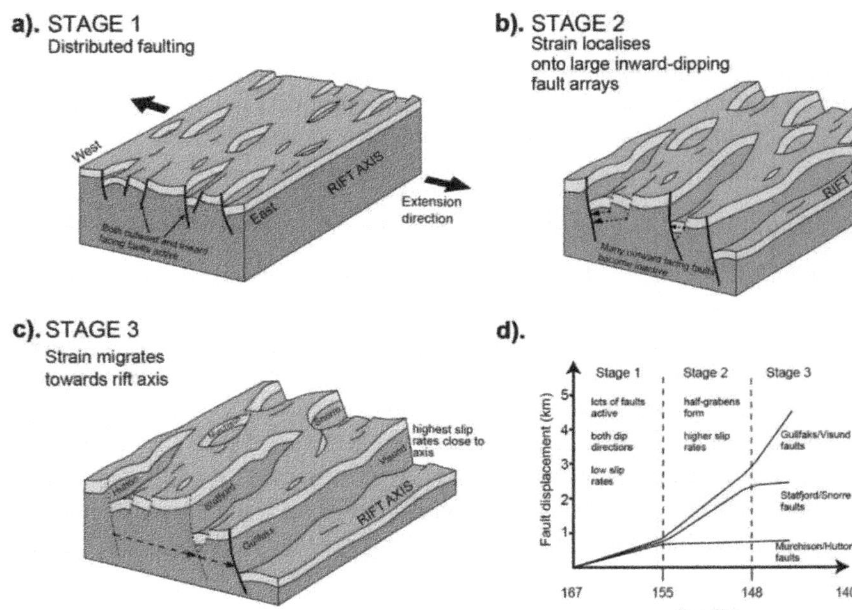

Figure 1.27. *Illustration of the possible complex geometries developed during extension and fault/basin migration/interaction/linkage. The cartoons are based on a study of the North Sea East Shetland Basin. The graph in d) summarizes fault throw versus time for the three main fault systems (source: from Cowie et al. 2005)*

The way faults are organized and linked in an extensional setting is related to the way normal faults grow (see section 3.2) (Childs *et al.* 2017; Rotevatn *et al.* 2019).

Further reading.– The above descriptions are abbreviated and often simplified. If interested in reading and learning further, the reader is referred to the following list of publications and references:

– General: (Kingston *et al.* 1983; Leeder and Gawthorpe 1987; Leeder and Jackson 1993; Gawthorpe and Leeder 2000; Cowie *et al.* 2005; Childs *et al.* 2017; Rotevatn *et al.* 2019).

– For detailed explanations of basin analysis, we recommend the necessary (Allen and Allen 2013).

1.4. Subsidence

Subsidence is the stage that leads to the progressive deepening of the basin floor and hence allows the accumulation of sediments in rift basins. It is the process by which the lithosphere regains isostatic equilibrium. **Accommodation** or **accommodation space** refers to the amount of space available for sediments and/or magmatic material to accumulate.

On rifted continental margins, three main types of mechanisms can contribute to subsidence:

1) **Extension**: the continental crust stretching leads to the formation of faults whose activity generates topographic depressions by the downward movement of the hanging walls. This is the **initial tectonic subsidence** and is a syn-rift stage.

2) **Thermal cooling**: with time, rifted margins get wider. As the rift axis moves away, the thinned lithosphere also becomes colder and denser, promoting further subsidence. Additionally, during lithospheric stretching, the underlying asthenosphere naturally rises underneath the rift axis. With time, the warm asthenosphere progressively cools into mantle lithosphere, a process occurring over many tens of millions of years. From these two processes, the lithospheric plate thickens until the thermally defined lithosphere–asthenosphere boundary regains its initial depth. Because lithospheric mantle is denser than asthenospheric mantle, this cooling process also induces subsidence in the rifted margin. The entire process is called **thermal subsidence**.

3) **Loading**, sedimentary or volcanic: additional subsidence is caused by the sediment/volcanic load and/or the weight of the water that fills the accommodation space of the rift zone. This is a process that is present at all rifting stages. The emplacement of large volumes of hot magma in the extending system can influence the uplift–subsidence curve of the basin, depending on the volume and the rate of their emplacement (Royden and Keen 1980).

Densification processes of the crust can also influence subsidence, such as increases in density driven by changing pressure/temperature conditions and/or the emplacement of higher density melts into the lower density upper crust, as mentioned above.

Additional mechanisms may also influence subsidence curves, like asthenospheric mantle flow. The asthenospheric mantle is in constant motion, due to convection caused by thermal or density gradients: hot mantle material rises, cools and then sinks in convection cells. Based on this, strictly speaking, very few places on Earth are actually in perfect "isostatic equilibrium" at geological scales (Watts 2001). Mantle convection is known to produce large-wavelength dynamic topography in the lithosphere, but even small-scale convection cells can impact local basin subsidence (Braun 2010; Petersen *et al.* 2010).

Based on the relationships of subsidence patterns and the sources causing subsidence, subsidence curves should be diagnostic of each type of sedimentary basin (McKenzie 1978; Bott 1979; Xie and Heller 2009; Allen and Allen 2013). In fact, Xie and Heller (2009) compared subsidence curves from various basin-types around the world and established a relationship between the shape of subsidence curves and different extensional settings (see Figure 1.28). In general, rift basins in extensional tectonic settings experience two distinct phases in their subsidence curves: an initial steep gradient, related to the tectonic subsidence that accompanies the rift tectonic episode, followed by a gentler but longer-lasting curve, corresponding to lithospheric thermal re-equilibration. On the opposite spectrum of rift-type basins, foreland basins typically have subsidence curves with an initial shallow curve before a steep one, and strike-slip basins are often observed as having very fast subsidence. In practice, however, subsidence curves are usually more complicated than envisaged by this basic approach. Rift evolution is often multiphase, with sometimes long-lasting periods of tectonic quiescence, leading to uplift–subsidence patterns that may be very different from the ideal case (see Chapter 2).

Further reading.– The above descriptions are abbreviated and often simplified. If interested in reading and learning further, the reader is referred to the following list of publications and references:

– General: (McKenzie 1978; Bott 1979; White *et al.* 1987; White and McKenzie 1989; Kusznir and Ziegler 1992; Xie and Heller 2009; Allen and Allen 2013).

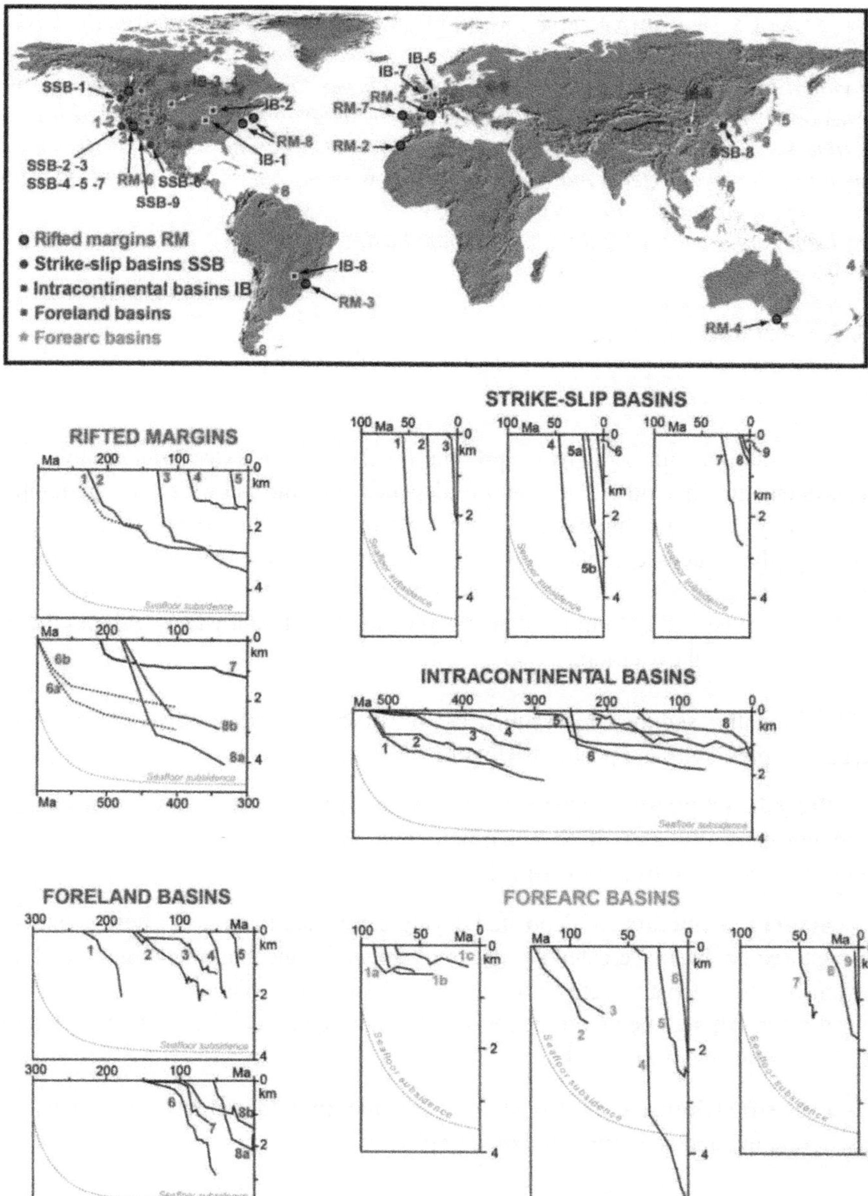

Figure 1.28. *Tectonic subsidence curves for various types of basins around the world (source: Xie and Heller 2009)*

CONTINUATION OF CAPTION FOR FIGURE 1.28.– *Rifted margins, strike-slip basins, intracontinental basins, foreland basins and forearc basins. Locations of each individual basin are shown on the map. In the curves for rifted margins, black solid curves correspond to the time scale at the top of the graph and black dotted lines to the time scale at the bottom of the graph. Note the different shapes of the curves depending on the extensional setting: the two-step curves for rifted margins (first steep, then gentle), the very steep curves for the strike-slip basins and the very long-term gentle curves for the intracratonic basins.*

1.5. Rift-related materials

1.5.1. *Sediments*

Sedimentation, although always present in rift basins, is extremely variable from one system to the other. It mainly depends on the surrounding geography (morphology) and lithological and climatic parameters. For detailed information regarding the processes related to these parameters, the reader is referred to contributions specifically dedicated to sedimentary processes. Here we will only list the basic architectural definitions that are considered pertinent when studying the rift space and temporal evolution.

Traditionally, sedimentary sequences of rift basins are subdivided into three categories: pre-, syn- and post-rifts (Figure 1.29):

– **Pre-rift sediments** are deposited before the rifting phase. They are considered to be mechanically part of the basement and they are deformed and displaced together during the subsequent rift-related tectonic phases.

– **Syn-rift sediments** are deposited during extensional activity. These sediments are captured in basins created by the movement of the basement blocks. If the relationship between the activity of the faults and the sediment supply correlates well, the syn-rift sediments are deposited in a fan geometry created by the rotation and the subsidence of the blocks.

– **Post-rift sediments** are deposited on the margin, after extensional activity has ceased, draping and smoothing the topography.

In theory, discontinuities mark the separation between pre-, syn- and post-rift sediments. For instance, the **breakup unconformity** is generated by an uplift event that is related to the final rupture of the continental lithosphere and emplacement of the first oceanic crust. The uplift generates a level of erosion in the stratigraphic record and/or an absence of record.

These basic definitions are extremely useful and often allow for a clear and effective subdivision of the rift history. However, they have been deeply challenged over the last few decades. Based on recently available high-resolution geophysical datasets, it appears that these strict definitions may not encompass the multiphase structural evolution of some extensional systems, and/or the complex geometries of certain faults: for instance, the tectonic movement accommodated by a high-angle normal fault (top left in Figure 1.29) is very different from the displacement generated on a low-angle detachment fault (bottom case in Figure 1.29). The sedimentary geometries that are potentially generated in these two opposing cases will contain very different characteristics. Additionally, it is now recognized that most rift systems evolve through distinct phases of successive tectonic activity that often propagate oceanwards. The result of these multiple phases is that a sedimentary unit that is interpreted as syn-rift in one basin may correspond to a pre-rift or a post-rift setting in an adjacent basin. The terms pre-, syn- and post-rifts are therefore misleading when used at margin-scale studies. Therefore, new definitions have been proposed to classify sediments in rift systems: **pre-tectonic, syn-tectonic and post-tectonic sediments** (see the discussion in Chapter 2 for further explanation). These distinctions allow sediments to be classified depending on whether they have been deposited before, during or after phases of tectonic activity. These new terms reflect the multiphase evolution characteristic of most extensional systems better than pre-rift, syn-rift and post-rift sediments.

The processes governing the production, transport and deposition of sediments in extensional areas are complex. Many parameters can influence the sediments' lifecycle and are, moreover, often inter-dependent. These influencers can be categorized into two categories: non-tectonic, such as climate conditions (arid, tropical), source rock composition, position relative to lake or sea level and eustatic fluctuations (Prosser 1993); and tectonic, such as the type of sedimentary basin, characteristics of the multiphase activity of the fault(s) bounding the basin and subsidence. All of these tectonic influences directly impact the source location, transport route and deposition location.

The fault type and fault growth pattern directly control the subsidence rate, depocenter geometry and position, and facies distribution in the basin (see the schematic comparison in Figure 1.30). In a half-graben with HANFs, the basin's asymmetry naturally imposes a strong influence on the syn-tectonic deposition patterns, and the footwall uplift, which can be very pronounced in some settings, strongly affects the drainage system (Leeder and Jackson 1993). The footwall provides coarse materials in the form of alluvial fans, whereas the material from the hanging wall is finer-grained on average (e.g. Gawthorpe and Leeder 2000) (Figure 1.31). In another case, basins sitting above low-angle detachment faults mostly collect fault products that are derived from the footwall breakaway, and the sedimentary geometries are very different from more standard half-graben-type basins (Figure 1.30). Therefore, special attention should be paid to the structural context in which the basin has been formed when studying the architecture and geometric relationship of a sedimentary basin. Additionally, fault propagation, growth, death and linkage are important key parameters controlling the final sedimentary architecture. For example, the presence of relay ramps between basins will cause deviations in the source and routing system.

The sediment composition also changes during rifting evolution because of changes in the source and routing system, as seen in changes from continental to marine environments. In general, most of the sedimentation histories begin with fluvial, lacustrine and subaerial deposits and evolve with time to more marine conditions ranging from near shore to distal offshore conditions.

The detailed study of sedimentary successions related to the geologic time scale is called **stratigraphy**. **Sequence stratigraphy** addresses how depositional architecture forms in response to (changes in) accommodation and sediment supply. Readers interested in these are referred to contributions specifically dedicated to these topics.

> Further reading.– The above descriptions are abbreviated and often simplified. If interested in reading and learning further, the reader is referred to the following list of publications and references:
> – General: (Leeder and Gawthorpe 1987; Prosser 1993; Gawthorpe and Leeder 2000; Wilson *et al.* 2001; Masini *et al.* 2013; Ribes *et al.* 2019).

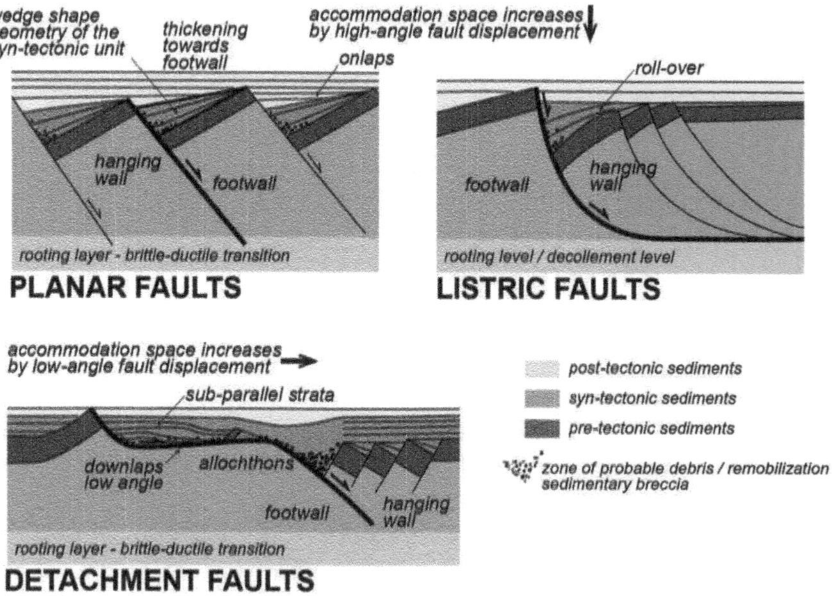

Figure 1.29. *Schematic representation of the definitions of sedimentary subdivision into pre-, syn- and post-tectonics, illustrated in different extensional contexts*

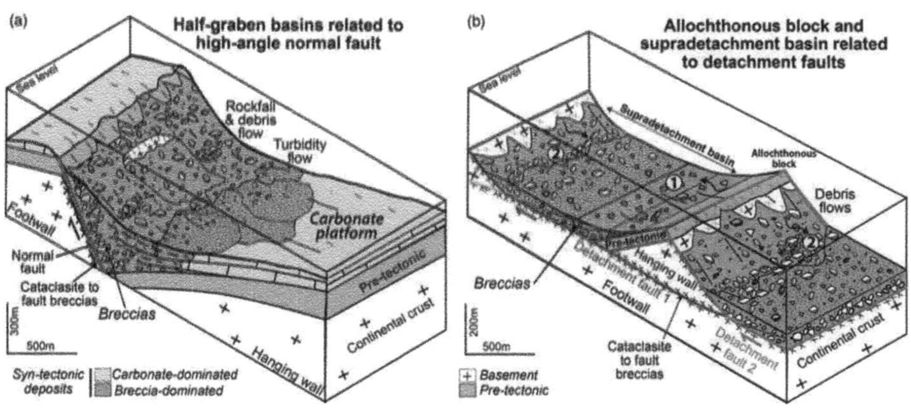

Figure 1.30. *Schematic diagrams summarizing the major structural and stratigraphic characteristics of (a) a high-angle normal fault bounding a half-graben-type basin inspired by the Il Motto half-graben in the Ortler Nappe (Northern Adriatic proximal margin) and (b) an extensional detachment fault system with extensional allochthonous blocks and a supra-detachment basin inspired by the Samedan basin in the Err nappe (Northern Adriatic margin) (source: from Ribes* et al. *2019, modified after Wilson* et al. *2001 and Masini* et al. *2011)*

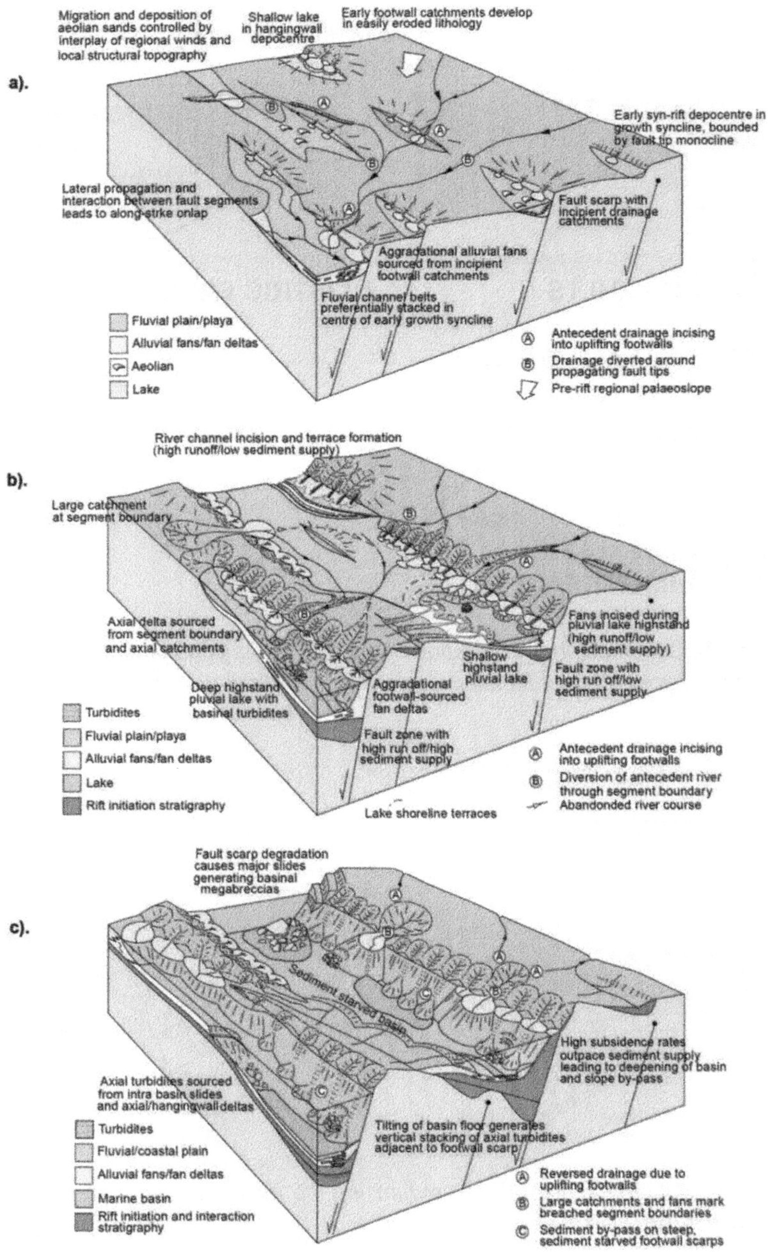

Figure 1.31. *Series of block diagrams illustrating the tectono-sedimentary evolution of a normal fault array (continental environments) (source: modified from Gawthorpe and Leeder 2000)*

CONTINUATION OF CAPTION FOR FIGURE 1.31.– *a) Initiation stage; b) interaction and linkage stage and c) through-going fault zone stage. The diagrams show how sediment transport pathways are influenced by fault geometry and activity and by antecedent drainage networks that are locally modified by surface topography. See Gawthorpe and Leeder (2000) for detailed explanations of the model that includes five more stages not shown here.*

1.5.2. *Salt*

Section written in collaboration with Mark Rowan.

In geology, the term "**salt**" refers to an evaporitic mineral, a rock (**halite**, also known as **rocksalt**) and an interbedded sedimentary sequence composed mostly or largely of evaporitic material. **Evaporites** designate rocks formed by the precipitation of minerals in arid, isolated basins where the loss of water by evaporation exceeds the inflow. There are more than 80 evaporite minerals (Warren 2016), with the most common being **halite** and **anhydrite/gypsum**. They are usually interbedded with carbonate and siliciclastic rocks in layered evaporite sequences (LES) that can be up to 5+ km thick.

Halite, anhydrite and other evaporite minerals are distinguished based on their composition and crystal structure: for example, halite is a sodium-chloride (NaCl) isometric crystal, anhydrite is a sulfate (CaSo4) with orthorhombic crystals, and gypsum is a hydrous variation of anhydrite. The minerals precipitate at different stages during water evaporation, in the reverse order of their solubilities (gypsum/anhydrite precipitates before halite, and bittern salts form in extremely evaporative conditions – see Warren (2016)). Halite is typically the most abundant constituent of LES.

In rift/rifted margin settings, similarly to sediments, salt is often classified as pre-rift, syn-rift or post-rift (e.g. Jackson and Vendeville 1994) according to the time of its deposition relative to the tectonic activity. Syn-rift salt can further be subdivided by adhering to the multistage evolution proposed for rifted margins (see Chapter 2) because the relative timing of rift evolution and evaporite deposition controls the spatial and thickness distribution of the salt (Rowan 2014). If salt is deposited early in margin development (syn-stretching or early syn-rift salt) or in rift basins that never evolve into rifted margins, it forms near sea level in elongate basins that are variably isolated or connected with significant thickness variations (Figure 1.32a). If salt is deposited later in the history (syn-thinning or mid syn-rift salt), the salt will be concentrated in the distal domain, with the base salt offset by large-displacement, low-angle faults. Finally, if salt is shortly deposited before the onset of spreading (syn-exhumation or late syn-rift salt), it will be widespread with little/no relief on the base salt except in the outer domain, where it is typically thickest in an outer trough bounded by oceanic crust (Rowan 2014) (see Figure 1.32b).

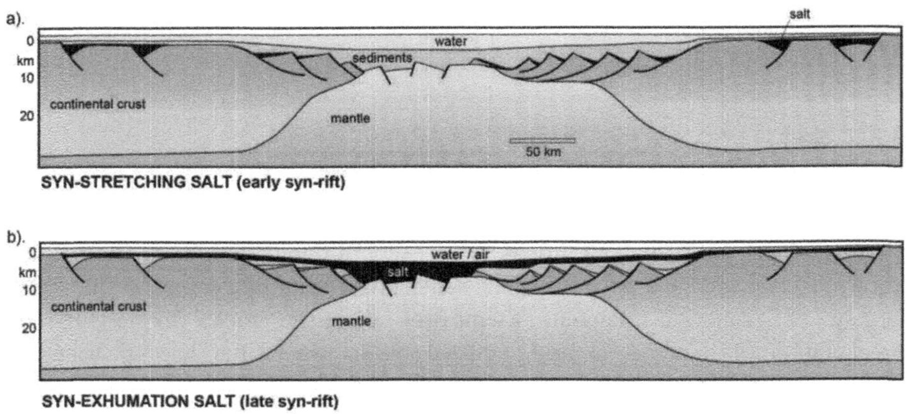

Figure 1.32. *Evaporite deposition at different stages of rifting, shown just prior to the onset of sea-floor spreading: a) syn-stretching (early syn-rift) salt; b) syn-exhumation (late syn-rift) salt. Rift template modified from Ribes* et al. *(2019); vertical exaggeration 3:1*

Salt has physical characteristics and mechanical properties (Jackson and Hudec 2017) that make it special in rift evolution and thus a critical consideration when interpreting subsurface geometries. First, it has a constant low density (2.160 g/cm^3 for pure halite) and is thus incompressible. Other sediments become denser with burial and compaction/cementation, so there will be a density inversion for the deeply buried salt. Much more important, however, is that halite has a low viscosity that allows it to easily flow, compared to surrounding rocks. Where the encasing rocks behave as relatively strong, brittle materials, salt is best viewed as a pressurized fluid that flows in a viscous manner (Vendeville and Jackson 1992). The drivers for salt movement are extension, contraction and differential loading, and the consequent way in which salt deforms is known as **salt tectonics**, **halokinesis** or **halotectonics**.

In a rift/rifted margin context, the style of halokinesis largely depends on whether the salt is pre-rift, early syn-rift or late syn-rift (Rowan 2014). In the first two cases, the deformation is thick-skinned in that the basement, salt and its cover are all extended. The salt tends to decouple the supra- and subsalt deformation to some degree, with a greater degree of coupling for thin salt and/or large-displacement faults (Withjack and Callaway 2000). In cases of extreme extension and mantle exhumation, salt can completely decouple low-angle detachment faults in the presalt from thin-skinned detachment faults in the suprasalt (Jammes *et al.* 2010). In less extended settings, typical structural features (Figure 1.33a) include listric or conjugate faults that are detached on the salt, drape folds and faulted drape folds over basement faults, and diapirs located over or slightly in the footwall of subsalt faults. A salt diapir is a body of salt with a truncating relationship with its overburden; in rifted settings, they are usually triggered by extension, going through reactive, active and passive stages (Vendeville and Jackson 1992). Passive diapirs continue to grow at or just beneath the surface as sediment is deposited around them.

In the case of late syn-rift salt, salt tectonics is dominated by post-rift gravitational failure of the salt and its overburden (see Rowan 2020). The primary drivers are the basinward tilt of the margin, caused by the thermal subsidence of oceanic crust, and proximal sediment loading. The deformation is thin-skinned, comprising of linked systems of proximal extension, translation and distal contraction (Figure 1.33b). Extension is accommodated by both basinward- and landward-dipping normal faults, contraction is manifested by both salt-cored folds and thrust faults as well as the squeezing of diapirs and the translational province has both symmetric and asymmetric salt-evacuation structures. Relief on the base salt can generate ramp-syncline basins. Diapirs are triggered by various processes, and the salt may break out and laterally flow to form salt sheets and canopies.

Salt has commonly been associated with geophysical imaging problems. Firstly, salt bodies may be non-reflective or have complex internal structures depending on the original layering and mode of deformation. Secondly, salt structures such as diapirs, typical of many rifted margins worldwide, usually have steep sub-vertical geometries that are often poorly imaged and that often also generate geophysical noise, making their exact shape and edge-diapir location difficult to define. Thirdly, the contrast between shallow high-velocity salt and low-velocity sediment causes migration problems and artifacts. Finally, there are many cases where little orno energy penetrates salt sheets/canopies into subsalt domains and then back to the surface receivers. However, new acquisition and processing techniques now allow much better imaging (Figure 1.34).

Further reading.– The above descriptions are abbreviated and often simplified. If interested in reading and learning further, the reader is referred to the following list of publications and references:

– General: (Vendeville and Jackson 1992; Jackson and Vendeville 1994; Withjack and Callaway 2000; Jammes *et al.* 2010; Rowan 2014, 2020; Warren 2016; Jackson *et al.* 2020).

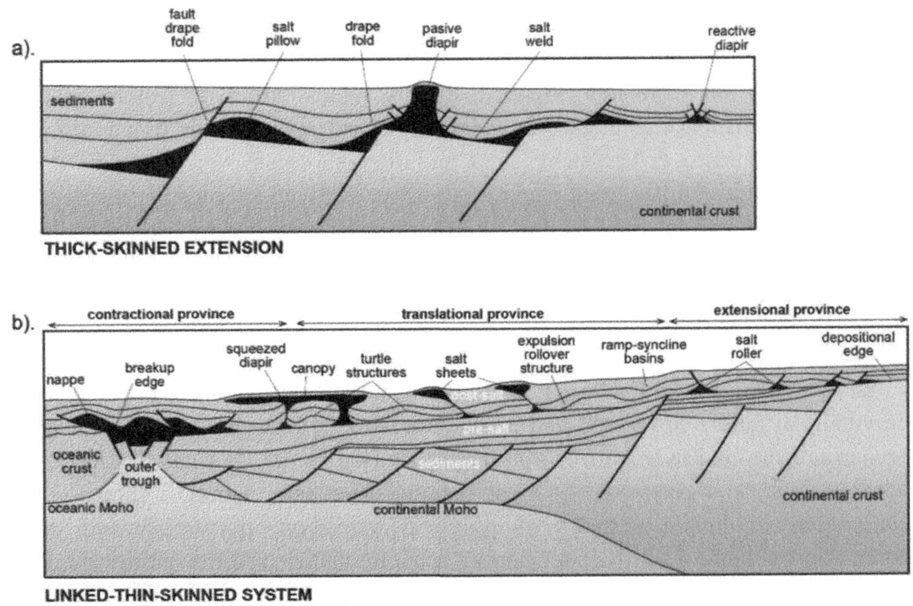

Figure 1.33. *Styles and features of rift-related salt tectonics: a) thick-skinned extension; b) gravity-driven, linked thin-skinned extension, translation and contraction. Not to scale, and a) and b) are at different scales*

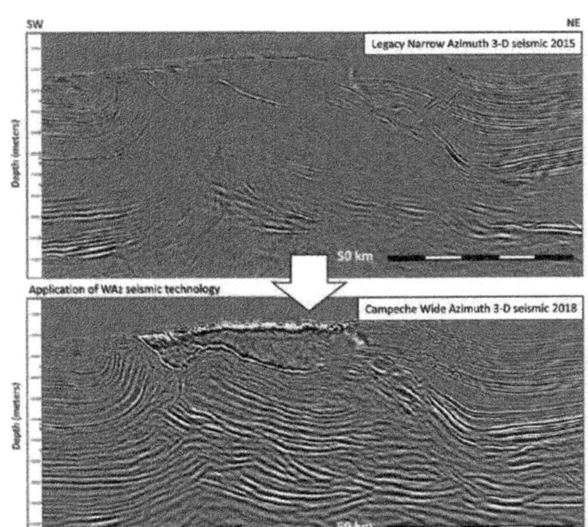

Figure 1.34. *Improvement of seismic imaging, especially in the subsalt domain (example from the southern Gulf of Mexico, source: Shann* et al. *(2020))*

1.5.3. *Magma*

Magmatism contributes to rift evolution and final margin architecture to various degrees, and depends on various parameters, such as extension rate and different characteristics of the lithosphere and asthenosphere (White and McKenzie 1989; White 1992). A series of typical volcanic geometries often observed, interpreted and/or modeled in rifted margins are summarized below and exemplified in Figure 1.35. These are traditionally distinguished into three main types: intrusives, extrusives and underplates.

Intrusive magmatic features refer to magmatic rocks emplaced in rocks that were previously in place. Depending on the emplacement geometry relative to the previous rocks, these intrusions can be concordant or discordant. **Sills** are tabular intrusions, emplaced between older layers, in a concordant intrusive sheet, while **dikes** are defined as discordant intrusive sheets that will crosscut older rocks. Sills are commonly fed by dikes. Additional intrusive magmatic bodies include the following: **pluton**, which refers to large intrusive bodies; **batholiths**, which are intrusive complexes of several magmatic bodies of large dimension (typically several plutons) and **laccoliths**, which are concordant plutons and form when the magmatic rocks are intruded between rock layers with high pressure, forcing an upward doming/folding of the overlying strata and giving rise to an overall mushroom-like form with a generally planar base.

Extrusive magmatic features are magmatic rocks emplaced above the Earth's surface, at the seafloor or at the surface. In rifted margins, these include a long series of possible features such as **lava flows**, **lava deltas** and **volcanoes**. Secondary geometries can also be generated, such as mass wasting events and volcanic-derived sediments. Their seismic reflection facies are often very well defined with typically high-amplitude reflectors; however, the conditions leading to their emplacement are still very much debated.

Underplates refer to rocks that are emplaced as uniform bodies at the base of pre-existing rocks – usually below the lower crust in extended rifted margin settings. Field outcrops and deep-sea drilling observation prove the existence of underplated magmatic bodies. In modern rifted margins, however, these structures cannot be directly observed. Their presence has been proposed based on the interpretation of geophysical models, making them indirect observations. **HVLC** and **LCB** are widely used acronyms that refer to these particular bodies. HVLC is an abbreviation for "high velocity lower crust" and "LCB" for "lower crustal body" (see Chapter 2 for further explanation and discussion around these terms).

Given the wide range of geometries and contexts of emplacement, the study of magmatic features in rifts and rifted margins is a specialized research topic: **seismic volcanostratigraphy**, a concept inherited from Mitchum *et al.* (1977) and Symonds *et al.* (1998). Seismic volcanostratigraphy uses seismostratigraphy and sequence stratigraphy concepts to identify and map magmatic sequences, based on their seismic facies. The study of seismic reflection geometries associated with volcanic successions is used to constrain the emplacement context, depositional environment and potential subsequent remobilization and transport. For many decades, these geometries remained unclear and unconstrained due to seismic imaging issues. However, the recent advances in seismic processing have greatly improved imaging and resolution. Some deep-sea drilling campaigns that are specifically dedicated to the sampling of the volcanic successions of the NE Atlantic rift have been able to place crucial constraints on their lithologies and ages (DSDP Leg 38, 104 and 152, 163; Eldholm *et al.* 1987; Saunders *et al.* 1998; Larsen *et al.* 1999). These modern advances allow for highly detailed studies of the temporal and spatial evolution of the magmatic activity of the rifted margin, notably in terms of paleogeography and morphology at breakup time.

We briefly list below the major magmatic geometries often encountered in distal rifted margin settings (Figure 1.35). We will follow the classification built on the West Indian margin case (Calvès *et al.* 2011) and descriptions of the Vøring mid-Norwegian margin (Abdelmalak *et al.* 2016). Both are based on the early descriptions made by Symonds *et al.* (1998) and Planke *et al.* (2000):

– **SDRs (seaward dipping reflectors)**: these are a specific feature identified on seismic reflection profiles by a wedge-shape/fan-like geometry, opening and dipping oceanward with medium- to high-amplitude seismic reflectors (Mutter *et al.* 1982). The top continentward termination has been drilled (e.g. in the NE Atlantic; Eldholm *et al.* 1989) and the related strong reflectors have been proven to correspond to subaerially-erupted flood basalts. Thus, when identified in distal rifted margin settings, SDR-like seismic geometries are often interpreted as fully basaltic structures. However, no constraints exist on their internal lithologies, notably the ratio between sediments and magma. Based on their position, the packages are often distinguished into inner SDRs and outer SDRs, depending on whether they are located landward or oceanward from the outer high, respectively:

- The **outer high** is a mound-like feature located between the inner and outer SDR packages. It is usually characterized by high-amplitude top reflectors and a rather chaotic internal reflection pattern.

– **Lava flows**: these are traditionally categorized into subaerial and submarine (hyaloclastite). They can also be distinguished as inner and landward flows: the inner flows are structurally located below the landward flows and lava delta and the landward flows are located landward from the inner SDRs. These are the flows that supply lava to the lava delta (Figure 1.35): the landward flows are subaerial, while the lava delta is subaqueous, marking the limit from where the lava enters water.

– The **escarpment** is a structural cliff observed along some distal margins. It is related to depositional/erosional processes, marking the transition from the subaerial to the submarine environment. The escarpment is interpreted to record the shoreline at the time of delta progradation (Wright *et al.* 2012). It is thus an important geometry that can give crucial information on the paleogeography of the margin and its topographic evolution. Structurally, when the lava enters water, it undergoes fragmentation into hyaloclastic beccias that can be transported downslope by gravitational processes to form the overall progradational foreset pattern with the escarpment.

– The **lava delta** corresponds to the structure built by the progradation of the lava flows outboard. The delta grows by the addition and stacking of new flows and hyaloclastite debris/breccias, resulting in the progradation of the shoreline oceanwards. The thickness of the delta may be used as a parameter to estimate the paleo-water depth, as it gives insights into the accommodation space available at the time of deposition.

Additional features can be defined and mapped, such as hyaloclastic mounds, various types of sills (e.g. saucer-shape), plugs, plutons, vents, dikes and volcanoes, volcanic-derived sediments, slumps and mass wasting. For more information, the reader is referred to the recommended publications below (on p. 58).

The source of the magma is very poorly understood, both locally and regionally. Locally, it is often proposed that dikes (called "feeder dikes") bring the magma to the sills and lava flows to build the various magmatic features. However, these dikes can be near-vertical structures and thus are extremely difficult to image and identify on seismic reflection data. Additionally, it has been proven that magma can be transported over very significant distances (hundreds to thousands of km) within widespread sill complexes, promoting the development of magmatic features that do not overlie the melt source (Magee *et al.* 2016). Therefore, the source location can be a great distance from the observed structures, adding huge uncertainties on the identification and characterization of the physical sources.

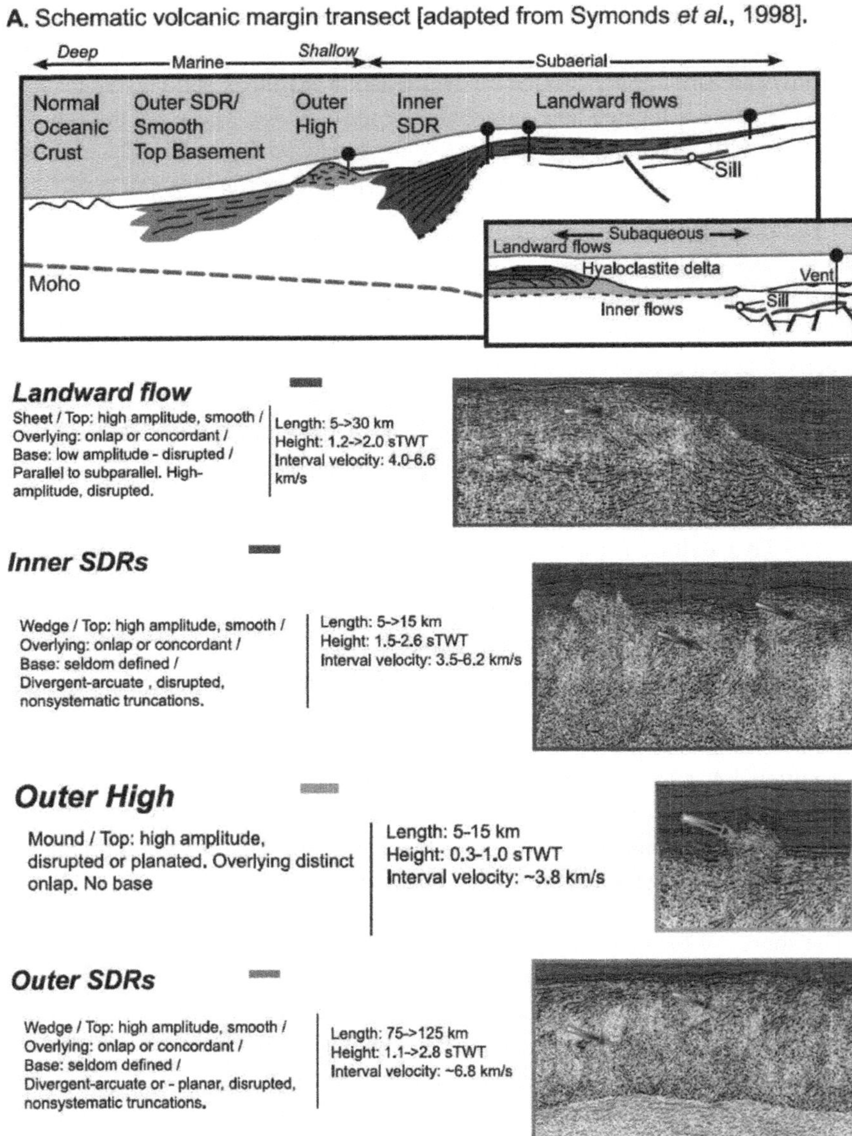

Figure 1.35a. *Summary of the main geometries related to magmatic activity encountered in rifts and rifted margin studies. The figure is from Calvès* et al. *(2011), based on the study of the West Indian margin, and adapted from Symonds* et al. *(1998). Schematic volcanic margin transect at the top panel illustrating seismic facies units associated with extrusive volcanic deposits*

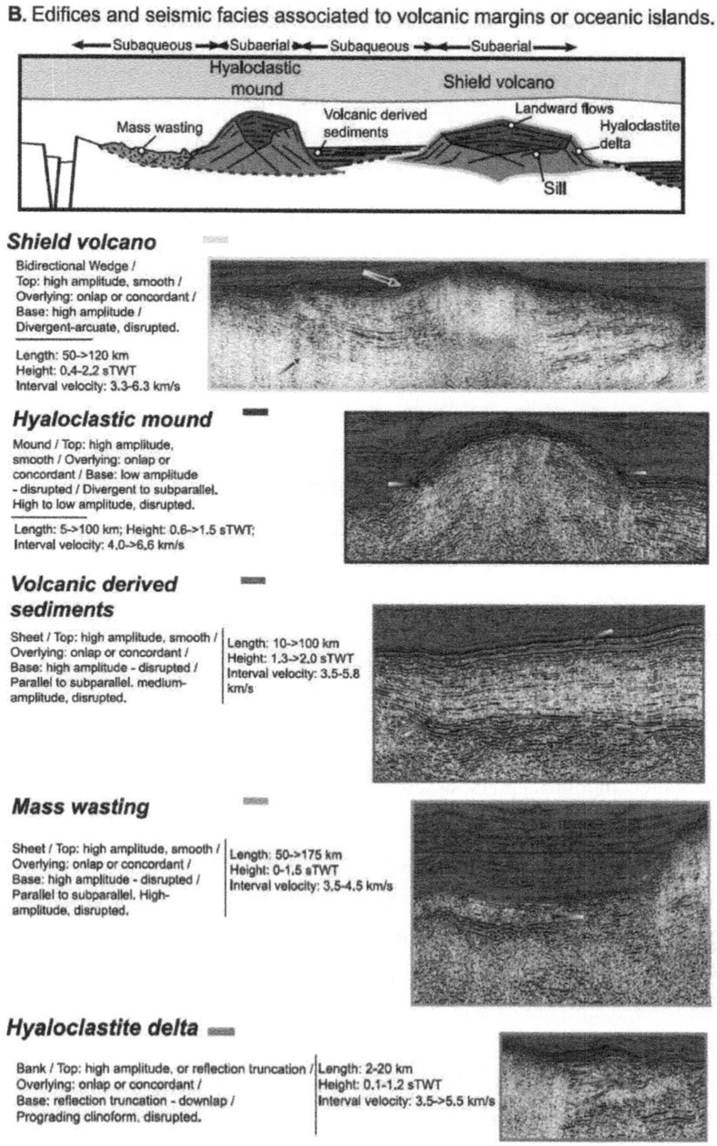

Figure 1.35b. *Seismic facies chart, morphometric data and interpretation of the volcanostratigraphic elements identified on the West Indian margin (source: from Calvès* et al. *2011)*

Further reading.– The above descriptions are abbreviated and often simplified. If interested in reading and learning further, the reader is referred to the following list of publications and references:
– General: (Mitchum *et al.* 1977; White *et al.* 1987; White and McKenzie 1989; Eldholm 1991; Saunders *et al.* 1997; Symonds *et al.* 1998; Planke *et al.* 2000; Calvès *et al.* 2011; Magee *et al.* 2016).

1.6. References

Abdelmalak, M.M., Planke, S., Faleide, J.I., Jerram, D.A., Zastrozhnov, D., Eide, S., Myklebust, R. (2016). The development of volcanic sequences at rifted margins: New insights from the structure and morphology of the Vøring Escarpment, mid-Norwegian Margin. *Journal of Geophysical Research: Solid Earth*, 121(7), 5212–5236.

Allen, J. and Allen, P. (2013). *Basin Analysis: Principles and Application to Petroleum Play Assessment*, 3rd edition. Wiley-Blackwell, Oxford.

Amante, C. and Eakins, B.W. (2009). ETOPO1 1 arc-minute global relief model: Procedures, data sources and analysis. Report, NGDC.

Anderson, E.M. (1905). The dynamics of faulting. *Transactions of the Edinburgh Geological Society*, 8(3), 387–402.

Armitage, J.J. and Allen, P.A. (2010). Cratonic basins and the long-term subsidence history of continental interiors. *Journal of the Geological Society*, 167(1), 61–70.

Artemieva, I.M. and Mooney, W.D. (2002). On the relations between cratonic lithosphere thickness, plate motions, and basal drag. *Tectonophysics*, 358(1), 211–231.

Atwater, T. (1970). Implications of plate tectonics for the Cenozoic tectonic evolution of western North America. *GSA Bulletin*, 81(12), 3513–3536.

Axen, G.J. (1992). Pore pressure, stress increase, and fault weakening in low-angle normal faulting. *Journal of Geophysical Research: Solid Earth*, 97(B6), 8979–8991.

Axen, G.J. (2004). Mechanics of low-angle normal faults. In *Rheology and Deformation in the Lithosphere at Continental Margins*, Karner, G.D., Taylor, B., Driscoll, N.W., Kohlstedt, D.L. (eds). Columbia University Press, New York.

Axen, G.J., Taylor, W.J., Bartley, J.M. (1993). Space-time patterns and tectonic controls of Tertiary extension and magmatism in the Great Basin of the western United States. *GSA Bulletin*, 105(1), 56–76.

Bastow, I.D., Nyblade, A.A., Stuart, G.W., Rooney, T.O., Benoit, M.H. (2008). Upper mantle seismic structure beneath the Ethiopian hot spot: Rifting at the edge of the African low-velocity anomaly. *Geochemistry, Geophysics, Geosystems*, 9(12).

Bird, P. (2003). An updated digital model of plate boundaries. *Geochemistry, Geophysics, Geosystems*, 4(3).

Bonatti, E. (1985). Punctiform initiation of seafloor spreading in the Red Sea during transition from a continental to an oceanic rift. *Nature*, 316(6023), 33–37.

Bosworth, W., Lučić, D., Stockli, D.F. (2021). North African Phanerozoic. In *Encyclopedia of Geology*, 2nd edition, Alderton, D. and Elias, S.A. (eds). Academic Press, Oxford.

Bott, M.H.P. (1979). Subsidence mechanisms at passive continental margins. *American Association of Petroleum Geologists Memoir*, 29, 8–19.

Braun, J. (2010). The many surface expressions of mantle dynamics. *Nature Geoscience*, 3(12), 825–833.

Braun, J. and Shaw, R. (2001). A thin-plate model of Palaeozoic deformation of the Australian lithosphere: Implications for understanding the dynamics of intracratonic deformation. *Geological Society, London, Special Publications*, 184(1), 165–193.

Brun, J.-P. and Choukroune, P. (1983). Normal faulting, block tilting, and décollement in a stretched crust. *Tectonics*, 2(4), 345–356.

Brun, J.P., Gutscher, M.A., Blum, R., Bois, C., Burg, J.P., Colletta, B., Dürbaum, H., Damotte, B., Durst, H., Fuchs, K., Grohmann, N., Hübner, M., Karcher, T., Kessler, G., Klöckner, M., Lucazeau, F., Lüschen, E., Marthelot, J.-M., Meier, L., Ravat, M., Reichert, C., Vernassa, S., Villemin, T., Wenzel, F., Wittlinger, G. (1992). Deep crustal structure of the Rhine Graben from DEKORP-ECORS seismic reflection data: A summary. *Tectonophysics*, 208(1), 139–147.

Brun, J.-P., Sokoutis, D., Tirel, C., Gueydan, F., Van Den Driessche, J., Beslier, M.-O. (2018). Crustal versus mantle core complexes. *Tectonophysics*, 746, 22–45.

Buck, W.R. (1988). Flexural rotation of normal faults. *Tectonics*, 7, 959–973.

Buck, W.R. (1991). Modes of continental lithospheric extension. *Journal of Geophysical Research*, 96(B12), 20161–120178.

Buck, W.R. (2004). Consequences of asthenospheric variability on continental rifting. In *Rheology and Deformation of the Lithosphere at Continental Margins*, Karner, G.D., Taylor, B., Driscoll, N.W., Kohlstedt, D.L. (eds). Columbia University Press, New York.

Bullard, E.C. (1936). Gravity measurements in East Africa. *Philosophical Transactions of the Royal Society A*, 235, 445–531.

Burke, K. (1977). Aulacogens and continental breakup. *Annual Review of Earth and Planetary Sciences*, 5, 371–396.

Burke, K. (1996). The African plate. *South African Journal of Geology*, 99, 341–409.

Burke, K. and Dewey, J.F. (1973). Plume-generated triple junctions: Key indicators in applying plate tectonics to old rocks. *The Journal of Geology*, 81(4), 406–433.

Burov, E. and Cloetingh, S. (1997). Erosion and rift dynamics: New thermomechanical aspects of post-rift evolution of extensional basins. *Earth and Planetary Science Letters*, 150(1), 7–26.

Cacace, M. and Scheck-Wenderoth, M. (2016). Why intracontinental basins subside longer: 3-D feedback effects of lithospheric cooling and sedimentation on the flexural strength of the lithosphere. *Journal of Geophysical Research: Solid Earth*, 121(5), 3742–3761.

Calvès, G., Schwab, A.M., Huuse, M., Clift, P.D., Gaina, C., Jolley, D., Tabrez, A.R., Inam, A. (2011). Seismic volcanostratigraphy of the western Indian rifted margin: The pre-Deccan igneous province. *Journal of Geophysical Research*, 116(B01101).

Cartwright, J.A., Trudgill, B.D., Mansfield, C.S. (1995). Fault growth by segment linkage: An explanation for scatter in maximum displacement and trace length data from the Canyonlands Grabens of SE Utah. *Journal of Structural Geology*, 17(9), 1319–1326.

Childs, C., Holdsworth, R.E., Jackson, C.A.-L., Manzocchi, T., Walsh, J.J., Yielding, G. (2017). Introduction to the geometry and growth of normal faults. *Geological Society, London, Special Publications*, 439(1), 1–9.

Chorowicz, J. (2005). The East African rift system. *Journal of African Earth Sciences*, 43(1), 379–410.

Clerc, C., Ringenbach, J.-C., Jolivet, L., Ballard, J.-F. (2018). Rifted margins: Ductile deformation, boudinage, continentward-dipping normal faults and the role of the weak lower crust. *Gondwana Research*, 53, 20–40.

Corti, G. (2009). Continental rift evolution: From rift initiation to incipient break-up in the Main Ethiopian Rift, East Africa. *Earth-Science Reviews*, 96(1), 1–53.

Corti, G., Bonini, M., Conticelli, S., Innocenti, F., Manetti, P., Sokoutis, D. (2003). Analogue modelling of continental extension: A review focused on the relations between the patterns of deformation and the presence of magma. *Earth-Science Reviews*, 63(3–4), 169–247.

Cowie, P.A., Underhill, J.R., Behn, M.D., Lin, J., Gill, C.E. (2005). Spatio-temporal evolution of strain accumulation derived from multi-scale observations of Late Jurassic rifting in the northern North Sea: A critical test of models for lithospheric extension. *Earth and Planetary Science Letters*, 234(3–4), 401–419.

Crosby, A.G., Fishwick, S., White, N. (2010). Structure and evolution of the intracratonic Congo Basin. *Geochemistry, Geophysics, Geosystems*, 11(6).

Daly, M.C., Lawrence, S.R., Diemu, K., Matouana, B. (1992). Tectonic evolution of the Cuvette Centrale, Zaire. *Journal of the Geological Society*, 149(4), 539–546.

Davis, M. and Kusznir, N.J. (2004). Depth-dependent lithospheric stretching at rifted continental margins. In *Rheology and Deformation of the Lithosphere at Continental Margins*, Karner, G.D., Taylor, B., Driscoll, N.W., Kohlsted, D.L. (eds). Columbia University Press, New York.

Davis, G.A. and Lister, G.A. (1988). Detachment faulting in continental extension; perspective from the southwestern U.S. Cordillera. In *Processes in Continental Lithospheric Deformation*, Clark, S.P., Burchfiel, B.C., Suppe, J. (eds). Geological Society of America, Boulder.

Dewey, J.F. (1988). Extensional collapse of orogens. *Tectonics*, 7, 1123–1139.

Dèzes, P., Schmid, S.M., Ziegler, P.A. (2004). Evolution of the European Cenozoic Rift System: Interaction of the Alpine and Pyrenean orogens with their foreland lithosphere. *Tectonophysics*, 389(1), 1–33.

Dickinson, W.R. (2002). The Basin and Range Province as a composite extensional domain. *International Geology Review*, 44(1), 1–38.

Ebinger, C.J. (2005). Continental break-up: The East African perspective. *Astronomy & Geophysics*, 46(2), 16–21.

Ebinger, C.J. and Sleep, N.H. (1998). Cenozoic magmatism throughout East Africa resulting from impact of a single plume. *Nature*, 395(6704), 788–791.

Eldholm, O. (1991). Magmatic tectonic evolution of a volcanic rifted margin. *Marine Geology*, 102(1–4), 43–61.

Eldholm, O., Thiede, J., Taylor, E. (1987). Norwegian Sea, sites 642–644. *Proceedings of the Ocean Drilling Program, Part A: Initial Reports*, Ocean Drilling Program, College Station.

Eldholm, O., Thiede, J., Taylor, B. (1989). Evolution of the Vøring volcanic margin. *Proceedings of the Ocean Drilling Program, Scientific Results*, Ocean Drilling Program, College Station.

Elliott, D., Ramsay, J.G., Wood, D.S. (1976). A discussion on natural strain and geological structure – The energy balance and deformation mechanisms of thrust sheets. *Philosophical Transactions of the Royal Society of London. Series A, Mathematical and Physical Sciences*, 283(1312), 289–312.

Fossen, H. (2010). *Structural Geology*. Cambridge University Press, Cambridge.

Foster, A. and Nimmo, F. (1996). Comparisons between the rift systems of East Africa, Earth and Beta Regio, Venus. *Earth and Planetary Science Letters*, 143(1), 183–195.

Gac, S., Huismans, R.S., Podladchikov, Y.Y., Faleide, J.I. (2012). On the origin of the ultradeep East Barents Sea basin. *Journal of Geophysical Research: Solid Earth*, 117(B4), B04401.

Gawthorpe, R.L. and Leeder, M.R. (2000). Tectono-sedimentary evolution of active extensional basins. *Basin Research*, 12(3–4), 195–218.

Gilbert, G.K. (1928). Studies of basin-range structure. *U.S. Geological Survey Professional Paper*, 153, 1–92.

Gilbert, H. (2012). Crustal structure and signatures of recent tectonism as influenced by ancient terranes in the western United States. *Geosphere*, 8(1), 141–157.

Gillard, M., Autin, J., Manatschal, G. (2016). Fault systems at hyper-extended rifted margins and embryonic oceanic crust: Structural style, evolution and relation to magma. *Marine and Petroleum Geology*, 76, 51–67.

Giresse, P. (2005). Mesozoic–Cenozoic history of the Congo Basin. *Journal of African Earth Sciences*, 43(1), 301–315.

Gregory, J.W. (1896). *The Great Rift Valley: Being the Narrative of a Journey to Mount Kenya and Lake Baringo: With Some Account of the Geology, Natural History, Anthropology and Future Prospects of British East Africa*. John Murray, London.

Gregory, J.W. (1923). The structure of the Great Rift Valley. *Nature*, 112, 514–516.

Hansen, S.E., Nyblade, A.A., Benoit, M.H. (2012). Mantle structure beneath Africa and Arabia from adaptively parameterized P-wave tomography: Implications for the origin of Cenozoic Afro-Arabian tectonism. *Earth and Planetary Science Letters*, 319–320, 23–34.

Hartley, R. and Allen, P. (1994). Interior cratonic basins of Africa: Relation to continental break-up and role of mantle convection. *Basin Research*, 6, 95–113.

Heine, C., Dietmar Müller, R., Steinberger, B., Torsvik, T.H. (2008). Subsidence in intracontinental basins due to dynamic topography. *Physics of the Earth and Planetary Interiors*, 171(1), 252–264.

Hellinger, S.J. and Sclater, J.G. (1983). Some comments on two-layer extensional models for the evolution of sedimentary basins. *Journal of Geophysical Research: Solid Earth*, 88(B10), 8251–8269.

Holbrook, W.S. (1990). The crustal structure of the Northwestern Basin and Range Province, Nevada, from wide-angle seismic data. *Journal of Geophysical Research: Solid Earth*, 95(B13), 21843–21869.

Huismans, R.S., Podladchikov, Y.Y., Cloetingh, S. (2001). Transition from passive to active rifting: Relative importance of asthenospheric doming and passive extension of the lithosphere. *Journal of Geophysical Research: Solid Earth*, 106(B6), 11271–11291.

Jackson, M.P. and Hudec, M.R. (2017). *Salt Tectonics: Principles and Practice*. Cambridge University Press, Cambridge.

Jackson, C.A.-L. and Rotevatn, A. (2013). 3D seismic analysis of the structure and evolution of a salt-influenced normal fault zone: A test of competing fault growth models. *Journal of Structural Geology*, 54, 215–234.

Jackson, M.P.A. and Vendeville, B.C. (1994). Regional extension as a geologic trigger for diapirism. *GSA Bulletin*, 106(1), 57–73.

Jackson, C.A.-L., Bell, R.E., Rotevatn, A., Tvedt, A.B.M. (2017). Techniques to determine the kinematics of synsedimentary normal faults and implications for fault growth models. *Geological Society, London, Special Publications*, 439(1), 187–217.

Jammes, S., Lavier, L., Manatschal, G. (2010). Extreme crustal thinning in the Bay of Biscay and the Western Pyrenees: From observations to modeling. *Geochemistry, Geophysics, Geosystems*, 11.

Kadima, E., Delvaux, D., Sebagenzi, S.N., Tack, L., Kabeya, S.M. (2011). Structure and geological history of the Congo Basin: An integrated interpretation of gravity, magnetic and reflection seismic data. *Basin Research*, 23(5), 499–527.

Keen, C.E. (1985). The dynamics of rifting: Deformation of the lithosphere by active and passive driving forces. *Geophysical Journal of the Astronomical Society*, 80, 95–120.

Keranen, K., Klemperer, S.L., Gloaguen, R., Group, E.W. (2004). Three-dimensional seismic imaging of a protoridge axis in the Main Ethiopian Rift. *Geology*, 32(11), 949–952.

King, S.D. and Ritsema, J. (2000). African hot spot volcanism: Small-scale convection in the upper mantle beneath cratons. *Science*, 290(5494), 1137–1140.

Kingston, D.R., Dishroon, C.P., Williams, P.A. (1983). Global Basin classification system. *American Association of Petroleum Geologists Bulletin*, 67, 2157–2193.

Klein, G.D. and Hsui, A.T. (1987). Origin of intracratonic basins. *Geology*, 15, 1094–10998.

Kusznir, N.J. (1987). The strength of the continental lithosphere. *Geophysical Journal of the Royal Astronomical Society*, 89(1), 451–451.

Kusznir, N.J. and Ziegler, P.A. (1992). The mechanics of continental extension and sedimentary basin formation – A simple-shear pure-shear flexural cantilever model. *Tectonophysics*, 215(1–2), 117–131.

Lăpădat, A., Imber, J., Yielding, G., Iacopini, D., Mccaffrey, K.J.W., Long, J.J., Jones, R.R. (2017). Occurrence and development of folding related to normal faulting within a mechanically heterogeneous sedimentary sequence: A case study from Inner Moray Firth, UK. *Geological Society, London, Special Publications*, 439(1), 373–394.

de Lapparent, A. (1898). Soulevements et affaissements. *Rev. Questions Sci.*, 14,5–33.

Larsen, H.C., Duncan, R.A., Allan, J.F., Brooks, K. (1999). Southeast Greenland Margin, sites 988–990. *Proceedings of the Ocean Drilling Program, Scientific Results*, Ocean Drilling Program, College Station.

Lavier, L.L., Buck, W.R., Poliakov, A.N.B. (1999). Selfconsistent rolling-hinge model for the evolution of large-offset, low-angle normal faults. *Geology*, 27, 1127–1130.

Leeder, M.R. and Gawthorpe, R.L. (1987). Sedimentary models for extensional tilt-block/half-graben basins. *Geological Society, London, Special Publications*, 28(1), 139–152.

Leeder, M.R. and Jackson, J.A. (1993). The interaction between normal faulting and drainage in active extensional basins, with examples from the western United States and central Greece. *Basin Research*, 5(2), 79–102.

Lister, G.S. and Davis, G.A. (1989). The origin of metamorphic core complexes and detachment faults formed during Tertiary continental extension in the northern Colorado River region, U.S.A. *Geology*, 11(2), 65–94.

Lister, G.S., Etheridge, M.A., Symonds, P.A. (1986). Detachment faulting and the evolution of passive continental margins. *Geology*, 14, 246–250.

Lister, G.S., Etheridge, M.A., Symonds, P.A. (1991). Detachment models for the formation of passive continental margins. *Tectonics*, 10(5), 1038–1064.

Longwell, C.R. (1945). Low-angle normal faults in the Basin and Range Province. *EOS Transactions American Geophysical Union*, 26, 107–118.

Magee, C., Muirhead, J.D., Karvelas, A., Holford, S.P., Jackson, C.A.-L., Bastow, I.D., Schofield, N., Stevenson, C.T.E., McLean, C., McCarthy, W., Shtukert, O. (2016). Lateral magma flow in mafic sill complexes. *Geosphere*, 12(3), 809–841.

Masini, E., Manatschal, G., Mohn, G., Ghienne, J.-F., Lafont, F. (2011). The tectono-sedimentary evolution of a supra-detachment rift basin at a deep-water magma-poor rifted margin: The example of the Samedan Basin preserved in the Err nappe in SE Switzerland. *Basin Research*, 23(6), 652–677.

Masini, E., Manatschal, G., Mohn, G. (2013). The Alpine Tethys rifted margins: Reconciling old and new ideas to understand the stratigraphic architecture of magma-poor rifted margins. *Sedimentology*, 60(1), 174–196.

Matthews, K.J., Müller, R.D., Wessel, P., Whittaker, J.M. (2011). The tectonic fabric of the ocean basins. *Journal of Geophysical Research: Solid Earth*, 116(B12), B12109.

McKenzie, D.P. (1978). Some remarks on the development of sedimentary basins. *Earth and Planetary Science Letters*, 40, 20–32.

Merle, O. (2011). A simple continental rift classification. *Tectonophysics*, 513(1), 88–95.

Middleton, M.F. (1989). A model for the formation of intracratonic sag basins. *Geophysical Journal International*, 99(3), 665–676.

Miller, E.L., Dumitru, T.A., Brown, R.W., Gans, P.B. (1999). Rapid Miocene slip on the Snake Range–Deep Creek Range fault system, east-central Nevada. *GSA Bulletin*, 111(6), 886–905.

Mitchum, R.M., Vail, P.R., Sangree, J.B. (1977). Stratigraphic interpretation of seismic reflection patterns in depositional sequences. In *Seismic Stratigraphy: Application to Hydrocarbon Exploration*, Payton, C.E. (ed.). American Association of Petroleum Geologists, Tulsa.

Molnar, P. and Atwater, T. (1973). Relative motion of hot spots in the mantle. *Nature*, 246, 288–191.

Montelli, R., Nolet, G., Dahlen, F.A., Masters, G. (2006). A catalogue of deep mantle plumes: New results from finite-frequency tomography. *Geochemistry, Geophysics, Geosystems*, 7(11).

Müller, R.D., Cannon, J., Qin, X., Watson, R.J., Gurnis, M., Williams, S., Pfaffelmoser, T., Seton, M., Russell, S.H.J., Zahirovic, S. (2018). GPlates: Building a virtual earth through deep time. *Geochemistry, Geophysics, Geosystems*, 19(7), 2243–2261.

Mutter, J.C., Talwani, M., Stoffa, P.L. (1982). Origin of seaward-dipping reflectors in oceanic crust off the Norwegian margin by "subaerial sea-floor spreading". *Geology*, 10, 3–12.

Neumann, E.R. and Ramberg, I.B. (1978). *Petrology and Geochemistry of Continental Rifts*. D. Reidel Publishing Company, Dordrecht and Boston.

Olade, M.A. (1975). Evolution of Nigeria's Benue Trough (Aulacogen): A tectonic model. *Geological Magazine*, 112(6), 575–583.

Osmundsen, P.T. and Andersen, T.B. (2001). The middle Devonian basins of western Norway: Sedimentary response to large-scale transtensional tectonics? *Tectonophysics*, 332(1–2), 51–68.

Osmundsen, P.T., Braathen, A., Nordgulen, Ø., Roberts, D., Meyer, G.B., Eide, E. (2003). The Devonian Nesna shear zone and adjacent gneiss-cored culminations, North-Central Norwegian Caledonides. *Journal of the Geological Society*, 160, 137–150.

Parsons, T. (2006). The Basin and Range Province. In *Developments in Geotectonics*, Olsen, K.H. (ed.). Elsevier Science, Amsterdam.

Peron-Pinvidic, G. and Osmundsen, P.T. (2020). From orogeny to rifting: Insights from the Norwegian "reactivation phase". *Scientific Reports*, 10(1), 14860.

Petersen, K.D., Nielsen, S.B., Clausen, O.R., Stephenson, R., Gerya, T. (2010). Small-scale mantle convection produces stratigraphic sequences in sedimentary basins. *Science*, 329(5993), 827–830.

Planke, S., Symonds, P.A., Alvestad, E., Skogseid, J. (2000). Seismic volcanostratigraphy of large-volume basaltic extrusive complexes on rifted margins. *Journal of Geophysical Research-Solid Earth*, 105(B8), 19335–19351.

Prosser, S. (1993). Rift-related linked depositional systems and their seismic expression. *Geological Society, London, Special Publications*, 71(1), 35–66.

Ragab, A.I. and El-Kaliouby, B.A. (1992). Geodynamic of the Gulf of Suez-Red Sea rifting and origin of within plate magmatism. *Journal of African Earth Sciences (and the Middle East)*, 14(3), 351–360.

Reston, T.J. (2007). Extension discrepancy at North Atlantic nonvolcanic rifted margins: Depth-dependent stretching or unrecognized faulting? *Geology*, 35(4), 367–370.

Reston, T.J. (2009). The extension discrepancy and syn-rift subsidence deficit at rifted margins. *Petroleum Geoscience*, 15(3), 217–237.

Reston, T.J. and McDermott, K.G. (2011). Successive detachment faults and mantle unroofing at magma-poor rifted margins. *Geology*, 39(11), 1071–1074.

Reston, T.J. and McDermott, K.G. (2014). An assessment of the cause of the "extension discrepancy" with reference to the west Galicia margin. *Basin Research*, 26(1), 135–153.

Rey, P., Vanderhaeghe, O., Teyssier, C. (2001). Gravitational collapse of the continental crust: Definition, regimes and modes. *Tectonophysics*, 342(3), 435–449.

Ribes, C., Ghienne, J.-F., Manatschal, G., Decarlis, A., Karner, G.D., Figueredo, P.H., Johnson, C.A. (2019). Long-lived mega fault-scarps and related breccias at distal rifted margins: Insights from present-day and fossil analogues. *Journal of the Geological Society*, 176(5), 801–816.

Ricketts, J.W., Karlstrom, K.E., Kelley, S.A. (2015). Embryonic core complexes in narrow continental rifts: The importance of low-angle normal faults in the Rio Grande rift of central New Mexico. *Geosphere*, 11(2), 425–444.

Robert, C.M. (ed.) (2008). Aulacogens. In *Developments in Marine Geology: An Interplay between Geodynamics and Paleoenvironment*. Elsevier Science, Amsterdam.

Rooney, T.O. (2017). The Cenozoic magmatism of East Africa: Part I – Flood basalts and pulsed magmatism. *Lithos*, 286–287, 264–301.

Rooney, T.O. (2019). The Cenozoic magmatism of East Africa: Part II – Rifting of the mobile belt. *Lithos*, 360–361, 105291.

Rooney, T.O. (2020a). The Cenozoic magmatism of East Africa: Part III – Rifting of the craton. *Lithos*, 360–361, 105390.

Rooney, T.O. (2020b). The Cenozoic magmatism of East Africa: Part IV – The terminal stages of rifting preserved in the Northern East African Rift System. *Lithos*, 360–361, 105381.

Rooney, T.O. (2020c). The Cenozoic magmatism of East Africa: Part V – Magma sources and processes in the East African Rift. *Lithos*, 360–361, 105296.

Rooney, T.O., Furman, T., Yirgu, G., Ayalew, D. (2005). Structure of the Ethiopian lithosphere: Xenolith evidence in the Main Ethiopian Rift. *Geochimica et Cosmochimica Acta*, 69(15), 3889–3910.

Rotevatn, A., Jackson, C.A.-L., Tvedt, A.B.M., Bell, R.E., Blækkan, I. (2019). How do normal faults grow? *Journal of Structural Geology*, 125, 174–184.

Rowan, M.G. (2014). Passive-margin salt basins: Hyperextension, evaporite deposition, and salt tectonics. *Basin Research*, 26(1), 154–182.

Rowan, M.G. (2020). The South Atlantic and Gulf of Mexico salt basins: Crustal thinning, subsidence and accommodation for salt and presalt strata. *Geological Society, London, Special Publications*, 476(1), 333–363.

Royden, L. and Keen, C.E. (1980). Rifting process and thermal evolution of the continental margin of Eastern Canada determined from subsidence curves. *Earth and Planetary Science Letters*, 51(2), 343–361.

Ruppel, C. (1995). Extensional processes in continental lithosphere. *Journal of Geophysical Research-Solid Earth*, 100(B12), 24187–24215.

Saunders, A.D., Fitton, J.G., Kerr, A.C., Norry, M.J., Kent, R.W. (1997). The North Atlantic igneous provinces. In *Large Igneous Provinces: Continental, Oceanic, and Planetary Flood Volcanism*, Mahoney, J.J. and Coffin, M.F. (eds). American Geophysical Union and Wiley, Washington, D.C.

Saunders, A.D., Larsen, H.C., Wise Jr., S.W. (1998). East Greenland Margin, sites 914–919. *Proceedings of the Ocean Drilling Program, Scientific Results*, Ocean Drilling Program, College Station.

Scholz, C.H. (2019). *The Mechanics of Earthquakes and Faulting*, 3rd edition. Cambridge University Press, Cambridge.

Scott, R.J. and Lister, G.S. (1992). Detachment faults: Evidence for a low-angle origin. *Geology*, 20(9), 833–836.

Şengör, A.M.C. and Burke, K. (1978). Relative timing of rifting and volcanism on Earth and its tectonic implications. *Geophysical Research Letters*, 5, 419–421.

Şengör, A.M.C., Tüysüz, O., İmren, C., Sakınç, M., Eyidoğan, H., Görür, N., Pichon, X.L., Rangin, C. (2005). The North Anatolian Fault: A new look. *Annual Review of Earth and Planetary Sciences*, 33(1), 37–112.

Shann, M.V., Vazquez-Reyes, K., Ali, H.M., Horbury, A.D. (2020). The Sureste Super Basin of southern Mexico. *AAPG Bulletin*, 104(12), 2643–2700.

Sleep, N.H. (1971). Thermal effects of the formation of Atlantic continental margins by continental break up. *Geophysical Journal International*, 24(4), 325–350.

Sonder, L.J. and Jones, C.H. (1999). Western United States extension: How the west was widened. *Annual Review of Earth and Planetary Sciences*, 27(1), 417–462.

Spencer, J.E. and Reynolds, S.J. (1990). Relationship between Mesozoic and Cenozoic tectonic features in west central Arizona and adjacent southeastern California. *Journal of Geophysical Research: Solid Earth*, 95(B1), 539–555.

Stern, R.J. (2002). Subduction zones. *Reviews of Geophysics*, 40(4), 3-1–3-38.

Suess, E. (1891). Die Brüche des östlichen Afrika. In *Beiträge zur geologischen Kenntniss des östlichen Afrika*, von Höhnel, L., Rosiwal, A.,; Toula, F., Eduard Suess, E. (eds). Denkschriften Kaiserlichen Akademie Wissenschaften, Vienna.

Symonds, P.A., Planke, S., Frey, O., Skogseid, J. (1998). Volcanic evolution of the Western Australian continental margin and its implications for basin development. In *The Sedimentary Basins of Western Australia 2*, Purcell, P.G. and Purcell, R.R. (eds). Petroleum Exploration Society of Australia, Perth.

Ten Brink, U.S. and Taylor, M.H. (2002). Crustal structure of central Lake Baikal: Insights into intracontinental rifting. *Journal of Geophysical Research: Solid Earth*, 107(B7), ETG 2-1–ETG 2-15.

Tucholke, B.E., Behn, M.D., Buck, W.R., Lin, J. (2008). Role of melt supply in oceanic detachment faulting and formation of megamullions. *Geology*, 36(6), 455–458.

Turcotte, D.L. and Emerman, S.H. (1983). Mechanisms of active and passive rifting. In *Developments in Geotectonics*, Morgan, P. and Baker, B.H. (eds). Elsevier Science, Amsterdam.

Turcotte, D.L. and Oxburgh, E.R. (1973). Mid-plate tectonics. *Nature*, 244(5415), 337–339.

Van Wijk, J., Koning, D., Axen, G., Coblentz, D., Gragg, E., Sion, B. (2018). Tectonic subsidence, geoid analysis, and the Miocene–Pliocene unconformity in the Rio Grande rift, southwestern United States: Implications for mantle upwelling as a driving force for rift opening. *Geosphere*, 14(2), 684–709.

Vendeville, B.C. and Jackson, M.P. (1992). The rise of diapirs during thin-skinned extension. *Marine and Petroleum Geology*, 9(4), 331–354.

Vening Meinesz, E. (1950). Les grabens africains, résultat de compression ou de tension dans la crofite terrestre ? *Bull. Inst. R. Colonial Belge*, 21, 539–552.

Walsh, J.J., Bailey, W.R., Childs, C., Nicol, A., Bonson, C.G. (2003). Formation of segmented normal faults: A 3-D perspective. *Journal of Structural Geology*, 25(8), 1251–1262.

Warren, J.K. (2016). *Evaporites: A Geological Compendium*. Springer, Berlin.

Watterson, J. (1986). Fault dimensions, displacements and growth. *Pure and Applied Geophysics*, 124(1), 365–373.

Watts, A.B. (2001). *Isostasy and Flexure of the Lithosphere*. Cambridge University Press, Cambridge.

Watts, A.B., Tozer, B., Daly, M.C., Smith, J. (2018). A comparative study of the Parnaíba, Michigan and Congo cratonic basins. *Geological Society, London, Special Publications*, 472(1), 45–66.

Wernicke, B. (1981). Low-angle normal faults in the Basin and Range Province: Nappe tectonics in an extending orogen. *Nature*, 291, 645–648.

Wernicke (1985). Uniform-sense normal simple shear of the continental lithosphere. *Canadian Journal of Earth Sciences*, 22, 108–125.

Wernicke, B. and Axen, G.J. (1988). On the role of isostasy in the evolution of normal fault systems. *Geology*, 16(9), 848–851.

White, R.S. (1992). Crustal structure and magmatism of North-Atlantic continental margins. *Journal of the Geological Society*, 149, 841–854.

White, R.S. and McKenzie, D. (1989). Magmatism at rift zones: The generation of volcanic continental margins and flood basalts. *Journal of Geophysical Research*, 94, 7685–7729.

White, R.S., Spence, G.D., Fowler, S.R., McKenzie, D.P., Westbrook, G.K., Bowen, A.N. (1987). Magmatism at rifted continental margins. *Nature*, 330(6147), 439–444.

Whitney, D.L., Teyssier, C., Rey, P., Buck, R. (2014). Continental and oceanic core complexes. *GSA Bulletin*, 125(3–4), 273–298.

Wiest, J.D., Osmundsen, P.T., Jacobs, J., Fossen, H. (2019). Deep crustal flow within postorogenic metamorphic core complexes: Insights from the southern western Gneiss Region of Norway. *Tectonics*, 38(12), 4267–4289.

Willis, B. (1928). The Dead Sea problem: Rift valley or ramp valley. *Geological Society of America Bulletin*, 39, 452–490.

Wilson, J.T. (1966). Did the Atlantic close and then re-open? *Nature*, 211(5050), 676–681.

Wilson, R.C.L., Manatschal, G., Wise, S. (2001). Rifting along non-volcanic passive margins: Stratigraphic and seismic evidence from the Mesozoic successions of the Alps and western Iberia. In *Non-Volcanic Rifting of Continental Margins: A Comparison of Evidence from Land and Sea*, Wilson, R.C.L., Whitmarsh, R.B., Taylor, B., Froitzheim, N. (eds). Geological Society, London.

Wilson, D., Aster, R., West, M., Ni, J., Grand, S., Gao, W., Baldridge, W.S., Semken, S., Patel, P. (2005). Lithospheric structure of the Rio Grande rift. *Nature*, 433(7028), 851–855.

Wilson, R.W., Houseman, G.A., Buiter, S.J.H., McCaffrey, K.J.W., Doré, A.G. (2019). Fifty years of the Wilson Cycle concept in plate tectonics: An overview. *Geological Society, London, Special Publications*, 470(1), 1–17.

Withjack, M.O. and Callaway, S. (2000). Active normal faulting beneath a salt layer: An experimental study of deformation patterns in the cover sequence. *AAPG Bulletin*, 84(5), 627–651.

Wright, K.A., Davies, R.J., Jerram, D.A., Morris, J., Fletcher, R. (2012). Application of seismic and sequence stratigraphic concepts to a lava-fed delta system in the Faroe-Shetland Basin, UK and Faroes. *Basin Research*, 24(1), 91–106.

Wu, J.E., Mcclay, K., Whitehouse, P., Dooley, T. (2009). 4D analogue modelling of transtensional pull-apart basins. *Marine and Petroleum Geology*, 26(8), 1608–1623.

Xie, X. and Heller, P.L. (2009). Plate tectonics and basin subsidence history. *GSA Bulletin*, 121(1–2), 55–64.

Yang, T., Chou, Y.-M., Ferré, E.C., Dekkers, M.J., Chen, J., Yeh, E.-C., Tanikawa, W. (2020). Faulting processes unveiled by magnetic properties of fault rocks. *Reviews of Geophysics*, 58(4), e2019RG000690.

Ziegler, P.A. (1992). European Cenozoic rift system. *Tectonophysics*, 208(1), 91–111.

2

What is a Rifted Margin? From the Early Models to Modern Views and Future Challenges

Gwenn PERON-PINVIDIC[1,2]

[1] Geological Survey of Norway (NGU), Trondheim, Norway
[2] Department of Geoscience and Petroleum,
Norwegian University of Science and Technology, Trondheim, Norway

Rifted margins are major geological objects that mark the transition between continents and oceans, the two first-order types of land masses present on Earth. They include specific types of tectonic structures, basement compositions and sedimentary basins, which record the thinning of continental crust and the birth of oceanic spreading centers.

Early interpretations of rifted margins assumed that continental crust uniformly deformed and gradually thinned oceanwards, ending in a sharp juxtaposition with a tripartite oceanic crust (Figure 2.1). The emergence of new data over the past 30 years challenged this early representation, and research in rifting and rifted margin formation subsequently underwent a major **paradigm shift**. The drilling of exhumed mantle and the following extensive work on the Iberia–Newfoundland system (Ocean Drilling Program results; for example, Tucholke *et al.*

For a color version of all the figures in this chapter, see www.iste.co.uk/peronpinvidic/margins1.zip.

Continental Rifted Margins 1,
coordinated by Gwenn PERON-PINVIDIC.

2007) resulted in the development of a **new understanding of rifted margin architecture**. The integration of field analogues (e.g. Manatschal 2004) and the increasing number of modern high-resolution seismic datasets (e.g. Zalán *et al.* 2011) showed that the evolution of rifted basins reflects composite geological processes that are very different from those defined in the pioneering models of rifting.

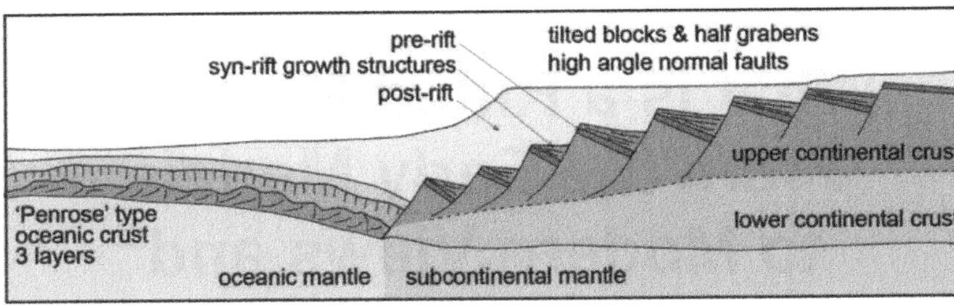

Figure 2.1. *An early "text book"-style illustration of a rifted margin, standard in the 1970s and 1980s. Note the sharp juxtaposition of continental and oceanic lithosphere (source: Peron-Pinvidic* et al. *2019)*

2.1. What is interesting about studying rifted margins?

Rifted margins record plate tectonic separation and thus constitute one of the fundamental tectonic components of the Earth, together with mountain belts, transform margins, oceanic ridges and subduction zones (Figure 2.2).

For various reasons, such as accessibility, assessment of seismicity and volcanic risks, the scientific community has long been focused on the study of mid-ocean ridges, subduction zones and mountain belts, rather than on rifted margins. However, **neither the onset of seafloor spreading, nor the formation of subduction zones and collisional mountain belts can be investigated without a proper understanding of rifted margins**. The architecture of an orogen is, at least partly, controlled by the rocks taking part in the collisional processes, their geometries and composition. And as orogens tend to form at the locus of pre-existing oceans (Wilson 1966), rifted margins can legitimately be considered as one of the primary structures that influences orogenic spatial and temporal evolution (Rouméjon *et al.* 2015).

It is essential to correctly constrain and understand the architecture and evolution of rifted margins for purposes of fundamental research, but equally for more applied questions.

Rifted margin research has historically been deeply linked to applied problems in **hydrocarbon exploration**. The oil and gas industry has undoubtedly been the main driver behind data acquisition and improvements in geophysical processing and imaging techniques. The sedimentary archives of rifted margin formation have proven extremely valuable for the petroleum industry: more than 60% of the so-called "giants" (i.e. the most voluminous oil/gas fields) are identified in rifts and rifted margins (Mann *et al.* 2001). But these archives can also help address some of the most pressing environmental issues currently faced by humanity. These include **landslides** and **tsunami hazards**, which are both of high concern to the onshore coastal areas that are often densely populated. Because of the tectonic and sedimentary processes caused by lithospheric thinning, rifts and rifted margins often contain large accumulations of sediments. These are sometimes associated with abrupt bathymetric slopes, which can enhance submarine slumps, debris flows and turbidity currents – all potential causes of tsunamis.

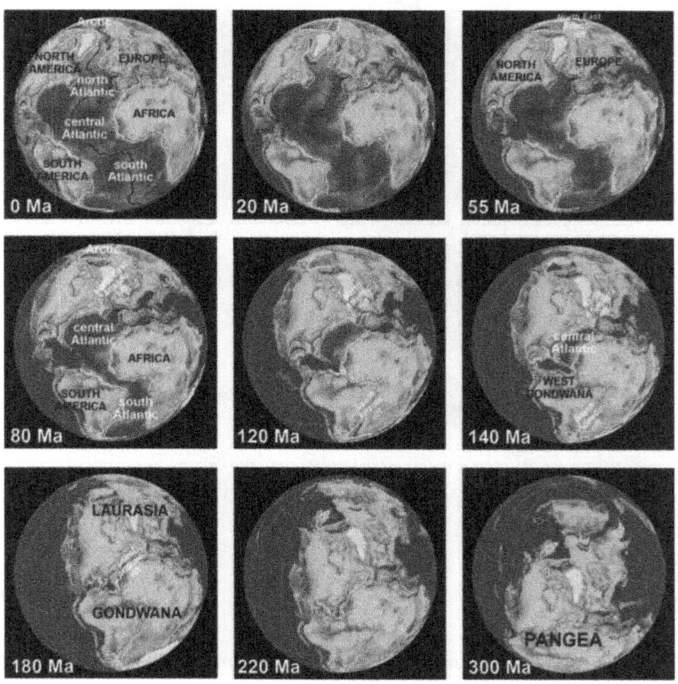

Figure 2.2. *Series of global plate reconstructions selected as a representative of the tectonic evolution of the Atlantic Ocean since the dispersal of the Pangea. Note that the present-day Atlantic Ocean continental masses are all bordered by rifts and rifted margins. The reconstructions were performed using Gplates and are based on the global plate motion poles, ridges (black lines) and continent–ocean boundaries (white lines) (source: Müller* et al. *2018)*

The northeastern Atlantic margins contain a very nice illustration of the hazardous potential of these events: the Storegga slide is a major scarp on the bathymetric slope offshore Norway (Figure 2.3). It has been estimated to have occurred 8,200 years ago, and involved 2,400–3,200 km^3 of displaced sediments, affecting a surface of 100,000 km^2 (Haflidason *et al.* 2004; Bondevik *et al.* 2005; Bryn *et al.* 2005). Tsunami deposits related to the slide have been identified in various places, including the Faroe Islands and the coast of Scotland, which are at very substantial distances from the initial instability. Other similar slide events have been identified around the northeast Atlantic Ocean, such as the Træna djupet (500 km^3; Laberg *et al.* 2002), the Andøya (900 km^3; Laberg *et al.* 2000) and the Bear Island (1,100 km^3; Leynaud *et al.* 2009) events, illustrating the potentially high hazard for such events.

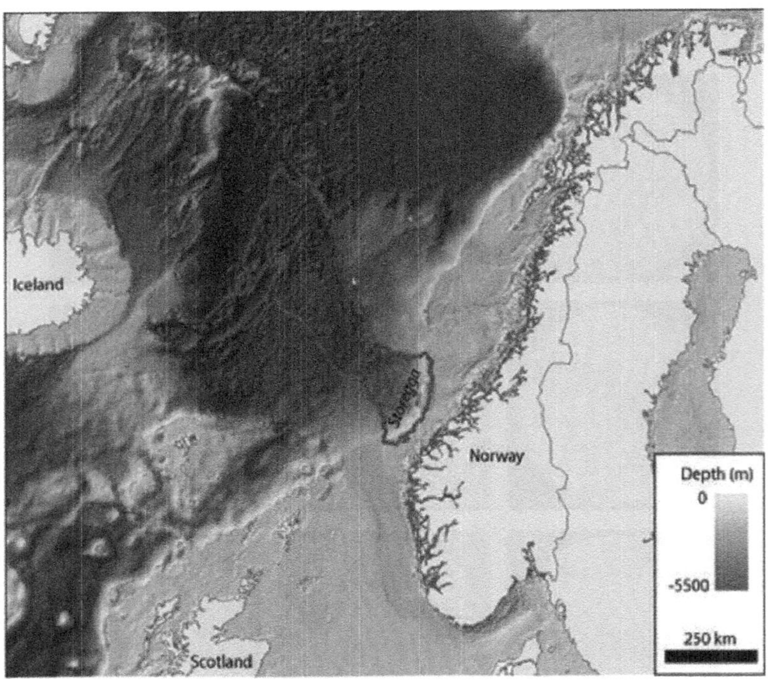

Figure 2.3. *Location map showing the extent of the Storegga slide (red polygon). The slide deposits are interpreted to extend 810 km offshore from the upper edge of the slide (source: Haflidason* et al. *2004)*

Rifting is also often associated with **seismicity** and **magmatic activity** (Ebinger *et al.* 2010). The East African Rift is a very significant example illustrating the need for a better understanding of these processes: several earthquake events of a magnitude greater than M6 have been identified or registered in the area (Zielke and Strecker 2009); many major volcanoes are active and potentially highly destructive for human activities (e.g. the Nyragongo near the city of Goma), and secondary processes such as the emanation of lethal gases (e.g. Nyos lake) are poorly constrained. Local landslides are also of very high concern in these highly unstable volcanic regions.

Recently, it has been shown that rifted margin evolution may aid our understanding of **climate changes**. Magma production during final rifting and breakup events has also been related to massive CO2 production, which could explain the natural excursions of methane and CO2 in Earth's history (e.g. Escartín *et al.* 2008; Crosby *et al.* 2010; the Paleocene–Eocene Thermal Maximum (PETM), Svensen *et al.* 2004). The discovery of exhumed mantle domains and related serpentinization processes can account for substantial mass transfers of elements from the mantle into the hydrosphere, as well as the production of high volumes of methane and other gases (Buck *et al.* 2005). And, at a different time scale, as well as of high importance for planetary evolution, is the formation of oceanic gateways during continental rifting, which exert a strong influence on the circulation of oceanic currents, and hence on the evolution and distribution of life.

Rifted margins and offshore rift basins are of high interest for many various reasons. A correct understanding of their architecture as well as their temporal and spatial evolution is of high importance for many applied geological questions for humanity.

2.2. Morphological definition

By definition, **a rifted margin marks the transition from the continent to the ocean, where there is no active plate boundary** (i.e. subduction zones). The principal identification of rifted margins can be easily made on bathymetric maps where these specific areas are clearly identifiable (Figure 2.4). Rifted margins include several morphologically distinct entities; from the shore to the ocean, these are:

– the **continental shelf**, which is a relatively flat area, slightly inclined towards the sea, 80 km wide and under 100 m deep on average;

– the **continental slope**, which is ~45 km wide, with water depths varying from 200 to 4,000 m on average, and a typical slope of 5°;

– the **continental rise** with very gentle slopes (<1°), which extends from 4,000 to 5,000 m depth and is connected to the ocean basin;

– the **abyssal plain** corresponds to the deep ocean floor, usually found at depths >3,000 m, between the continental rise and the mid-ocean ridge. The abyssal plain represents more than 50% of the Earth's surface.

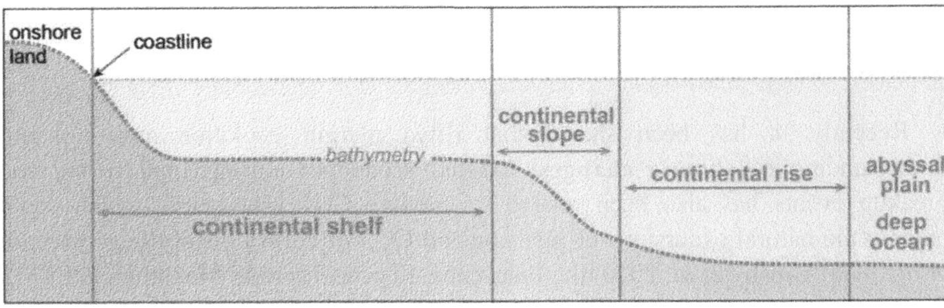

Figure 2.4. *Diagram illustrating the morphology of offshore rifted margins based on bathymetric information*

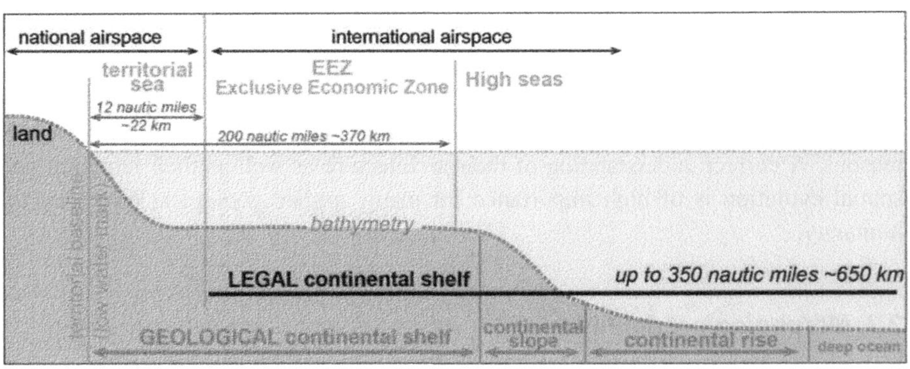

Figure 2.5. *Illustration of the coastal legal zones, including the territorial sea, the exclusive economic zone (EEZ), the legal continental shelf and the high seas. No country can claim any zone as per their wish; it must be as per UNCLOS (The United Nations Convention on the Law of the Sea). Note the possible mismatch between the defined areas of legal zones and geological domains*

This morphological definition is a primary feature based on the present-day bathymetry, and hence relates to the youngest surficial deposits and today's topography. It therefore does not provide any information on the underlying architecture or spatial and temporal evolution of the rifted margin (that is covered in the next chapter). However, this geographical definition is fundamental, notably because of high pertinence to the "**law of the sea**" – the group of laws which codify and regulate the rights and duties of countries in the offshore areas, notably for the navigation and fishing rights. The law of the sea also helps countries define their territorial offshore limits, which has a huge economical impact on their potential for natural resource exploitation (seabed minerals, hydrocarbons). The definitions of the continental shelf, continental slope and continental rise directly impact the definition of the **exclusive economic zone** (EEZ), which identifies the surfaces claimable by the various countries (Figure 2.5).

2.3. Classification schemes for rifted margins

Rifted margins are extremely diverse, and encompass a variety of geometries, basement compositions and sedimentary architectures that, in turn, indicate differences in temporal and spatial evolution. To facilitate studies of individual margins as well as comparative analyses, **different classification types have been proposed in the literature. The most commonly used classification schemes are based on the opening geometry** (see section 2.3.1), **on the amount of magma** (see section 2.3.2) **or on the thickness of the sediment infill sediments** (see section 2.3.3). For practical reasons, studies often focus on the evolution of the end-member types of rifted continental margins, thus allowing for an easy outline of the main characteristics. However, rifted margins cover a large spectrum of architectural possibilities, so even though a classification is always handy, we must bear in mind that natural cases are rarely limited to end-members.

Below, we present each classification scheme in two steps: one that outlines its theoretical justification (the "Theory") and one that discusses case examples and related issues (the "Evaluation").

2.3.1. *Classifications based on the extensional context: the divergent, sheared and transtensional rifted margins*

Theory

Continental rifted margins can be classified as divergent, sheared, transtensional. These classifications describe the **obliquity** of the rift opening. Rift obliquity is defined as the angle between the local rift axis or line of breakup and the regional relative plate motion (Withjack and Jamison 1986; Basile and Braun 2016). Within that framework, a **divergent margin** has extension occurring at ~70–90° to the line of breakup, an **oblique margin** an extension at 30–70° and a **transform margin** extension at 0–30° to the line of breakup (Figure 2.6).

Evaluation

Although these definitions appear straightforward, there are major issues with this approach which point to its oversimplification. First, it can be extremely challenging to determine the exact angle of obliquity in a rift system for various reasons: lack of information or constraints on the regional plate movement, poor data coverage and/or resolution can prohibit accurate identification of the rift axis; and along-strike segmentation of the rift system can produce very different kinematic contexts from one area to the next, even at short distances, preventing any coherent definition of a regional rift axis. Second, the angle of opening can considerably change during the rifting history, from one phase of deformation to the other, and from one margin segment to the other.

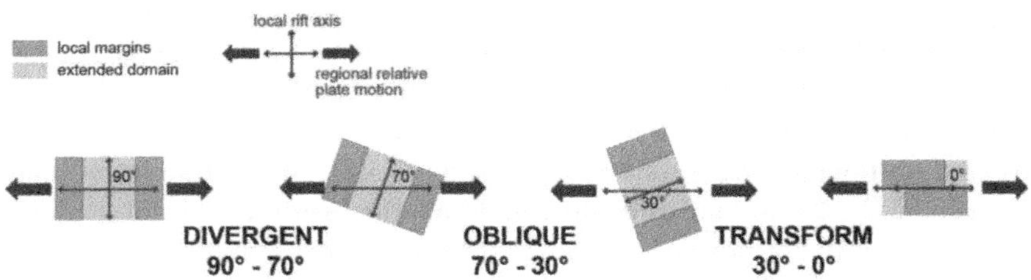

Figure 2.6. *Sketches defining rift obliquity. Continental margins with an obliquity of more than 70° are defined as divergent, with less than 30° as transform, and between 30° and 70° as oblique margins*

However, it is observed that this angle of opening does impact rift architecture (see below), which points to the relevance of the classification. Globally, there is a tendency towards generating narrower and more abrupt margins with higher angles of obliquity. In that sense, transform margins are typically described to differ from divergent passive margins with a marked narrow continental shelf (generally less than 30 km wide) and a sharp crustal thinning over a distance of less than 10–40 km (Figure 2.7; Bird 2001; Basile 2015; Mercier de Lépinay *et al.* 2016).

The Falkland Plateau–Agulhas Fracture Zone (Figure 2.8) is a typical example of a transform margin with a very high opening obliquity (<10°) during the final continental separation and breakup phase (Moulin *et al.* 2010). The transition from continental basement to oceanic crust has been derived from gravity modeling and constrained by seismic reflection and magnetic data. It is defined as a very abrupt transition, operating over an area of less than 50 km (Becker *et al.* 2012). Within the same margin category, at the northern end of the South Atlantic Ocean, the Ghana transform margin is situated at the continental prolongation of the prominent Romanche Fracture Zone (Figure 2.8). While the entire Equatorial Atlantic Ocean has been modeled with an obliquity of 60° (Heine and Brune 2014), rift obliquity in the Romanche Fracture Zone has extremely low values from 5° to 20° (Basile and Braun 2016). The continental crust approaching the transform margin dramatically thins over a distance of 20 km, from about 23 to 10 km (Bird 2001). Similarly, the Gulf of California (Figure 2.8) is a highly oblique rift system with an active system of en echelon basins, connected by long transform faults, and hosting the transition from the spreading of new oceanic crust in the south to the rifting of continental crust in the north (Aragón-Arreola *et al.* 2005). The obliquity in the Gulf of California varies between 20° and 25° (Withjack and Jamison 1986); the Gulf thus may be considered as being intermediate between oblique and transform margins. With an obliquity of about 50°, the Gulf of Aden is close to an "ideal" oblique rift system, based on the rift obliquity definition (Leroy *et al.* 2004; Autin *et al.* 2010). Fracture zones and transfer fault zones offset and strongly segment the margin (see section 2.6.5). On the other side of the spectrum, the Iberia–Newfoundland conjugate margins (Figure 2.8) may be considered as an archetype of divergent margins (see Chapter 3 for case examples).

Based on the key examples of divergent, oblique and transform margins listed above, it appears that a systematic crustal-scale asymmetry is observed along the conjugate sections despite the angle of obliquity applied to the rifted system. This relates to the distinction of upper-plate and lower-plate settings in rifted margins (see section 2.4.2.1).

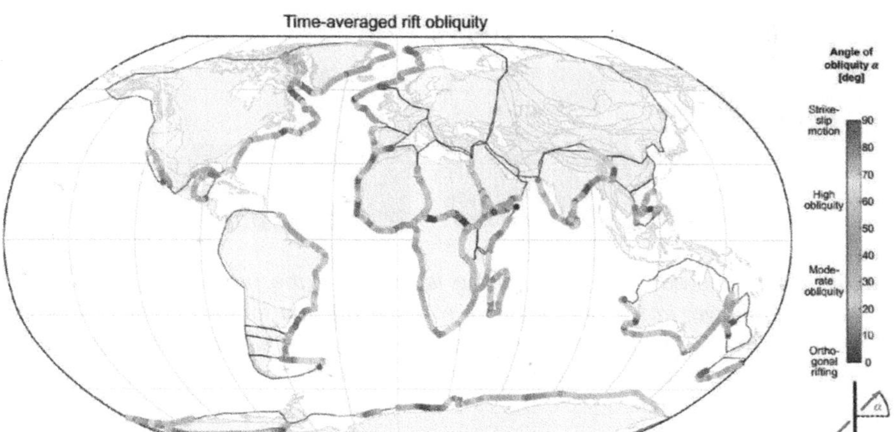

Figure 2.7. *Worldwide map with rifts and rifted margin segments color-coded based on their opening obliquity (averaged over time). The map illustrates the prevalence of oblique rifting since Pangea fragmentation, but that temporal changes in rift obliquity cannot be visualized in this representation (source: Brune* et al. *2018)*

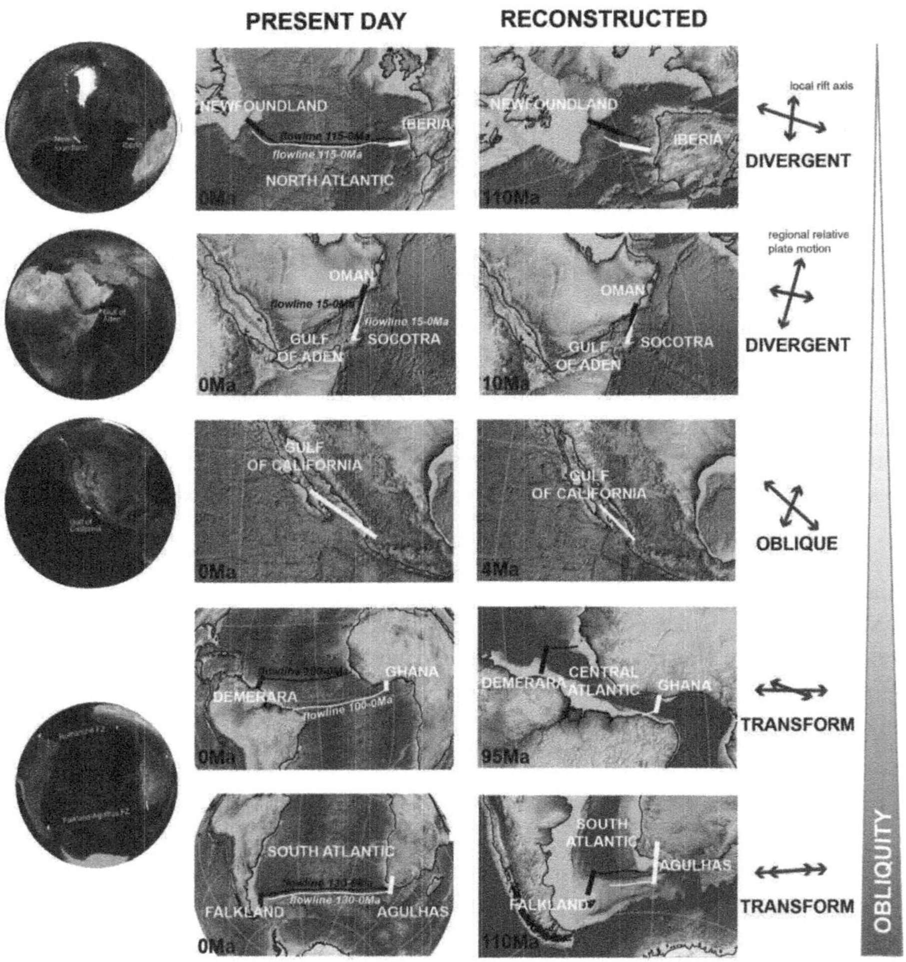

Figure 2.8. *Worldwide examples: in the left column, the present-day location and flowline are shown; in the right column, a reconstruction to shortly after breakup is presented. The reconstructions were performed using Gplates and are based on the global plate motion poles (source: Seton* et al. *2012)*

2.3.2. Classifications based on the amount of magma involved: the volcanic and non-volcanic rifted margins

Theory

Rifted margins can also be subdivided into **volcanic** and **non-volcanic** margins, depending on the amount of magmatic material observed and/or modeled

(Whitmarsh and Sawyer 1996). **Magmatic** and **amagmatic** are also widely used terms for these groupings.

Volcanic margins are characterized by significant magmatic activity, while non-volcanic margins are characterized by little or no magmatism – a status which is theorized to be directly related to the thermal state and fertility of the underlying upper mantle at the time of the continental extension.

In early studies, a non-volcanic margin was theoretically characterized by (1) a thinned continental crust deformed into a series of tilted blocks flanked by high-angle normal faults, (2) a transitional zone of unconstrained composition, neither oceanic nor continental, and (3) an abnormally thin oceanic crust (Figure 2.9). The Iberia–Newfoundland conjugate margins are traditionally referred to as the classic example of an amagmatic rifted system, with wide areas of exhumed serpentinized mantle and little evidence of magmatic activity (Whitmarsh *et al.* 2001; see Chapter 3).

On the other hand, the Norwegian–Greenland Sea conjugate margins are usually placed in the other end-member group – magmatic or volcanic margins – because they display classic magmatic features, such as seaward dipping reflectors (SDRs), outer high complexes, intrusions and lower crustal bodies (LCBs; usually interpreted in terms of magmatic underplatings), all together attesting to a pronounced magmatic activity associated with the construction of the marginal high area (Planke *et al.* 2000; Berndt *et al.* 2001; for more information on the magmatic geometries, see section 1.5.3). The East and West Greenland rift system or the Hatton–Rockall basins are also regularly cited as key examples of magmatic margins.

Within this magmatic versus amagmatic approach, volcanic margins are also classified as **Large Igneous Provinces**, often abbreviated as **LIPs** (Coffin Eldholm 1994; see Figure 2.10). LIPs correspond to areas associated with volumes of igneous rocks that are considered as anomalously large and emplaced over short time intervals (typically a few millions of years). These include underplatings, intrusives (sills, dikes) and extrusives (lava flows). Types of LIPs include continental flood basalts, oceanic plateaus, large dike swarms, ocean basin flood basalts or volcanic rifted margins (Figure 2.10). The formation of LIPs is attributed to mantle plumes or plate tectonics (Coffin Eldholm 1994; Foulger 2002).

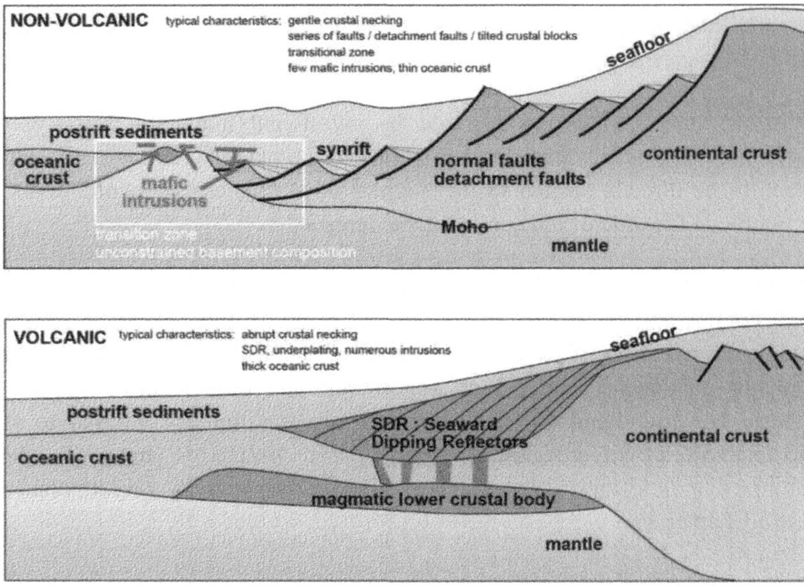

Figure 2.9. *Schematic representation of theoretical end-member categories of rifted margins: volcanic and non-volcanic (source: adapted from Franke (2013))*

Evaluation

Although the definition is very intuitive, modern datasets coupled with improved modeling techniques and new interpretation protocols now highlight the oversimplification of these two categories. Recent observations show that rifted margins correspond to a spectrum of possibilities rather than fit into an end-member scenario. Various recent studies have begun to revise these partitioned classifications. We list several key observations below:

1) Based on structural mapping, it has been shown that in general, rifted margins share numerous structural similarities, including the partition into proximal, necking and distal domains and the systematic presence of comparable entities, such as platform areas, necking domains, series of tilted crustal blocks, regions of "transitional" basement and deep sag/supra-detachment-type basins (Franke 2013; Peron-Pinvidic *et al.* 2013; see section 2.5). Comparable observations have been reported from many settings worldwide, such as the Armorican margin (Thinon *et al.* 2003), the Nova Scotia margin (Funck *et al.* 2004), the South China Sea (Franke *et al.* 2011), the Gulf of Aden (Autin *et al.* 2010), the Argentinian margin (Blaich *et al.* 2009), all of which were previously classified as magmatic or

non-magmatic. Thus, from the structural point of view, volcanic and non-volcanic margins appear very similar for most of them.

2) The volumes of magmatic material present in many rifted margins have been dramatically revised, which also weakens the end-member-type argument. For instance, the Iberia–Newfoundland conjugates, which are traditionally referred to as the archetype example of an amagmatic system, have numerous features usually associated with the volcanic settings, such as widespread sill complexes (the U-reflector sill system; Peron-Pinvidic *et al.* 2010) and outer high structures with magmatic underplatings (the J-anomaly; Bronner *et al.* 2011). Peron-Pinvidic *et al.* (2010) showed that the volume of magma in the distal Newfoundland margin is largely underestimated. Based on a conservative approach, the authors mapped the extent of the distal margin sill system and estimated a corresponding magmatic volume which was far greater than previously estimated and in the same order of magnitude as the sill complexes described at the volcanic Møre margin (Planke *et al.* 2005). Additionally, ODP samples show that the wide windows of exhumed mantle in between Iberia and Newfoundland potentially contain very significant volumes of magmatic material (Müntener and Manatschal 2006; Müntener *et al.* 2010). Likewise, the origin of the magmatic material in the mid-Norwegian margin (plume- vs. non-plume-related) has been questioned (Van Wijk *et al.* 2001), and its volume has been significantly lowered compared to previous estimations (e.g. Wangen *et al.* 2011).

3) Refraction models are often used to argue for abnormally high volumes of magma in magma-rich margins. Based on the modeling of OBS measurements, Mjelde *et al.* (2003, 2007, 2009, 2012) proposed the presence of significant volumes of underplated magmatic material in the distal mid-Norwegian margin, based on the interpretation of the "LCB" or "HVLC" (lower crustal body and high-velocity lower crust, respectively; see Chapter 1). These bodies are modeled at lower crustal depths, with velocities and densities higher than expected for "normal" lower crustal material. They can then be interpreted in terms of mafic material underplated at a certain rift stage under the extended crust. This interpretation proposes the addition of significant volumes of magmatic material to the margin, which argues in favor of the "volcanic/magma-rich" classification. The Icelandic plume often serves as explanation for the anomalous magma volume in these NE Atlantic regions. However, Rouzo *et al.* (2006) showed that the "OBS Vøring-2000" raw dataset used by Mjelde *et al.* (2003, 2007, 2009, 2012) can be modeled (with another method) without the presence of LCB or HVLC at depth. These results question the presence of such abnormal melt volumes in the margin and can open the door to alternative interpretations. In the last decade, several different models have been proposed and the interpretation of the HVLC or LCB bodies has been challenged at multiple locations. Alternative interpretations propose the existence of bodies not solely

made of magmatic material, but containing other types of lithologies such as lower crustal and mantle rocks, potentially altered and deformed at various degrees.

Figure 2.10. *Outlines (in pink) of the main LIPs (Large Igneous Provinces) on a worldwide elevation and bathymetric map. The LIP outlines are after Coffin, and include continental flood basalt provinces, volcanic rifted margins, oceanic plateaus and ocean basin flood basalt provinces. Note that these outlines are dependent on the definition used for LIPs (source: NOAA Etopo1; Amante and Eakins 2009)*

Based on the observations listed above, **it can be concluded that the amount of magma in many rifted margins is still highly uncertain, questioning the applicability of the end-member magmatic/amagmatic rifted margin classification**.

Even more interestingly, from the examples of the mid-Norwegian and Iberia–Newfoundland margins, it appears that the differences between the "magmatic" and "amagmatic" archetypes are only found in their outer domains (Figure 2.11), where one system presents well-developed SDRs and sill complexes (mid-Norway), which are lacking in the other (Iberia–Newfoundland). Although magmatic volumes are difficult to estimate, Peron-Pinvidic *et al.* (2016) suggested that it is the expression of the magma that differs from one setting to the other: volcanic extrusions such as SDRs and sill intrusions, clearly observed on the mid-Norwegian margin (e.g. Berndt *et al.* 2001), are obviously much more striking and seismically better imaged than melt fluids trapped in wide zones of exhumed and/or shallow mantle (Müntener *et al.* 2010). The volumes are certainly more significant in some margins than in others, as is the case with other rift parameters (e.g. the extension rate, the sedimentation rate). However, the amount of magma may not be as dichotomized as usually implied in the volcanic versus non-volcanic partition. Based on this, Peron-Pinvidic and Osmundsen (2016) recommended restricting the use of "magma-rich" and "magma-poor" to the breakup phase *sensu stricto*, which is the specific deformation phase that governs the construction of the outer margin and related marginal high: most rifted margins share a similar multiphase rift evolution ending with similar sets of structural entities. It is the magmatic input in the breakup phase that differs from one setting to the other.

The timing of magma genesis and emplacement are equally crucial. For instance, Tugend *et al.* (2018) emphasized the temporal dependence of the extensional setting in the overall magmatic evolution.

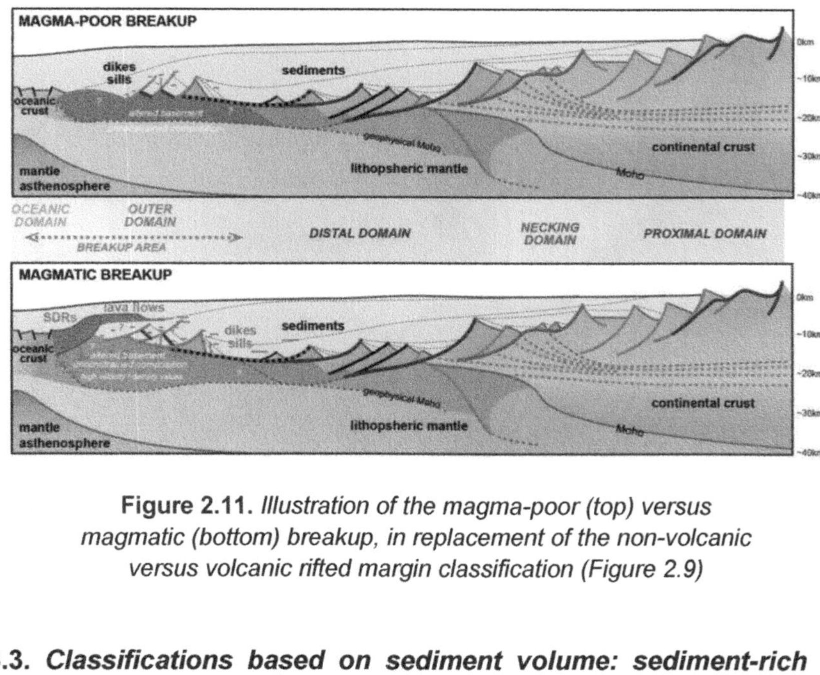

Figure 2.11. *Illustration of the magma-poor (top) versus magmatic (bottom) breakup, in replacement of the non-volcanic versus volcanic rifted margin classification (Figure 2.9)*

2.3.3. *Classifications based on sediment volume: sediment-rich and sediment-starved rifted margins*

Theory

One of the other striking characteristics of rifted margins is the volume of sediments captured in associated basins. Some margins show very thin sedimentary accumulations (e.g. Iberia, South Australia), while others display impressive (>15 km thick) sedimentary basins (e.g. Norway; Figures 2.12 and 2.13). These are defined as **sediment-starved** and **sediment-rich** rifted margins, respectively.

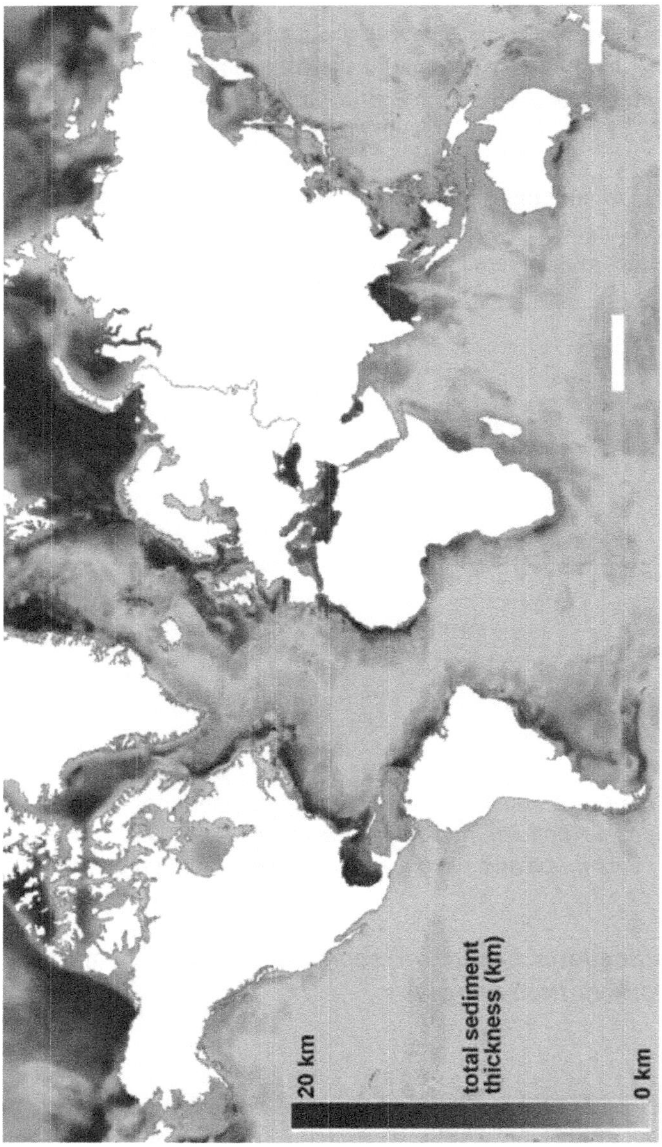

Figure 2.12. *Offshore total sediment thickness map (source: Straume et al. 2019)*

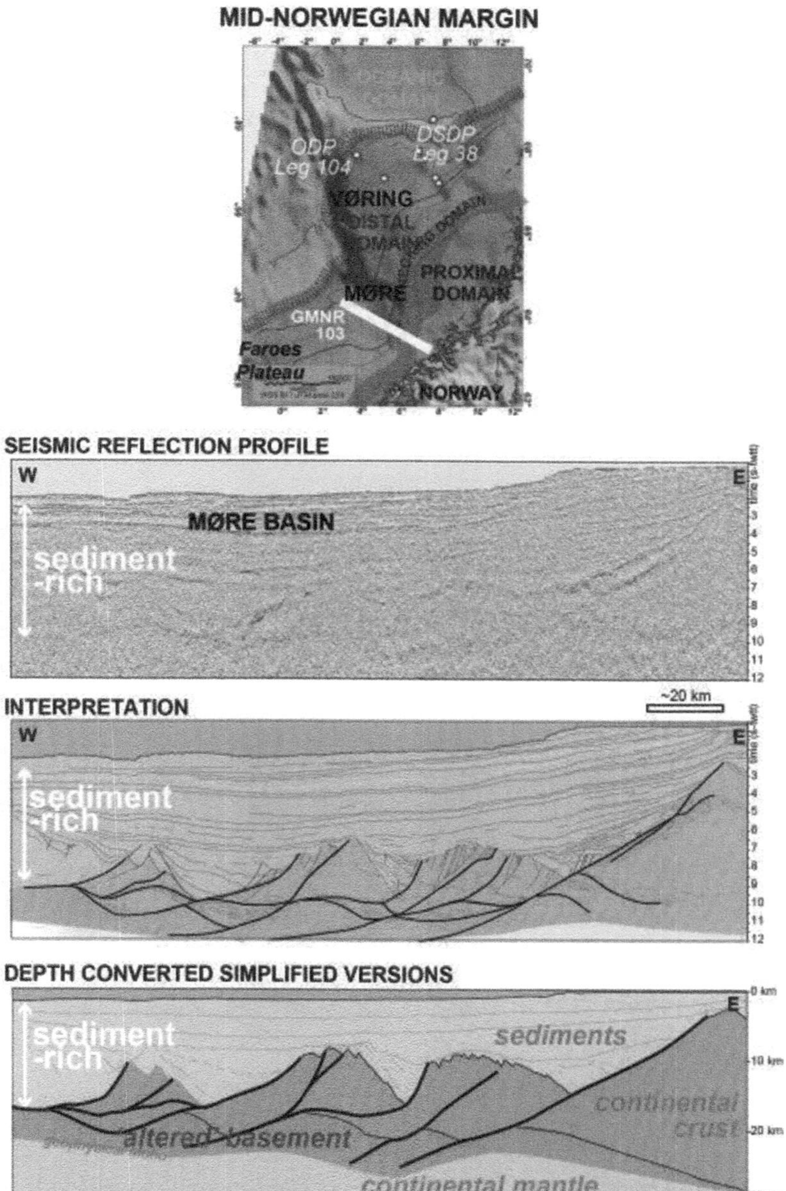

Figure 2.13a. *Ilustration of a sediment-rich rifted margin, based on examples from the mid-Norwegian margin, with the seismic reflection profile GMNR-103 imaging the Møre margin (source: adapted from Peron-Pinvidic and Osmundsen 2018)*

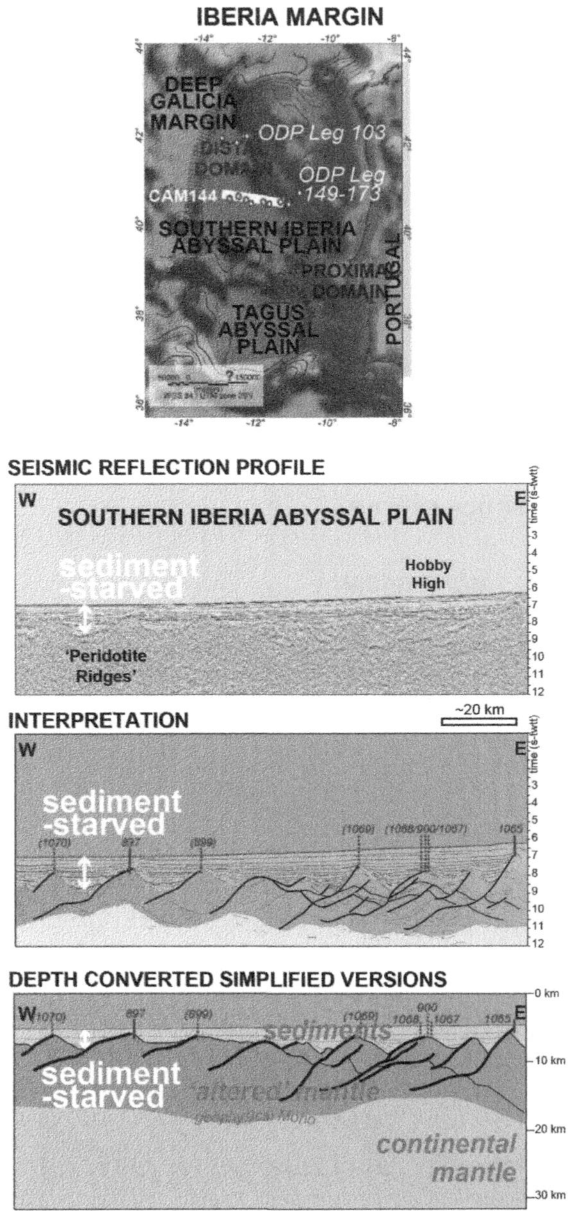

Figure 2.13b. *Ilustration of a sediment-starved rifted margin, based on examples from the Iberia margin, with the seismic reflection profile CAM-144 imaging the Southern Iberia Abyssal Plain (source: adapted from Peron-Pinvidic and Osmundsen 2018)*

Sedimentation occurs during the entire extensional process, and the volume and nature of the involved sediments can span a wide range of possibilities. The final thickness of a sedimentary basin is related to various parameters, the first being the space available for the sediments to accumulate at a specific location, which is determined by the tectonic activity, basement thinning and related subsidence. The sediment sourcing, sediment transport and storage conditions in the basinal area are similarly crucial parameters. Numerous studies exclusively focus on these patterns. The entire integrated process is often called "**source to sink**" (Figure 2.14). Its study consists of investigating the fate of sedimentary particles through a connected suite of geomorphic environments, from the onshore catchment to the area of deposition. The latter could be an adjacent flood plain or lake, a continental shelf or an abyssal deep-water sink. Within that framework, the basin is the **depositional area**, separate from the **erosional area** where the sediments originate from. The path sediments use during this journey constitutes the **sediment routing system**. The combined study of the various components of the sediment journey acknowledges the importance of the individual environments and the connectivity between them, from the mountains to the plains and offshore shelf or abyssal plains (e.g. Martinsen *et al.* 2010). It is now recognized that these environments are separated by dynamic boundaries that vary in time and space in response to variations in the sedimentary fluxes.

The sediment fluxes are, in turn, governed by various parameters that can additionally have feedback effects, such as the source rock composition, climate, landscape topography, catchment geography, tectonic parameters and sea-level fluctuations. Various studies have tried to investigate the impact of the sedimentary history and accumulation thickness on the overall rift evolution and final rifted margin architecture. Unfortunately, results and conclusions have been somehow rather contradictory. Analogue modeling shows that syn-rift sedimentation does not significantly influence the large-scale evolution of rift segments (Zwaan *et al.* 2018). The effect is identified at a more local scale on sedimentary wedge thicknesses and half-graben subsidence. Adversely, geophysical studies have suggested that sediments are a controlling parameter of the rifting process, triggering more subsidence and influencing rift width and magmatism (e.g. Andrés-Martínez *et al.* 2019).

Erosion and deposition have a noticeable impact on the **thermal regime** of the extending rift system, and thus on the maturation history of the hydrocarbon source rocks. These parameters have been extensively studied, notably by the oil and gas industry.

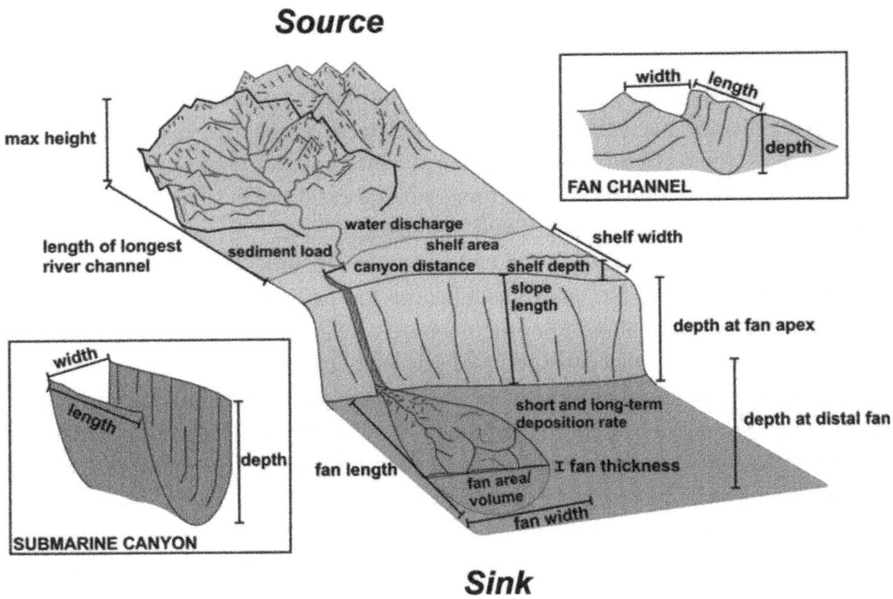

Figure 2.14. *Schematic representation of the key elements constituting a rifted margin source-to-sink system, with a distinct drainage basin–shelf–slope–basin floor profile (source: Martinsen et al. 2010)*

Evaluation

Contrary to the two other classifications presented above in this section, these sediment-rich and sediment-starved categories are used without much debate on the domination of a system under one category or the other. The amount of sediment in a rifted margin is generally easy to determine from seismic reflection imagery. But in some cases, it can be difficult to identify the actual top of basement, and thus the base of the overlying sediments. Thus, questions may remain on regarding the correct identification of the more deeply buried sediments which, because of compaction and diagenetic effects, may become undistinguishable from basement on standard geophysical datasets.

Further reading.– The above descriptions are abbreviated and often simplified. If interested in reading and learning further, the reader is referred to the following list of publications and references:

– General: (Coffin and Eldholm 1994; Whitmarsh and Sawyer 1996; Planke *et al.* 2000; Whitmarsh *et al.* 2001; Foulger 2002; Martinsen *et al.* 2010; Ebinger *et al.* 2010; Franke 2013; Peron-Pinvidic *et al.* 2013; Basile 2015; Mercier De Lépinay *et al.* 2016; Brune *et al.* 2018; Peron-Pinvidic and Manatschal 2019).

2.4. Historical review of how our understanding of rifted margins evolved

Although rifting is a fundamental plate tectonic process, the geological and geophysical characterization of basement geometries, stratigraphic architecture and magmatic products in rifted margins is still debated. Numerous models have been proposed for continental rifting, for individual study areas, as well as for the global understanding of rifting. Although some consensus has been reached on specific questions, there is still significant disagreement between researchers over models for continental rifting.

In this chapter, we review the evolution of knowledge on rift and rifted margin evolution and architecture, from the pioneering models that defined the fundamental physical rules governing lithospheric extension to modern conceptual models and today's consensus and debates.

2.4.1. *The pioneering models*

Two fundamental tectonic models cover the basics for studying rifts and rifted margins, the pure shear model by McKenzie (1978) and the simple shear model by Wernicke (1985). As more recent progress in geological and geophysical understanding has advanced our understanding of rifts, both models have been proven to be overly simplistic. However, our understanding of the extensional processes remains limited, and both of these pioneering models are surprisingly extremely robust.

2.4.1.1. *The pure shear model: McKenzie 1978*

To explain the development and evolution of sedimentary basins, McKenzie (1978) proposed a model which became famously known as the "**pure shear model**" or "**McKenzie model**". In this simple yet robust approach, the basin is envisioned as a simple vertical column of homogeneous continental lithosphere that is stretched uniformly and instantaneously (hence pure-shear deformation) to form a symmetrical basin, causing localized initial subsidence and passive upwelling of the asthenosphere with a thermal anomaly. The deformation is supposed to be uniform with depth across the entire rift.

When stretching ceases, the asthenosphere cools and becomes denser, provoking a slow subsidence of the basin, which is then progressively filled with sediments. The subsidence curve of pure-shear basins is typically divided into two stages: a first **tectonic-stage** with a rather short time scale (e.g. 10 Myrs) and relatively abrupt, followed by a **post-rift thermal stage**, very gentle and long-lasting (e.g. 40–100 Myrs; Figure 2.15). The amount of subsidence depends on the amount of initial stretching, which is estimated by the **stretching factor** (see Chapter 1).

The **pure shear model** is attractive in that it is a simple one-dimensional model, easy to implement mathematically, and can offer first-order estimates of thermal gradient evolution. The model is often used to determine an approximate estimate of subsidence and paleotemperature studies of rift basins.

Even though the theory is four decades old and uncomplicated, the concept still holds and remains very relevant to modeling the overall behavior of stretched lithosphere. When examining rifting in detail, many observations reveal that there is most likely no basin, rift or rifted margin developed by strict homogeneous stretching. Additionally, the pure shear model does not incorporate fundamental geodynamical processes, such as flexural isostasy and decompression melt, to correctly account for natural rift cases. Over the decades, the approach received various modifications to account for other rifting parameters, notably:

– the dependence of subsidence and heat flow histories with the rate of extension (Jarvis and McKenzie 1980);

– faulting and related flexural effects (Watts *et al.* 1982);

– small-scale convection cells in the asthenospheric mantle beneath the stretched lithosphere to account for more dynamic mantle behaviors, which can impact the amount of lithospheric thinning and width of the asthenospheric rise (Keen *et al.* 1987);

– different amounts of extension of the upper and lower lithosphere, which can cause a physical decoupling of the two layers and alter the balance between the stretching stage and the thermal subsidence stage (Royden and Keen 1980). This also relates to the simple shear model by Wernicke (1985), which assumes an offset between the stretched region of the upper crust and a ductile stretching of the lower crust/mantle (see section 2.4.1.2).

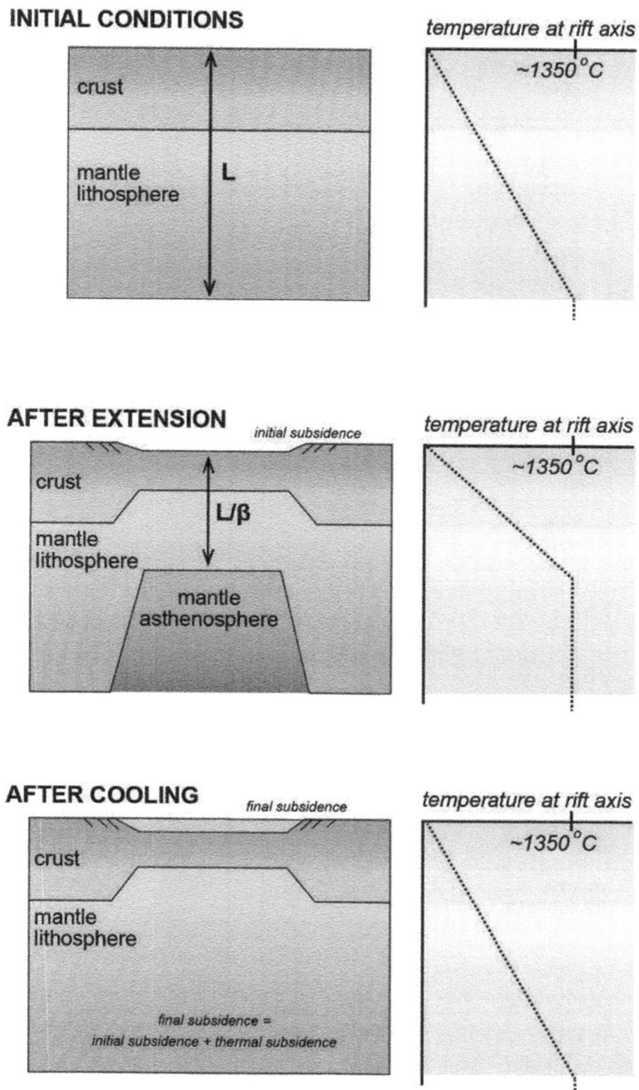

Figure 2.15. *Illustration of the McKenzie rift model (McKenzie 1978): the lithosphere is stretched; the lower continental crust plastically extends while the upper continental crust deforms by brittle faults that sole out on deeper ductile layers. The Moho rises at the rift axis and as the lithospheric mantle is thinned, it is replaced by hot asthenosphere at the rift axis (after extension). After cooling, the asthenosphere gradually cools over tens of millions of years and the lithosphere–asthenosphere boundary deepens while becoming cooler; the asthenosphere also becomes denser and participate to the subsidence of the above-lying sedimentary basin*

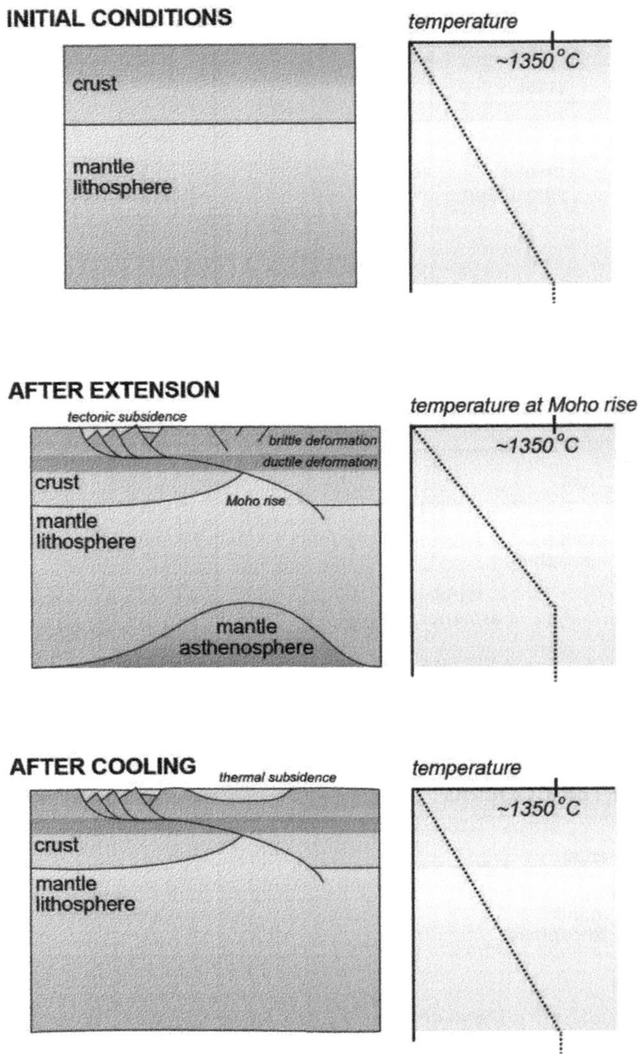

Figure 2.16. *Illustration of the Wernicke rift model (Wernicke 1985). Similarly to the McKenzie model, the continental lithosphere is stretched; the upper continental crust deforms by brittle faulting, the lower continental crust deforms ductily and the Moho rises, bringing up hot asthenosphere. The main difference is that stretching is partly accommodated by a low-angle detachment system that produces a strong asymmetry in the whole rift system, with a lateral shift between the thin-skinned basin developed initially due to faulting of the upper crust and the thermal basin developed at a second stage above the axis of the rising Moho and asthenosphere*

2.4.1.2. *The simple shear model: Wernicke 1985*

Later, Wernicke (1985) presented a rift model that emphasizes the role of faulting. Contrary to the McKenzie approach, the Wernicke model assumes a heterogeneous stretching of the lithosphere, where deformation is conducted in a simple shear regime. The rift basin is asymmetrically extended along a gently dipping large-scale detachment fault that traverses the lithosphere as a whole (Figure 2.16). Within this framework, the continental lithosphere is separated into two plates: a lower plate consisting mainly of lower crust at the rift axis and an upper plate. This model predicts an offset between the locus of maximum crustal thinning and the locus of maximum thinning of the lithospheric mantle (Figure 2.16). The effect is that the rising lithosphere–asthenosphere boundary, and thus the subsequent thermal subsidence, is laterally displaced from the area of maximum crustal thinning – which significantly differs from the pure shear model.

Various authors have extrapolated this model to rifted margins (Lister *et al.* 1986) or fossil margins (Lemoine *et al.* 1987). This is detailed in section 2.4.2.

2.4.2. *The evolution of knowledge*

Although rather simplistic in their approach, both the McKenzie and Wernicke concepts were considered to explain important aspects of rifting evolution. Following that line of thought, **composite models** emerged rapidly, which sought to reconcile the most important features of the two models. In the composite models, simple and pure shear act together to accommodate continental rifting. These mixed models often proposed that the major basement faults root at the brittle–ductile boundary at ~10–15 km depth or the middle crust, and that the rest of the lithosphere is deformed by distributed ductile pure shear extension (Le Pichon and Sibuet 1981; Sibuet 1992). Building upon this model, Kusznir *et al.* (1987) proposed the **flexural cantilever model**, where planar faults are prevalent in the upper crust and listric faults dominate at deeper crustal levels.

The following sections review some of the rifting models proposed in the "post- McKenzie/Wernicke" literature. The models summarized below are considered as milestone contributions, very often cited in conferences and peer-reviewed papers as key references because they advanced our knowledge and understanding of rifted margins.

2.4.2.1. *Rift models applied to rifted margins: the upper-plate/lower-plate distinction*

Simultaneous to the development of the McKenzie and Wernicke models and their successful application to intracontinental rifts, various derivative models rapidly emerged and inspired the rifted margin community. Lister *et al.* (1986) applied the detachment fault model (e.g. Wernicke 1985) to rifted margins by hypothesizing that detachment faults, which are related to the formation of continental core complexes, also exist in rifted continental margins, although they were not yet identified at the time. Following Wernicke's simple shear model, Lister *et al.* (1986) developed a rifting model based on the assumption of two major types of rifted margins: the "upper margins" and the "lower margins" or the **upper-plate** and **lower-plate margins**. The upper-plate margin corresponds to the hanging wall of the detachment system directing rifting; they consist of crustal rocks that were structurally above the detachment. The lower-plate margins are primarily formed of rocks located in the footwall of the detachment, with highly faulted remnants of the upper plate at the interface (Figure 2.17). These two fundamental categories of margins differ in their syn-rift structural configuration and subsidence pattern, and in their uplift and thermal histories. The location of the ocean–continent boundary, the geometries of the detachment faults and the rates of displacement along the major faults are used to constrain the characteristics of the margins. Lister *et al.* (1986) then proposed the existence of **transfer faults** that can accommodate the lateral structural shift from one setting to the other. This model led to a primary structural asymmetry, or a complementary geometry.

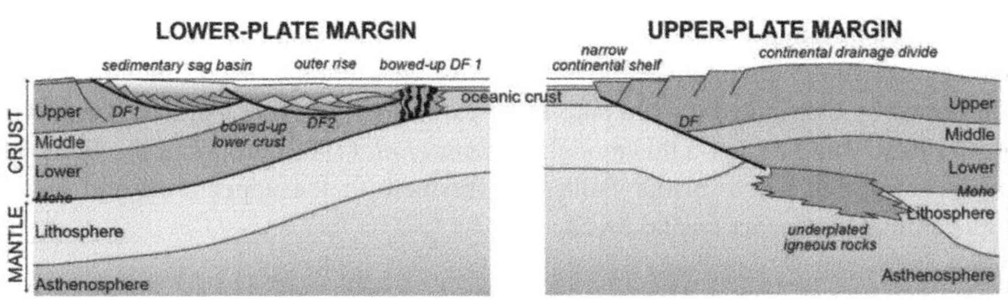

Figure 2.17. *Detachment fault (DF) model for continental passive margins showing the two types of margins formed: the "lower-plate margin" and the "upper-plate margin" (source: adapted from Lister* et al. *1986)*

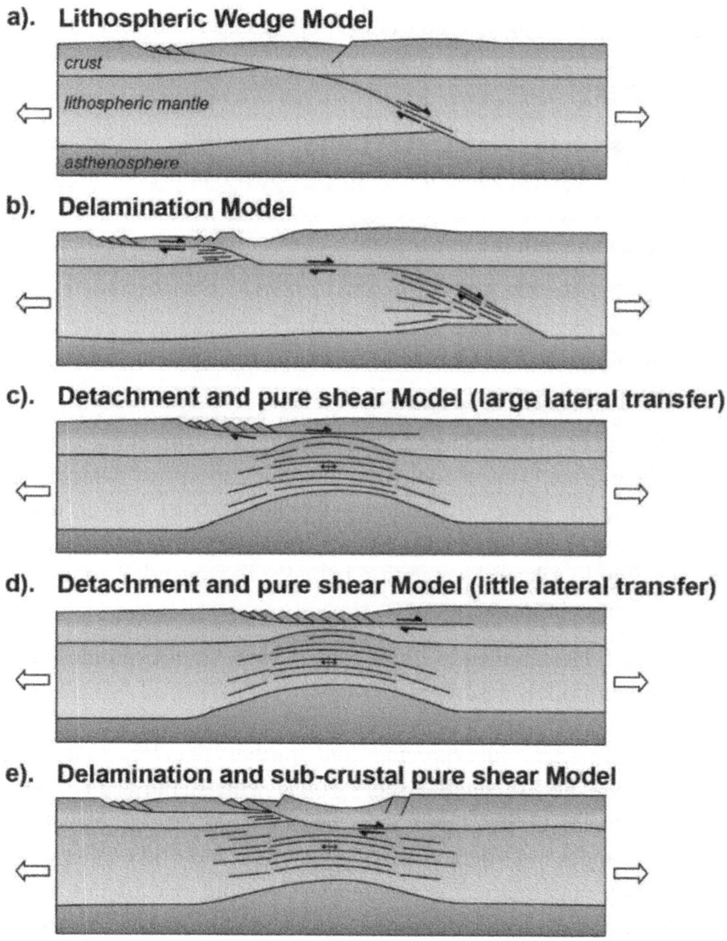

Figure 2.18. *Five extension models using the concept of detachments. a) Lithospheric wedge model, based on one or more generations of detachments cutting through the lithosphere. b) Delamination model, based on ramp geometries and at least two generations of detachments leading to the formation of marginal highs and ramp basins. c) Detachment and pure shear model with large lateral transfer, based on stretching of the lithosphere by pure shear, with the presence of a ductile zone potentially displaced from the brittle zone. d) Detachment and pure shear model with little lateral transfer, with pure shear stretching of the lithosphere directly located below the brittle detachment zone. e) Delamination and sub-crustal pure shear model, based on multiple detachments with composite geometries including ramps and flats, with ductile deformation directly located under the brittle detachment (source: adapted from Lister* et al. *1991)*

Following their 1986 study, Lister *et al.* (1991) described four main conceptual models intended to represent the main types of architectures observed at rifted margins (Figure 2.18):

– The "Lithospheric Wedge (Wernicke) Model", where extension is accommodated by extreme amounts of displacement along a single, large-scale detachment fault; the uplift and unroofing of the lower plate occur as it is pulled to the surface from underneath the upper plate.

– The "Delamination Model" corresponds to the delamination model initially introduced and described by Lister *et al.* (1986). The detachment develops in regions of rheological weaknesses throughout the lithosphere, displacing material over great distances at intracrustal depths.

– The "Detachment + Pure Shear Model" has ductile stretching occurring under a mid-crustal shear zone. This model combines the concept of detachment and flat shear zones with pure shear extension of the lower crust and/or upper mantle, resulting in a horizontal transfer of the lithospheric extension through the detachment, leading to a lateral separation between the extended zones in the upper crust and in the deeper lithosphere.

– The "Stepped Detachment + Pure Shear Model" incorporates most of the features of the other models described above.

Overall, these models incorporated and further developed the concept of detachment faults related to ductile shear zones with shallow dips. Modeling of subsidence and crustal uplift shows that variations in crustal elevations can be explained by changes in detachment geometry and lithospheric stretching. Significant igneous underplating is predicted if a rapid rise of hot asthenosphere takes place, which also creates additional lifting of the overlying structures. The location of breakup does not coincide with the deepest rift basin, but is instead located in regions more oceanwards, in areas where the asthenosphere is the shallowest (Figure 2.18).

2.4.2.2. *The upper-plate paradox and depth-dependent extension*

In parallel with the evolution of the upper-plate/lower-plate margin concepts and its derivatives being developed, crucial observations were made at various rift settings, such as the NW Australian shelf (Driscoll and Karner 1998), Norwegian margin (Roberts *et al.* 1997), Goban Spur, Galicia Bank or South China Sea (Davis and Kusznir 2004). The structural observations led to the conclusion that all rifted margins appear to correspond to an upper-plate setting, corresponding to what

Driscoll and Karner (1998) named the "**upper-plate paradox**". They argue that these cases illustrate a mismatch between stretching values determined from faulted geological surfaces and wholesale lithosphere thinning calculations. The extension of the upper crust is calculated by summing the fault throws measured on seismically imaged faults. Full crustal stretching is constrained by a combination of seismic refraction, seismic reflection and gravity inversion constraints. The stretching of the entire lithosphere is determined through post-breakup thermal subsidence analysis using flexural backstripping, decompaction and reverse thermal subsidence modeling (Davis and Kusznir 2004). The resulting values differ substantially, confirming the depth-dependent deformation pattern predicted by the simple-shear lithosphere extensional model. However, this analysis points to a failure in either the upper-plate/lower-plate model or the observations, measurements and modeling performed to estimate the amounts of stretching.

Later, Huismans and Beaumont (2011, 2014) further refined the concept of depth-dependent extension using numerical modeling, pointing to the crucial role of the depth-dependent extension pattern in explaining the end-member styles of rifted margins. They notably confirmed that margin subsidence characteristics can be explained by diachronous syn-rift extension.

Depth-dependent stretching is usually attributed to the rheological partitioning of the lithosphere, which, in turn, leads to the development of decoupling interfaces between layers of different strengths. These interfaces are favored sites for the initiation of major deformation structures, such as discrete or diffuse shear zones, detachments and decollement levels – all of which affect the extensional geometry of the margin. Additional parameters can also have a crucial role in the asymmetric development of rifted margins, such as the thermal lithospheric status, opening geometry, presence of melts and fluids or inherited structures (see section 2.5.3).

2.4.2.3. *The deformation mode approach*

Various key observations made in the 1990s regarding rifted margin architecture led to significant advancements in the understanding of the related rifting processes. In particular, research on the Iberia–Newfoundland conjugate rift system has been at the forefront for many years, and many of the following models refer to this iconic study site.

The Whitmarsh *et al.* (2001) model

Based on the Iberia margin and geological studies carried out on the alpine fossils of the Tethyan margins, Whitmarsh *et al.* (2001) proposed an innovative approach of rifting evolution. The authors showed that observations supported the existence of tens of km-wide zones of exhumed continental mantle lying between oceanic and continental crust – an observation that was not predicted by previous models. The model (Figure 2.19), designed for a magma-poor setting, assumes that rifting begins within a continental crust in isostatic equilibrium, about 30 km thick with a four-layer rheological partition. The crust can be locally thickened by pre-rift, underplated gabbro, as observed in the Alpine Tethys fossils and in the southern Iberia Abyssal Plain (Manatschal 2004). In the Whitmarsh *et al.* (2001) model, the onset of rifting is controlled by the ductile flow of the lower continental crust; the upper crust and mantle deform in the brittle field and the extension are dominated by simple shear at the entire system scale. Basins are bordered by listric normal faults that root in the rheologically weaker middle crust, which allows for the decoupling of the deformation between the upper crust and the mantle. The rifting is therefore distributed over a large area across the entire margin. The stretched crust cools slowly and magmatic activity is almost entirely absent. At that early rifting stage, the rift is symmetrical at the lithospheric scale. It is only later that the extension becomes asymmetric: this is the "**lithospheric necking**" **stage**. Mechanically or thermally induced weaknesses in the lithosphere control the initial location of lithospheric necking. Then, with the further development of fault activity, exhumation of the subcontinental mantle is accommodated by concave-downward detachment faults: this is the "**lithospheric detachment faulting**" **stage**. The geometrical change in the extensional structures most likely reflects a change in the distribution of the rheologically weaker layers, whereas the listric faults are flattened in the weak horizontal layers, the concave-downward faults are favored by sub-vertical zones of weakness, probably resulting from rising melts and stronger thermal gradients above the upwelling asthenosphere. The weaknesses guide the rise of the asthenosphere, which then controls the evolution of the final stages of rifting and the location of the continental breakup for the "**onset of seafloor spreading**" **stage**. The oceanwards increase of mid-ocean ridge basalt (MORB)-type magma volumes eventually generates a thin oceanic crust.

The novelty of the Whitmarsh *et al.* (2001) model resides in the consideration of a large number of constraints from modern rifted margins, as well as fossil onshore analogues. The remnants of rifted margins incorporated in orogens were attracting increased interest in the late 1990s and 2000s, leading to a serious revision of many of their geometries within the framework of new rift models (e.g. Manatschal Bernoulli 1998). This, in turn, triggered a growing interest from the rifted margin community, which was more traditionally focused on offshore cases.

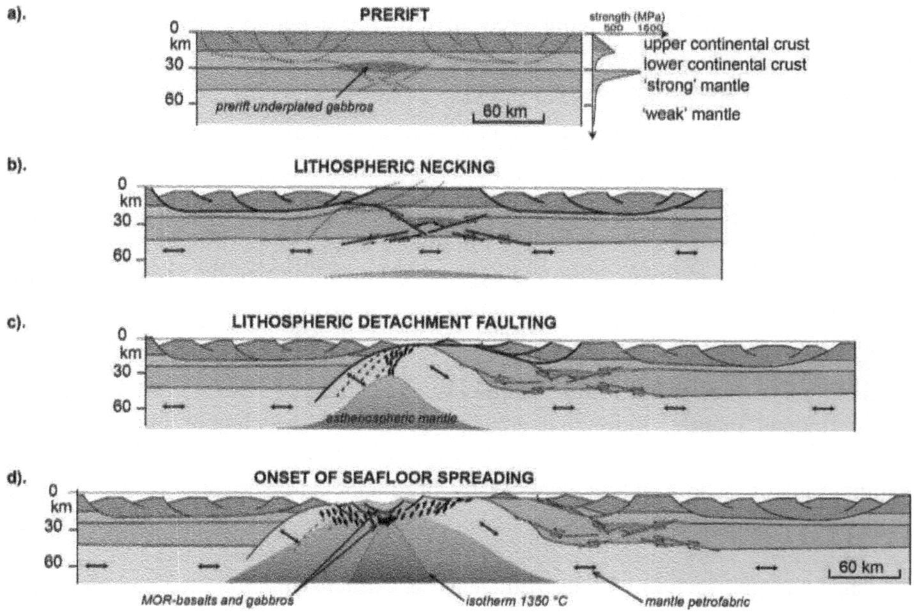

Figure 2.19. *Conceptual lithosphere-scale model of development of a rifted magma-poor margin relative to a fixed right-hand edge (source: Whitmarsh* et al. *2001)*

The Lavier and Manatschal (2006) model

The conceptual model by Whitmarsh *et al.* was soon refined by Lavier and Manatschal (2006), who used geodynamic numerical experiments to simulate continental rifting. The code (initiated on the PARAVOZ code; Poliakov *et al.* 1993) was run in a series of tests in order to respect the observations made at both the Iberia and Alpine Tethys margins. Initial lithospheric conditions, such as the crustal thickness, thermal structure and rheological profile of the continental crust had to be defined and the lithosphere was then modeled as a nonlinear visco-elasto-plastic deforming medium.

In their model, Lavier and Manatschal (2006) identified a succession of three major modes of deformation which can be directly correlated with the model by Whitmarsh *et al.* (2001; see Figure 2.20):

1) The **stretching mode** is characterized by the formation of multiple normal faults distributed throughout the entire extensional system. The faults have an overall listric geometry, cutting through the brittle upper continental crust and soling

out at more ductile levels encountered at mid-crustal depths. The accommodated displacements are relatively small, the total extension is limited and the accommodation space remains very minor.

2) The subsequent **thinning mode** is concentrated within the future distal margin. Shear zones develop at the crustal scale, thinning the continental crust to a thickness of less than 10 km. The ductile middle crust flows along the lower rigid crust and allows for delocalization of the deformation. The brittle upper crust and the upper brittle mantle are deformed by normal faults thinning the crust and mantle and exhuming the middle crust and mantle in an overall rolling-hinge-type fault system.

3) Finally, the **exhumation mode** is characterized by unroofing of the subcontinental mantle. Individual concave-downward detachment faults develop, probably influenced by serpentinization processes. These particular faults allow the exhumation of deep levels of mantle by producing large displacements without generating pronounced topography.

The margin section finally obtained is structurally highly complex and juxtaposes different levels of crust and mantle, both horizontally and vertically.

2.4.2.4. *The ductile deformation approach*

Many rifting models involve ductile deformation of some part of the lithosphere, often the lower crust (Huismans Beaumont 2011), with lateral extractions of the ductily deformed crust towards the ocean or the continent (Driscoll Karner 1998; Brun Beslier 1996) and boudinage structures (Reston 1988). However, in most cases, the rifting models only focus on brittle deformation, and ductile deformation is often envisioned as secondary deformation, mainly responsible for the partitioning of the deformation into distinct layers. Recently, however, some publications have proposed **interpretations and models that give a more significant role to the ductile regime in the overall rifted margin evolution and final geometry.**

For instance, Clerc and Lagabrielle (2014), followed by Clerc *et al.* (2016), used the fossilized Cretaceous margins of the North Pyrenean Zone to illustrate the fact that the local lithospheric thermal structure directly impacted the mode of deformation used to accommodate extension. Their alternative model implies a boudinage of the upper crust, allowed by a high geothermal regime combined with a blanketing effect of the sedimentary cover (see the case example in Chapter 3).

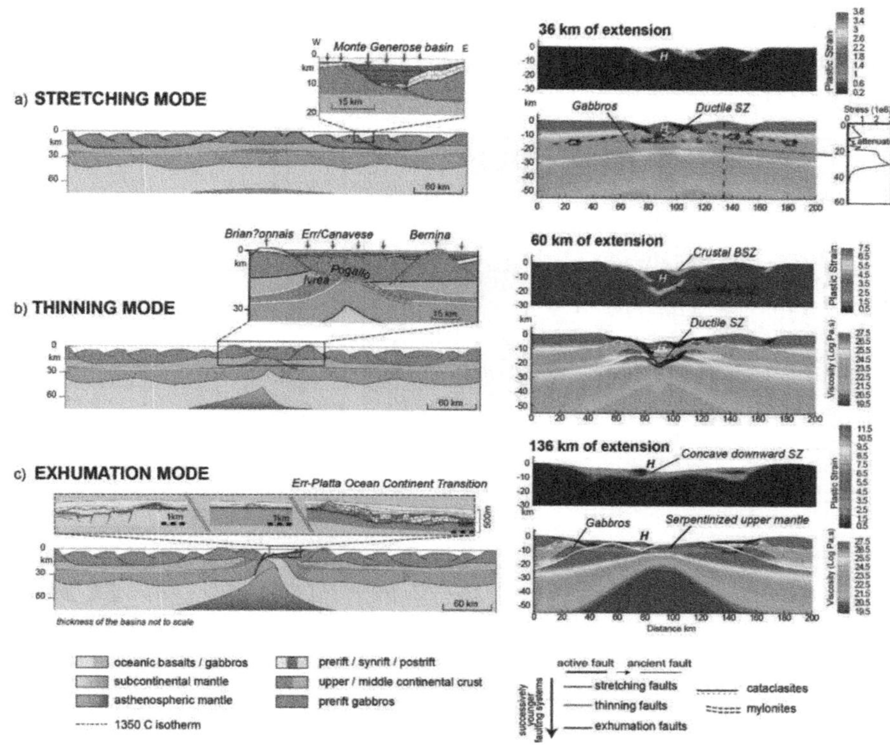

Figure 2.20. *Modes of extension leading to continental break-up. Left column: conceptual models for the different phases of extension based on observations from the Alpine and Iberia/Newfoundland margins. Right column: modeled evolution during lithospheric extension. The plastic strain (brittle deformation) and viscosity fields are plotted for the different phases of the modeled evolution of the lithosphere. Note the similarity between the observed and modeled structures. The stress envelope is plotted for a given depth profile (dashed lines for 36 km of extension). For each viscosity field, the Mohorovic boundary is shown by a white line. The letter H indicates the hanging block (source: Lavier and Manatschal 2006)*

Based on the increased number of high-resolution deep seismic reflection data, Clerc *et al.* (2015, 2018) were subsequently able to demonstrate that such boudinage fabric may be observed at other modern rift settings. In fact, before the recent access to high-resolution data, very few observations existed on the seismic facies and geometries of the lowermost parts of the basement. By using long-offset seismic profiles from the Uruguay, Namibia, Gabon and South China Sea margins, Clerc *et al.* (2015, 2018) showed that the lower basement deformation fabric may encompass greater amounts of ductile shear, boudinage and low-angle detachments than previously thought. Ductile, anastomosed shear zones delineate boudin-like remnants of upper crustal rocks on many regional seismic reflection profiles (Figure 2.21).

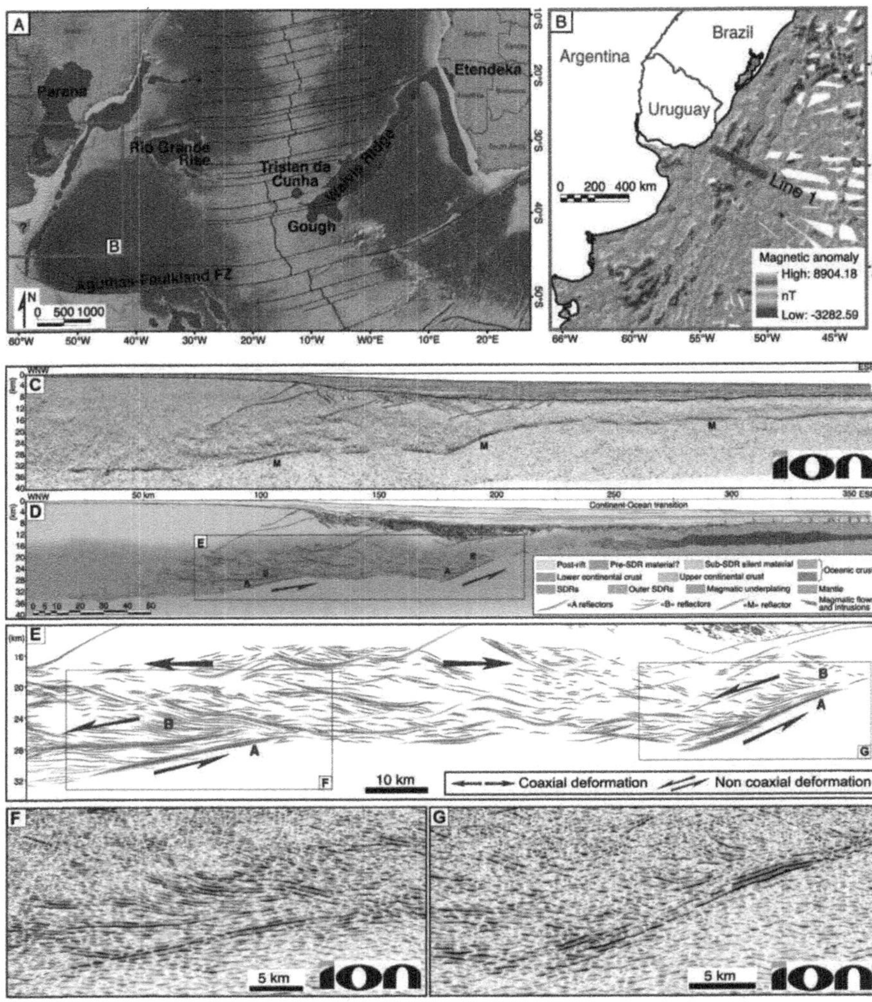

Figure 2.21. *Illustration of shear deformation based on seismic reflection profiles in the South Atlantic. a) Map of the South Atlantic Ocean around the Uruguay rift margin. b) Magnetic anomaly map of the Uruguayan rifted margin (Maus* et al. *2009). c) Seismic reflection profile and its interpretation d) and e) Interpreted close-up of the lower crust. f) and g) Details of the base of the continental crust presenting strong reflectors interpreted as typical shear patterns (source: adapted from Clerc* et al. *2015)*

Numerical experiments aiming at reproducing lithospheric extension also point to the fundamental role of the rheological state of the lower crust: the viscosity, which strongly depends on the lower crustal composition, thermal state and extension rate, has been identified as one of the key parameters controlling the margin evolution and final geometry (e.g. Brune *et al.* 2014).

2.4.2.5. *The faulting approach*

On the other side of the deformation spectrum, many models emphasize the role of normal faults and detachment faults, which are sometimes associated with shear zones and core complexes. However, when interpreting structural geometries in rifts and rifted margins, both onshore and offshore, it is often difficult to map the exact extent of a fault, and it is even more difficult to constrain its time–length evolution. For instance, the amount of extension measured from seismic reflection data is generally insufficient to explain the amount of crustal thinning and observed subsidence (Ziegler 1983). This problem is known as the **extension discrepancy**, which has been variably explained by sub-seismic faulting (Marrett Allmendinger 1992), unrecognized polyphase cross-cutting faulting (Reston 2005, 2007; Reston and McDermott 2014) or depth-dependent stretching/thinning with the lateral flow of lower crustal material (Davis Kusznir 2004). We briefly summarize some of these models below.

The polyphase faulting model

In order to solve this extensional discrepancy, and based on early studies such as Proffett (1977), Reston (2005) proposed that **polyphase faulting** may provide a simple explanation for the rifted margin cases. In this approach, the rift fault activity progressively evolves in successive phases with new faults cross-cutting, merging and/or displacing previous faults and fault segments.

Reston (2005) based his model on observations from the west Galicia margin and on Mohr–Coulomb theory, which predicts that faults become mechanically locked when rotated to low-angle values (<30°), thereby promoting the initiation and development of a new series of faults as extension progresses. The entire process results in intricate geometries with several generations of cross-cutting faults (Figure 2.22). The multiple generations of faults may be impossible to identify subsequently, both offshore and onshore. Reston (2007) was also able to apply this model to onshore transects, notably the Spanish Betics. Later, McDermott and Reston (2015) used seismic synthetic modeling to support the concept of polyphase faulting.

The sequential faulting model

To explain the asymmetry of conjugate margins and extension discrepancy, Ranero and Pérez-Gussinyé (2010) proposed an alternative to the polyphase model by Reston (2005, 2007). They used depth-converted seismic reflection profiles from the Iberian and conjugate Newfoundland margins to measure fault geometries, dip angles and displacements and compare the results to the interpreted crustal thinning.

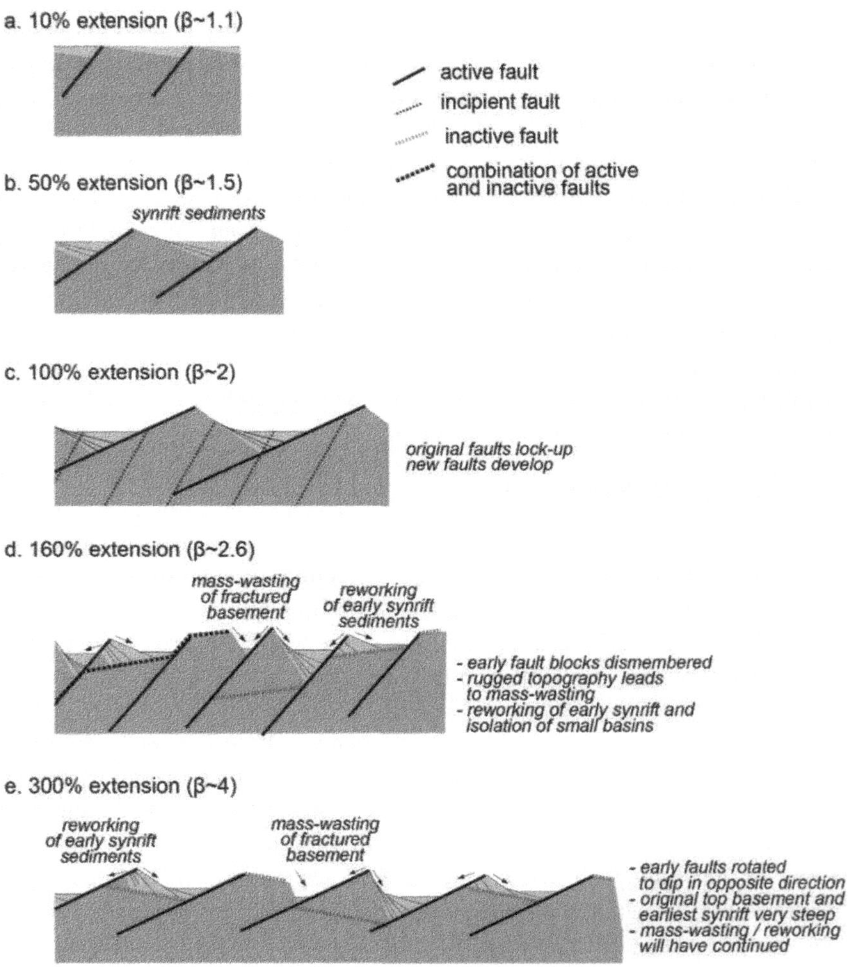

Figure 2.22. *Illustration of the variety of structures produced by polyphase faulting in a domino model (source: adapted from Reston 2005)*

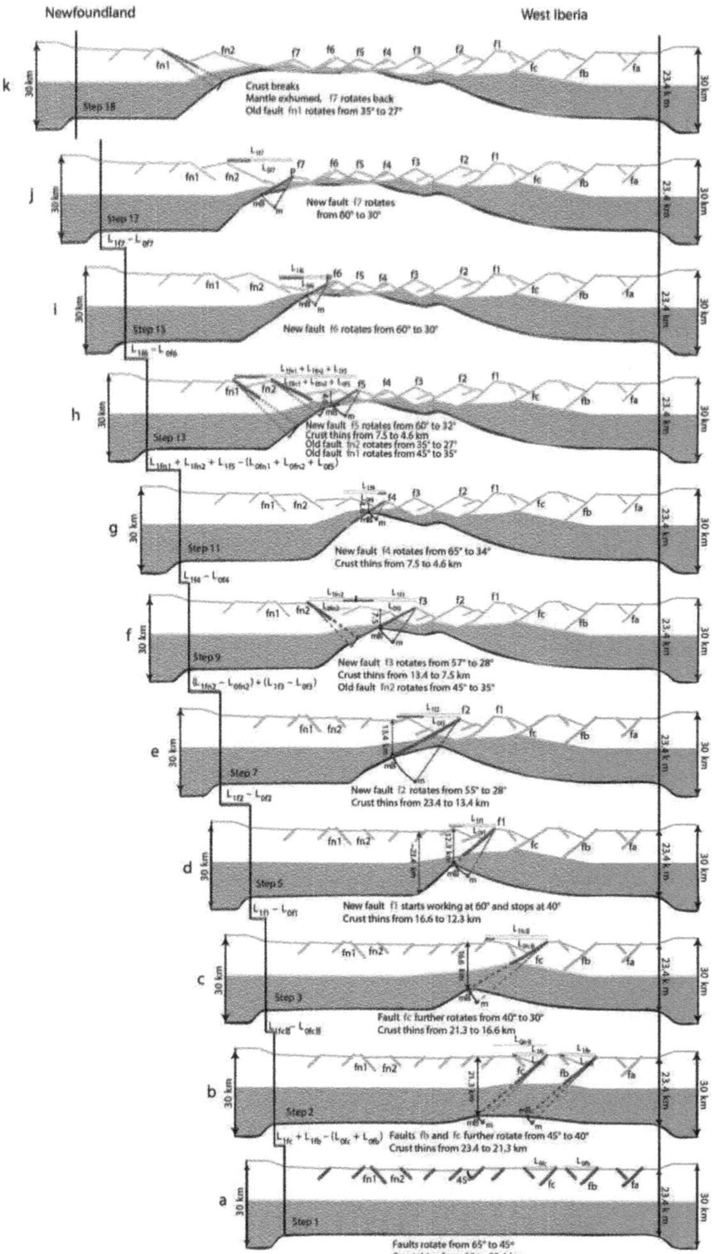

Figure 2.23. *Illustration of the sequential faulting model based on the Iberia–Newfoundland rift system. (source: adapted from Ranero and Pérez-Gussinyé 2010)*

CONTINUATION OF CAPTION FOR FIGURE 2.23. – *The model is fixed on the right-hand side. The basement is overlain by pre-rift sediments (in gray). The thick red and orange lines are active and inactive faults, respectively. The thin red lines mark the initial fault geometry. The black vertical arrows to the left of faults are crustal thicknesses. A blue line represents the Moho. The red, dashed lines are the projection of faults through ductile crust, used to geometrically estimate Moho movement. The pale orange shading represents the petrological lower crust, which switches from ductile (a–c) to progressively more brittle at the basin center (d–k). At each stage, horizontal extension equals the extension by active faults (indicated by the pink and blue horizontal lines). L0fi and L1fi are the horizontal lengths of a fault block before and after the extension of fault fi (i = a, b, c, 1, 2, 3, 4, 5, 6, 7, n1, n2). L0fc' and L1fc' are the horizontal lengths of the hanging-wall block to fault fc before and after the fault movement occurring between steps 2 and 3, respectively*

The fault mapping was used to create a kinematic-balanced profile along which rifting is driven by Andersonian-type normal faults that dissect the crust into tilted crustal blocks, with a progressive migration of the fault activity towards the center of the rift. The authors argue that the entire crustal thinning, up to mantle exhumation, can be explained by seismically visible normal faults forming in sequence (Figure 2.23).

Ranero and Pérez-Gussinyé (2010) explain that the asymmetry of the conjugate margins may arise from strain localization, promoting one direction of faulting from the early rifting stages. Furthermore, they suggest that slip along low-angle detachment structures is mechanically problematic (because they do not respect the Andersonian law) and may be related to misinterpretation of low-angle structures. They argue that low-angle faults can be created by the progressive rotation of inactive, originally steeper faults induced in their sequential faulting model (Figure 2.23).

The flipping fault model

The sampling of serpentinized mantle on offshore Iberia (Boillot *et al.* 1980), and the subsequent confirmation that rifting can lead to the exhumation of significant surfaces of subcontinental mantle on the seafloor (Tucholke and Sibuet 2007), led various authors to accept the existence of large-scale detachment faults in the distal margin. Similar to the rolling-hinge-type models developed for slow-spreading mid-ocean ridges (Tucholke *et al.* 1998; Lavier *et al.* 1999), concave-downward detachments rapidly became popular structures to explain the mantle unroofing in rifted margin settings.

Reston and McDermott (2011) proposed a model where the wide zones of exhumed mantle are generated through the successive formation of many generations of such detachment faults (Figure 2.24). Although each structure exhumes mantle in an asymmetric way, the detachments are supposed to initiate with successive opposite dips. Therefore, each fault cuts through the predecessor's fault's footwall, isolating segments of the previous detachment on one margin or the other, depending on the rift axis location. This elegant process allows the generation of roughly symmetric wide zones of exhumed mantle, as observed along the Iberia–Newfoundland conjugates.

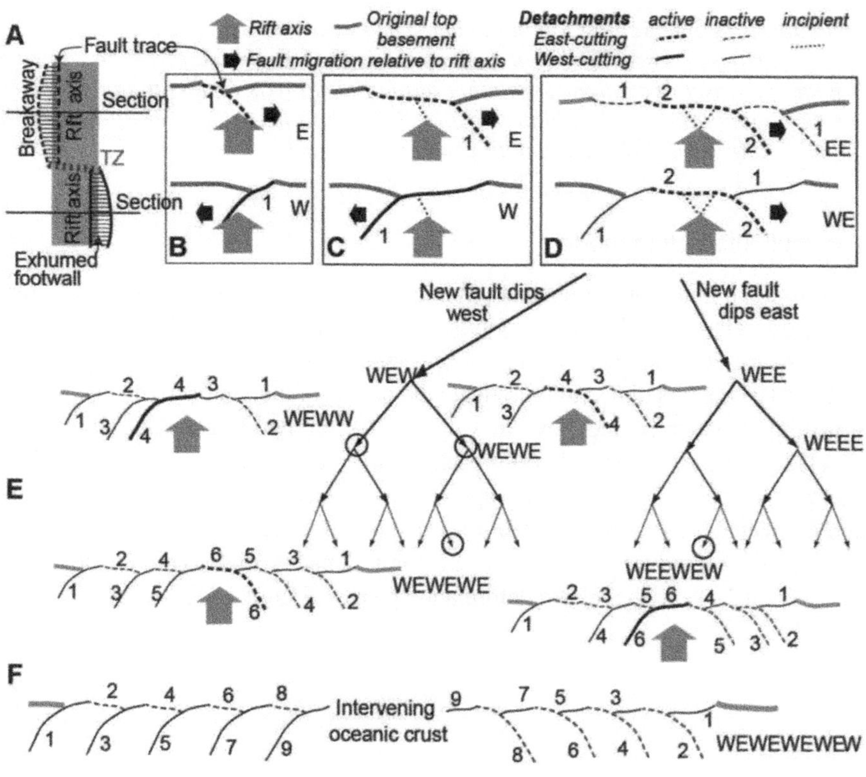

Figure 2.24. *Unroofing of the symmetric expanse of mantle by successive asymmetric rolling-hinge detachments (source: Reston and McDermott 2011)*

CONTINUATION OF CAPTION FOR FIGURE 2.24.– *a) Map view of two adjacent rift segments of opposite polarity separated by transfer zone (TZ). b) Amagmatic extension is taken up on faults originally rooted at the rift axis, dipping east (E) and west (W) in northern and southern segments, respectively. c) As the footwall is pulled out from beneath largely undeformed hanging wall, the active fault root moves (black block arrow) with the hanging wall over and beyond the rift axis and away from the root in the adjoining segment. A new fault cuts the rift axis through the exhumed footwall; the fault is shown here as the east-dipping structure in both rift segments, giving fault polarity sequences EE and WE in northern and southern segments. d) The new fault also moves with the hanging wall over and beyond the rift axis, until a third fault cuts up through footwall. e) Fault polarity tree showing how successive detachments cutting through footwalls of previous detachments always produce abandoned fault roots dipping away from the rift axis on both sides of the rift, whereas the exhumed slip surface of the rift-ward-cutting faults form the intervening top basement. The geometry resulting from four possible fault polarity sequences is shown, corresponding to the circled positions on the tree. f) Eventual onset of seafloor spreading along the rift axis results in abandoned root zones dipping landward on both margins. Note that the detachments, and therefore the generally unroofed mantle, both young oceanward*

The fault interaction model

Based on this consensus that fault activity is of primary importance for the evolution of the overall rifted margin geometry, Naliboff *et al.* (2017) used numerical experiments of continental rifting to understand the controlling parameters of the system. The authors aimed to investigate the details of the fault activity and associated structural geometries and how they evolve as rifting progresses. The numerical simulations are based on a thermo-mechanical finite element code (SULEC; refer to Buiter and Ellis (2012)), which approximates ductile flow and brittle failure in unprecedented high resolution (250 m) with a true free surface. This allows the models to correctly reproduce the distinct structural domains observed in rifted margins (see section 2.5) and the related deformation phases (Figure 2.25).

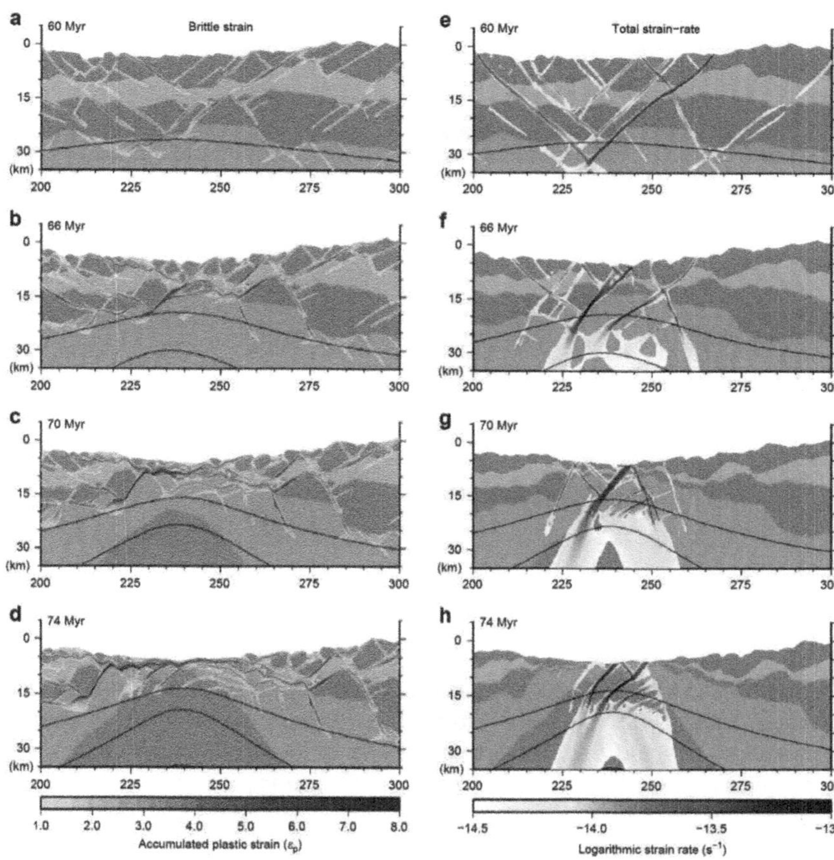

Figure 2.25a. *Evolution of finite strain and active deformation. Detailed evolution of the transition from stretching to hyperextension and exhumation in the numerical experiments. Each panel contains the compositional fields with the accumulated plastic (brittle) strain (a–d) or logarithmic strain-rate (e–h) superimposed. As extension proceeds, high-angle faults that formed in the early stretching phase are rotated sub-horizontally and stretched, while active shear zones remain at high angles*

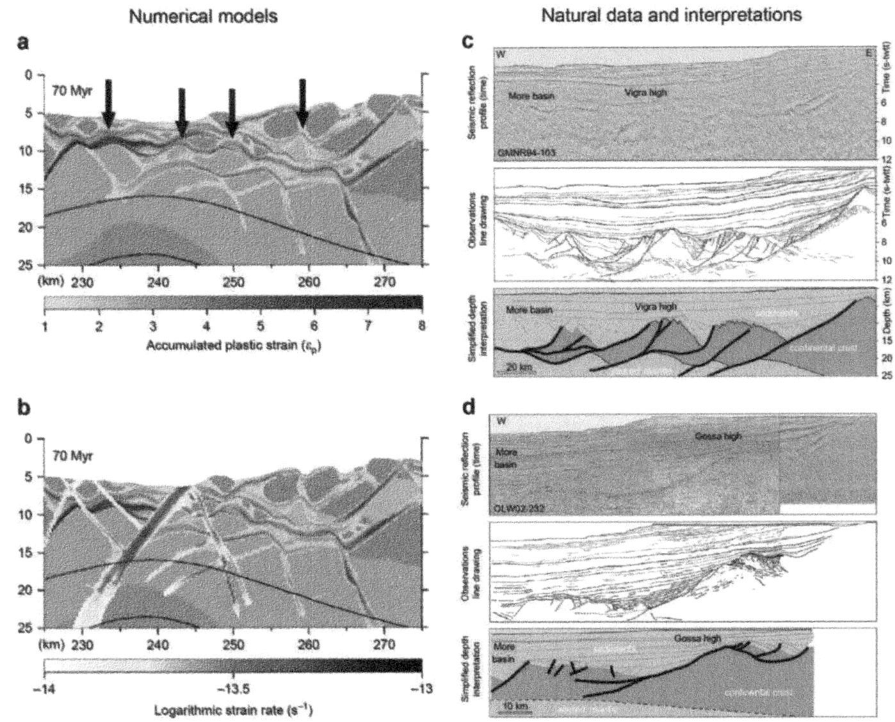

Figure 2.25b. *Comparison of the modeled and observed key structural features. a) Modeled lithospheric structure and accumulated deformation (plastic strain after 70 Myr). Arrows indicate the location of core complex structures. b) Strain-rate superimposed on the accumulated plastic deformation illustrates high-angle incision of active faulting on low-angle (rotated) structures. c) and d) Structure and interpretation, adapted from the previous work, of the distal margin structure along the mid-Norwegian margin. This region contains examples of both core complex geometries and incision of low-angle detachment faults by steeper faults penetrating into the mantle lithosphere (source: Naliboff* et al. *2017)*

The study indicates that the evolution of rifted margins is governed by multiphase rifting and inherited structural fabric, resulting in extremely complex geometries. The lifecycle of rift faults includes the definition of several fault subsegments through an intricate activity pattern with phases of fault-initiation, fault-activation, fault-linkage, fault-cross-cutting, fault-deactivation and possible fault-reactivation (Naliboff *et al.* 2017). The complex fault interactions are consistent with the theoretical models introduced by Reston (2005, 2007) and produce geometries resulting in complicated margin transects. **Standard seismic mapping protocols based on reflection data cannot yield the accurate identification of all of the polyphase structures, because their geometry and temporal complexity cannot be captured in geophysical datasets.** The numerical experiments revealed that what can be mapped is only a fraction of the true fault system and tectonic history, as proposed by McDermott and Reston (2015).

The fault breakaway model

The early studies of rifts and rifted margins identified the existence of distinct structural domains showing different evolutions and geometries; these were named the proximal and distal margins. With access to higher-resolution datasets, with increased areal coverage, these domains have been further characterized and key structures are identified. Section 2.5 summarizes these advances in detail.

Based on this domain-approach, Osmundsen and Peron-Pinvidic (2018) highlighted the key role performed by large-magnitude normal faults that bound the margin domains (Figure 2.26). Using the mid-Norwegian Vøring and Møre margins as examples, they showed that the evolution of margin domains and architecture are related to the linkage of such faults into regionally persistent "**breakaway complexes**". The strong along-strike geometrical variations observed in the Norwegian system is interpreted to reflect the interaction between these faults, and their interplay with fundamental parameters, such as variations in crustal thickness, rheology and inherited structures.

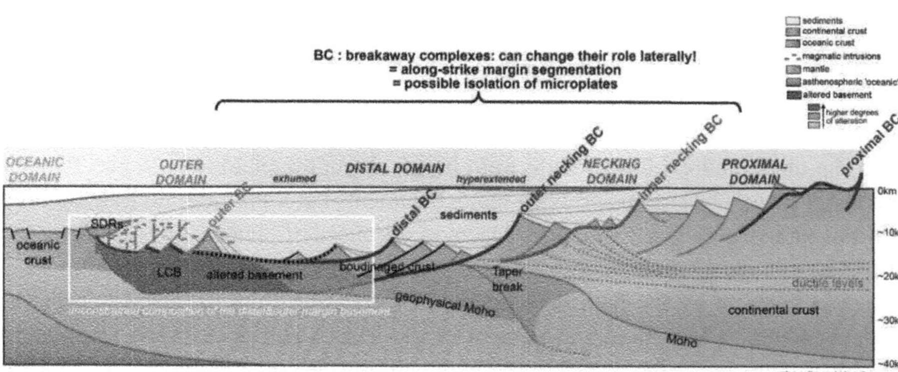

Figure 2.26. *Schematic representation of the structural setting of an idealized rifted margin, with margin domains and breakaway complexes (BCs) indicated (source: Peron-Pinvidic* et al. *2015 and Osmundsen and Peron-Pinvidic 2018)*

In the necking domain, where the crust is effectively thinned (see section 2.5.1.2), Osmundsen and Peron-Pinvidic (2018) introduced the notion of "**coupling efficiency**" to refer to the efficiency of the coupling processes between the crust and the mantle lithosphere.

The detachment factory model

Although Andersonian-type high-angle faults are favored in some settings, detachment faults are widely accepted as major structures that accommodate extension in rifts and rifted margins (see the previous sections). However, no information exists on where, when and how these detachments initiate. Therefore, Peron-Pinvidic and Naliboff (2020) used numerical experiments to further constrain the initiation and evolution of detachment faults during rifting, and proposed the detachment factory model.

The authors investigated the progressive formation and evolution of normal faults by mapping active and formerly active faults through short time steps (100 Kyr), by examining crustal structure, active deformation and finite deformation. In detail, the results of each time step contain active deformation (reflected in high strain rates) juxtaposed on accumulated plastic strain, which enables a detailed analysis of the fault activity through time (Figure 2.27).

The numerical simulations show that several detachment faults are simultaneously active at any given time, with some in the initiation stage and others in the process of deactivating. One, or at most, two detachments are actively exhuming mantle at a given time. These "detachment leaders" initiate at an ~15 km depth on the lower plate terminating against the weak zone termed the "nursery" (Figure 2.28). They gradually propagate upwards, extending in length and reaching higher dip angles in a fan-shape structure. They reach the seafloor surface with a high dip angle and exhume mantle from below the upper plate onto the seafloor of the lower plate. During this process, the detachments naturally migrate little by little off-axis and largely deactivate after reaching distances of about 10–15 km from the rift axis. The detachment footwall corresponds to an exposure of unroofed mantle, which is then cut by the next detachment leader arising from the nursery near the rift-axis center. The high-angle segment of the deactivated detachment remains on the upper plate and is the record of the fault activity. The overall geodynamic model is similar to the conceptual models proposed by Reston and McDermott (2011) for magma-poor rifted margins and Reston (2018) for slow-spreading oceanic ridges.

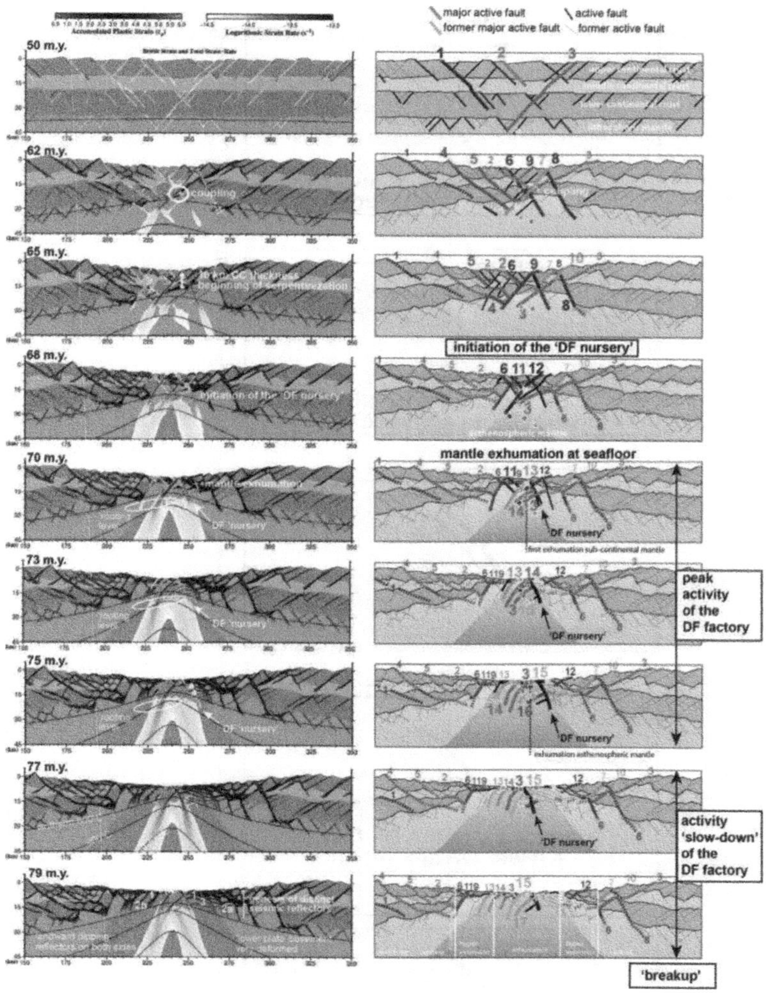

Figure 2.27. *Left column: numerical simulations with active deformation (strain-rate invariant) and accumulated plastic strain superimposed on crustal lithologies (see Naliboff* et al. *2017 for detailed description). Right column: interpretation of the simulations from mapping of the active and formerly active faults. The models allow for the identification of a tectonic system that accommodates mantle unroofing with multiple concave-downward detachment faults. We constrain a specific high-strain surface, the "detachment nursery", on which these detachments are born at depth and terminate while migrating upwards in an overall fan-shaped pattern. The detachments will cut the footwall of the previous detachments on seafloor at the rift axis to become the new detachment "leaders", and then migrate off-axis and deactivate when the hanging wall moves away (source: Peron-Pinvidic and Naliboff 2020)*

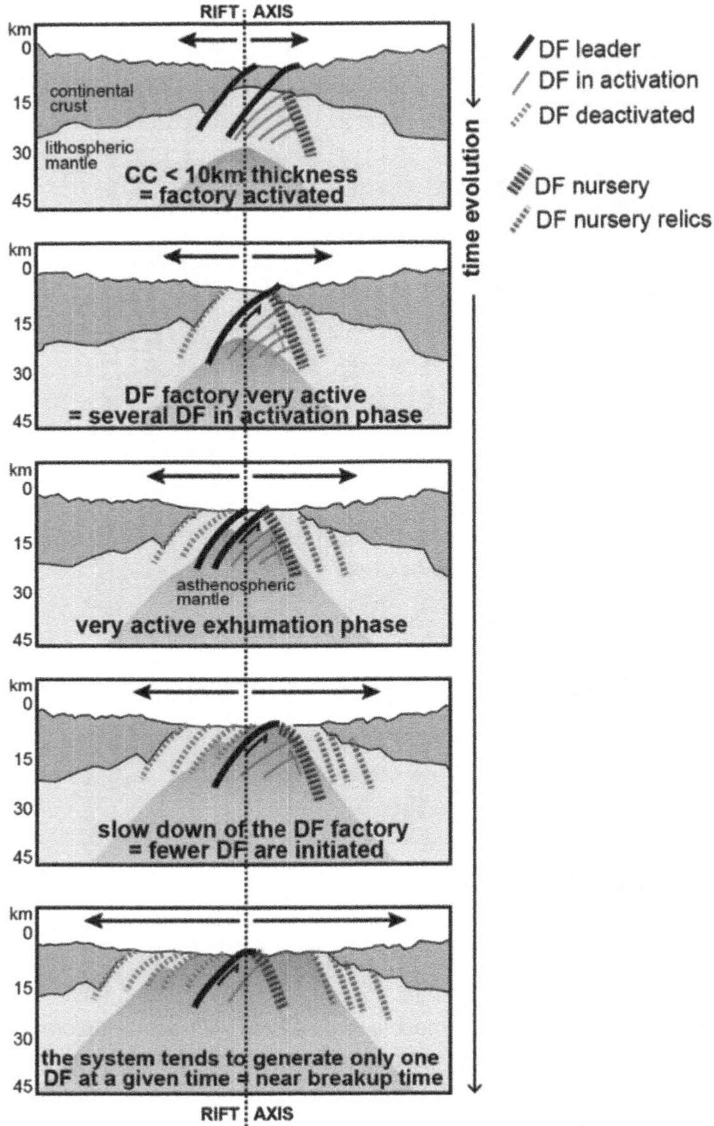

Figure 2.28a. *Schematic cartoon of the numerical simulations shown in Figure 2.27, highlighting the "detachment factory" activity through time (source: Peron-Pinvidic and Naliboff 2020)*

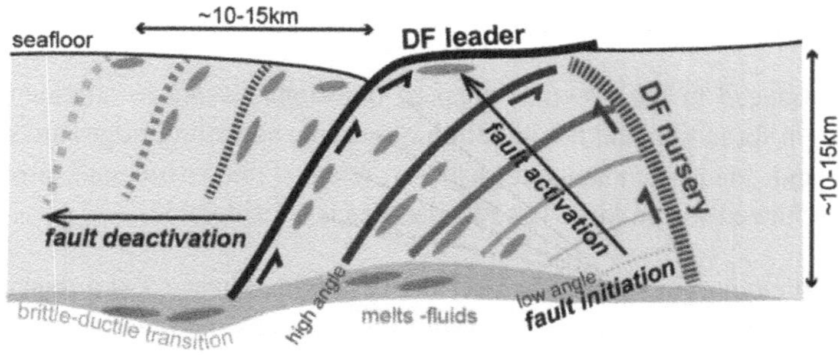

Figure 2.28b. *Cartoon illustrating the proposed exhumation detachment factory model: a weak surface, landward-dipping in the lower plate, is interpreted as a "detachment nursery". The exhumation detachment faults initiate and terminate on the nursery while propagating gradually upwards in an overall fan-shaped system. The faults activate while reaching higher angles and become the "leaders" when they reach the surface. The faults accumulate huge amounts of strain, which leads to rock damage that can favor fluid circulation, melt emplacement and serpentinization processes. The detachment fault leader then gradually migrates off-axis and is deactivated when the next detachment leader arises from the nursey at the rift axis. CC: continental crust. DF: detachment fault*

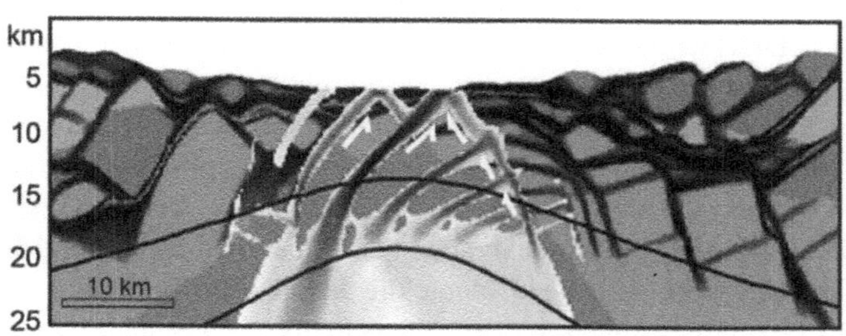

Figure 2.28c. *Close-up of the rift axis on the numerical simulation at the 73 Myrs time step*

Further reading.– The above descriptions are abbreviated and often simplified. If interested in reading and learning further, the reader is referred to the following list of publications and references:

– General: (Mckenzie 1978; Le Pichon and Sibuet 1981; Wernicke 1985; Lister *et al.* 1986, 1991; Kusznir 1987; Lemoine *et al.* 1987; Sibuet 1992; Driscoll and Karner 1998; Davis and Kusznir 2004; Reston 2005, 2009; Lavier and Manatschal 2006; Ranero and Pérez-Gussinyé 2010; Huismans and Beaumont 2011, 2014; Reston and McDermott 2011; Brune *et al.* 2014; Naliboff *et al.* 2017; Peron-Pinvidic and Naliboff 2020).

2.5. Continental rifted margins

As introduced in the previous chapter, questions related to continental rifting, rifted margin formation and the initiation of oceanic spreading ridges are still widely debated, and many fundamental aspects remain unconstrained. However, the progress achieved within these last two decades is considerable.

Early rifted margin studies focused on the proximal near-coastal domains, where data acquisition was concentrated in sedimentary basins with hydrocarbon potential. Little was known about the more distal settings or deeper crustal configuration, mainly due to lack of data and/or limitations of the imaging technology at that time. As a consequence, models applied to rifted margin formation assumed standard pure shear and/or simple shear deformation mechanisms (McKenzie 1978; Wernicke 1985) (see section 2.4.1). Although such models could partly explain the early evolution of rifted margins, including the formation of sedimentary basins and addition of extension-driven melt (i.e. magma), they were unable to predict the evolution of rift systems approaching lithospheric breakup. The discovery of unexpected structural settings, such as exhumed subcontinental mantle and/or hyperextended continental crust (Boillot *et al.* 1980) questioned the validity of the pioneering rifting models (see section 2.4.2) and set the stage for the future development of new concepts and approaches (e.g. Whitmarsh *et al.* 2001). Expeditions funded by international science programs such as ODP (Ocean Drilling Project) and IODP (International Ocean Drilling Project), combined with improved technology and onshore analogue studies, have provided the community with new observations that forced a revision of these pioneering models over the last three decades. **With access to newer and higher-resolution geophysical and geological datasets, new concepts and models emerged and the terminology used to describe rifted margins continuously changed. Most of these ground-breaking ideas and models were developed along the Iberia–Newfoundland conjugate margins and then later applied to other rift systems** (e.g. Osmundsen *et al.* 2002; Peron-Pinvidic *et al.* 2007; Zalán *et al.* 2011; Savva *et al.* 2013; Stica *et al.* 2014).

Numerous rift scenarios and structural geometries have been proposed to describe and explain the new observations (e.g. Lavier Manatschal 2006; Brune *et al.* 2014). The historical end-member partitioning between "volcanic" and "non-volcanic" settings (Eldholm *et al.* 1989) has been adjusted to "magma-poor" and "magma-rich", based on the finding that "non-volcanic" margins actually contain a fair amount of magma (e.g. Peron-Pinvidic *et al.* 2010) (see section 2.3.2). New terms and concepts have been introduced and definitions have been further developed to help better map these still obscure extensional systems (see section 2.4.2). These include morphological terms (e.g. taper break; Osmundsen and Redfield 2011), domain names (necking zone, transitional zone, ocean–continent transition, zone of exhumed continental mantle, ocean–continent boundary; Peron-Pinvidic and Manatschal 2009) and terms referring to geological processes (stretching, thinning, exhumation; Lavier Manatschal 2006). Recent high-resolution geophysical datasets have allowed similar structural observations to be made at many margins worldwide (e.g. Reston 2009), such as the Great Australian Bight (Ball *et al.* 2013), the South China Sea (Franke *et al.* 2011, 2013; Savva *et al.* 2013), the Gulf of Aden (d'Acremont *et al.* 2005; Autin *et al.* 2010), the Argentinean margin (Blaich *et al.* 2009), the Brazilian margin (Zalán *et al.* 2011), the Norwegian margin (Osmundsen *et al.* 2002; Faleide *et al.* 2010) and onshore analogues (Andersen *et al.* 2012; Masini *et al.* 2014).

Based on this, a consensus has been reached on the fact that rifting is fundamentally polyphase, and results from the interaction of crustal and mantle processes. A significant paradigm shift occurred, not only in our way of describing the strain evolution of rifted margins, but also in our understanding of the way that deformation is coupled with magmatic, thermal and isostatic processes.

2.5.1. *Structural domains*

The architectural similarities displayed by most of the rifted settings worldwide appear to reflect a **commonality of processes** with regard to rifted margin formation (e.g. Franke 2013). Within that framework, Peron-Pinvidic *et al.* (2013) proposed that rifted margins should be examined based on the definition of distinct **structural domains** that represent their geometrical foundations (Figures 2.29 and 2.30). These are the ***proximal***, ***necking***, ***distal***, ***outer*** and ***oceanic*** domains. Conceptually, each margin domain is associated with specific and corresponding modes of deformation and/or magmatic processes: ***stretching***, ***thinning***, ***hyperextension***, ***exhumation***, ***oceanization*** and ***spreading***. The phases of deformation migrate progressively oceanwards and each is characterized by a specific stratigraphic, isostatic, thermal and magmatic evolution. The deformation modes can overlap, overprint and interact in time and space, which may seriously complicate the resultant geometries.

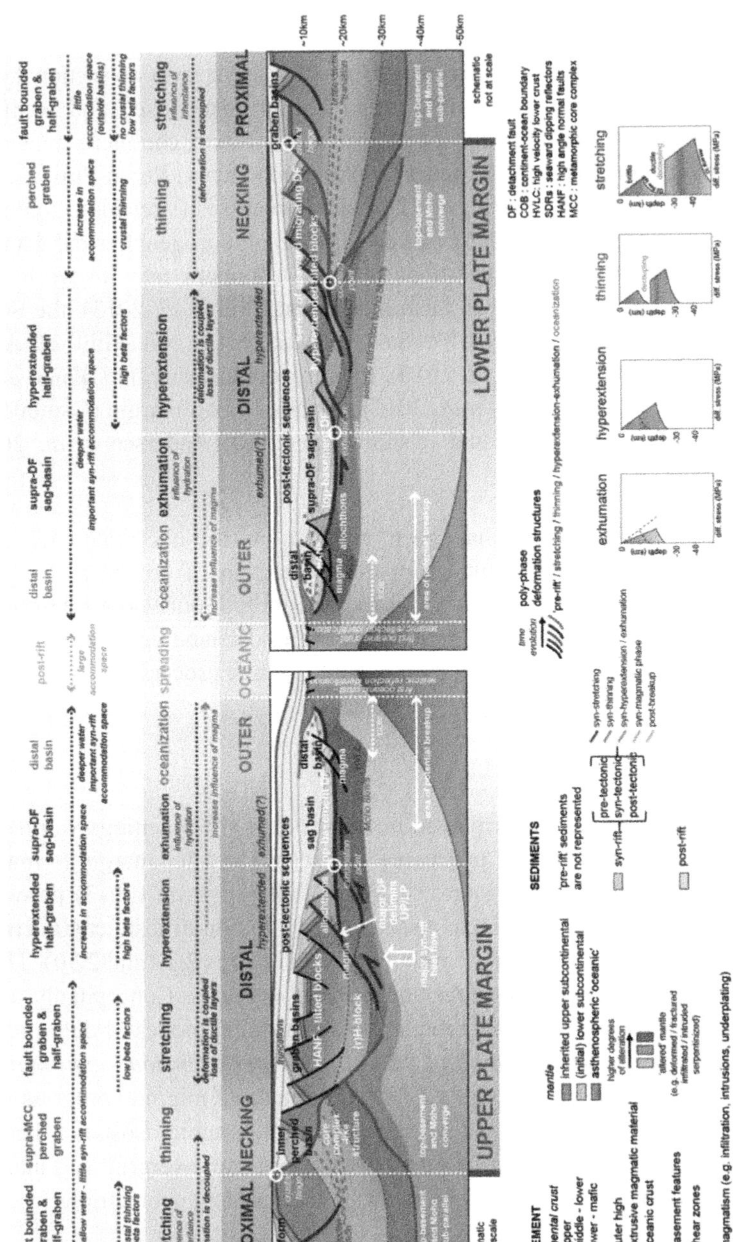

Figure 2.29. *Schematic cartoon summarizing the major structural and stratigraphic characteristics, dominant deformation mechanisms and structural domains of a conjugate set of lower-plate and upper-plate rifted margins (source: adapted from Peron-Pinvidic et al. 2015)*

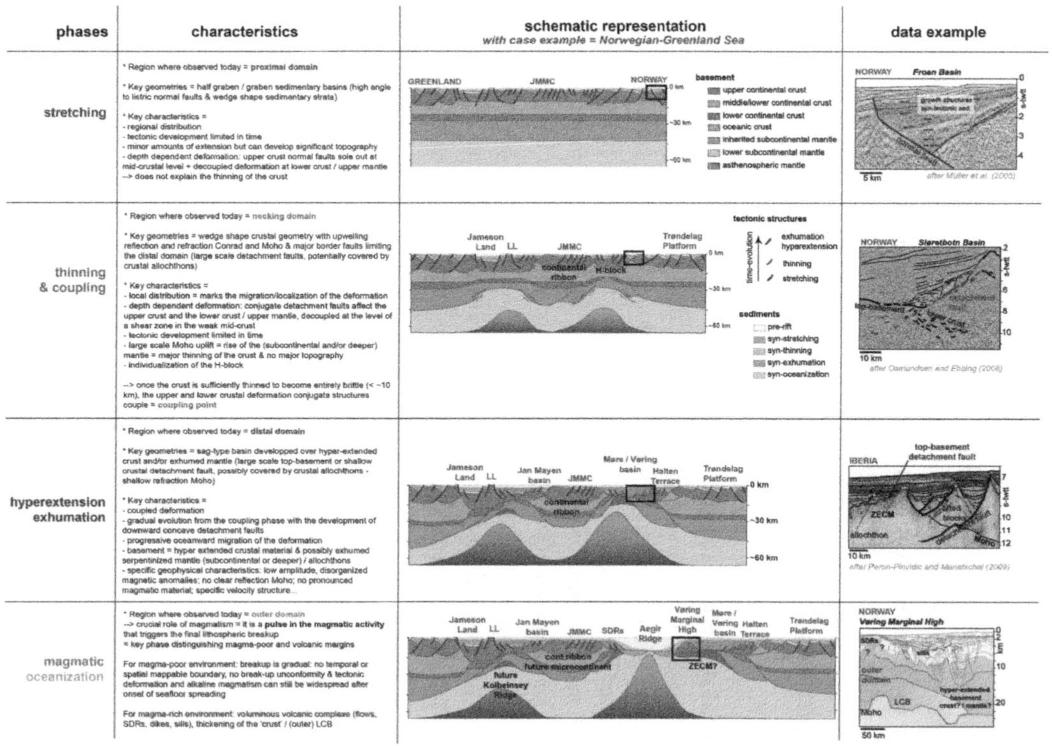

Figure 2.30. *Table summarizing the deformation modes interpreted to structure rifted margins, with their main characteristics, a schematic representation (based on the case example of the Norwegian–Greenland Sea rift system) and examples (source: Peron Pinvidic* et al. *2013)*

Pre-rift lithospheric status (structural, thermal and compositional inheritance) as well as opening obliquity and extension rates are parameters that are considered to directly impact basement and sedimentary geometries (e.g. Manatschal *et al.* 2014) and these will be detailed in the next section.

As the research is ongoing, the following sections are organized in the same way as previous sections in this chapter, with "**theory**" and "**evaluation**" sections in order to illustrate the distinction the research community is making between definitions, concepts and models, and the recent observations made possible with modern datasets. **Some of these observations will force a revision of the listed definitions in the coming years. This illustrates the vitality and necessity of research in the rifts and rifted margin community.**

2.5.1.1. *The proximal domain*

Theory

The **proximal domain** corresponds to the inboard continental crust that has been stretched at low values of extension (Peron-Pinvidic *et al.* 2013; Tugend *et al.* 2015) (Figure 2.30). The brittle upper crust is cut by high-angle normal faults that bound classical graben and half-graben basins. The faults have a listric shape, concave-upward and sole out at mid- to lower crustal levels (where the ductile deformation regime dominates), and hence do not affect the Moho (e.g. the Jeanne d'Arc basin and Murre fault; Welsink and Tankard 2012). At depth, other sets of faults are assumed to affect the rigid part of the upper lithospheric mantle and lower mafic crust (Fossen *et al.* 2014). The deformation pattern characterizing the proximal domain is therefore decoupled at the crustal/lithospheric scale: the faults affecting the upper crust and the ones affecting the upper mantle/lower crust are decoupled along ductile mid- to lower crustal levels. Since the bulk crust/lithosphere thicknesses are only slightly thinned, the subsidence is relatively small, and sediments are mainly restrained within the stretching basins that accommodate thick wedge-shaped syn-tectonic sedimentary units. The accommodation space outside of these basins is typically minimal. The sedimentary record includes continental to shallow water sedimentary systems often associated with subaerial exposure markers, and little to no aggradation of post-rift sequences.

Evaluation

Recent observations suggest that the above definition may not be representative of some rifted margins: based on the seismic reflection interpretation of the

Trøndelag Platform. Peron-Pinvidic *et al.* (2020) proposed a revised configuration of the proximal domain of the mid-Norwegian Vøring rifted margin (Figure 2.31). They show that the deep structure of the Trøndelag Platform is more complex than the homogeneously and slightly thinned continental crust assumed in the previously listed conceptual characteristics for proximal domains (op cit. and Osmundsen *et al.* 2021). The entire mid-Norwegian proximal area is floored by ductile crustal shearing and regional-scale core complexes that form a corrugated surface with basement culminations that are flanked and sometimes cut by detachment faults. The detachments are segmented, suggesting long-lived and multiphase tectonic activity. The observed geometries are directly comparable to the ones mapped onshore in and around the Devonian "Old Red Sandstone" basins in western and south-western Norway, and in the adjacent North Sea. The continental crust is, however, very much thinned, down to unexpectedly low values (~10 km) that are not predicted in standard rifting models. Thus, the observations demanded a revision of the conceptual approach of margins' proximal domain geometries – at least for rifted margins that developed on collapsed orogens. In this particular case, the combination of intensively sheared basement, core complexes, flanking and cutting detachment faults is interpreted to represent the offshore record of the final phases of the orogenic collapse and the initial phases of rifting. Figure 2.31(a–c) shows an illustration of new observations made on the mid-Norwegian margin that demand a revision of our approach of the geometries of the proximal domain.

A.

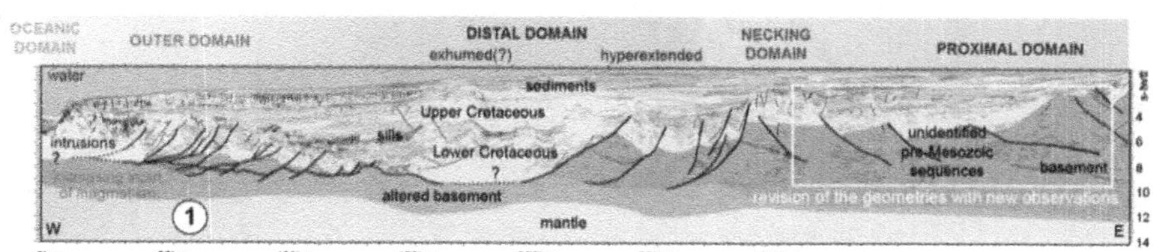

Figure 2.31a. *Cross-section illustrating the regional structural context of the mid-Norwegian rifted margin (location shown in Figure 2.31b). The interpretation is based on seismic reflection, seismic refraction and potential field data (source: adapted from Osmundsen* et al. *2016; Peron-Pinvidic and Osmundsen 2018)*

B.

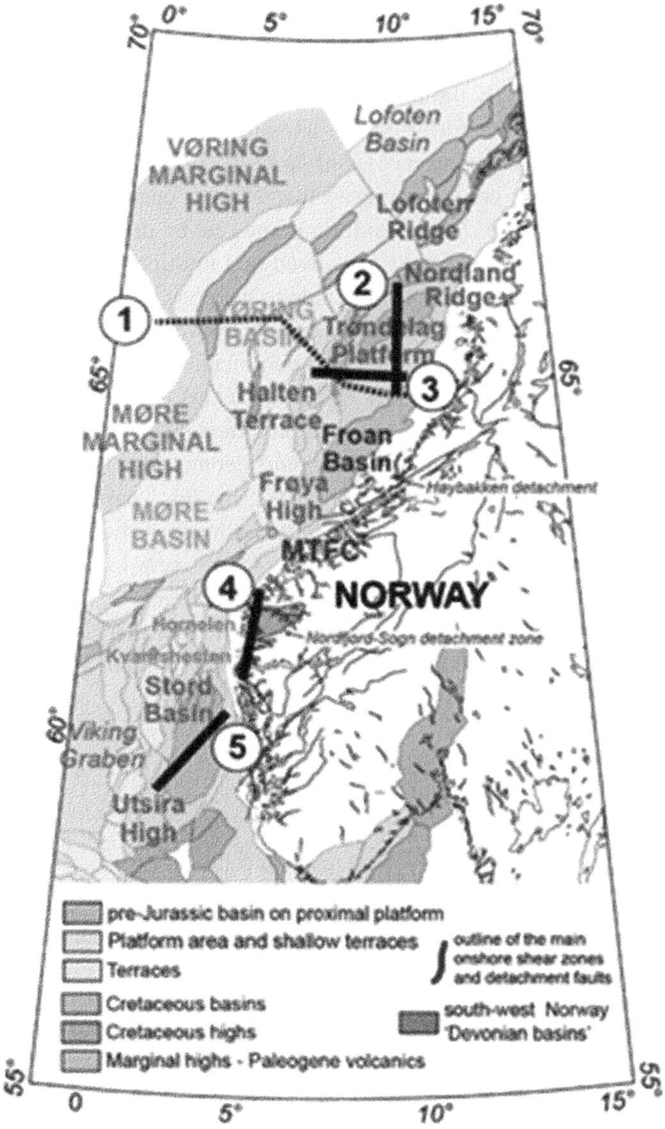

Figure 2.31b. *Simplified structural map of SW Norway. The onshore red lines outline the main mapped shear zones and detachment faults. MTFC: Møre-Trøndelag Fault Complex. Orange: outline of the "Devonian basins" (source: adapted from Blystad* et al. *1995)*

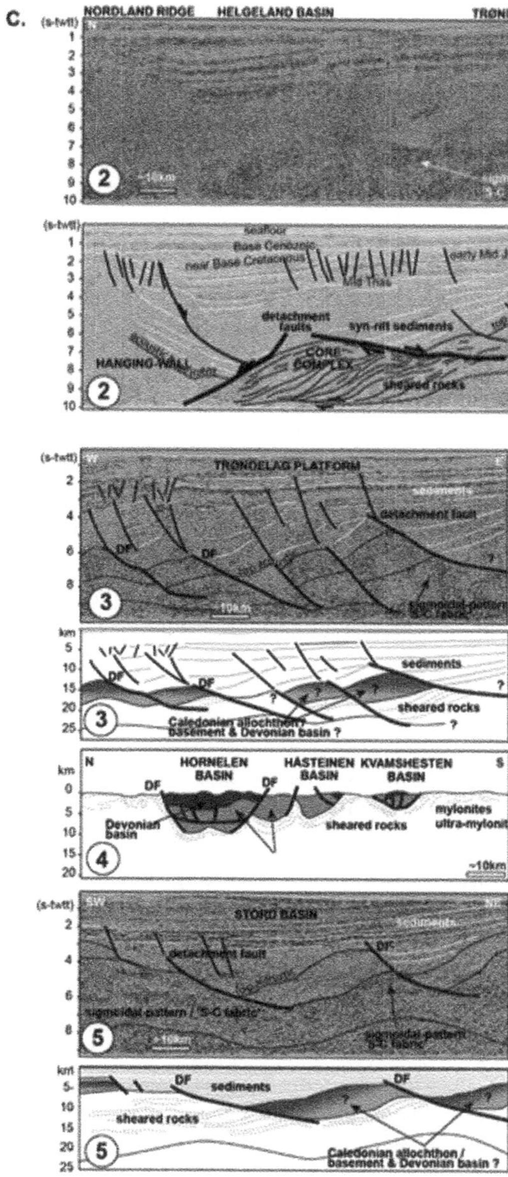

Figure 2.31c. *Seismic reflection profiles from the Trøndelag Platform and North Sea; in time version (s-twtt) and/or depth converted, without and/or with interpretation (location shown in Figure 2.31b)*

2.5.1.2. *The necking domain*

Theory

The **necking domain** is the domain where the lithosphere is efficiently thinned (Figure 2.30a–c). Similar to the proximal domain, deformation is still decoupled at the crustal scale. But in contrast to the proximal domain, the top basement and Moho are no longer sub-parallel. Seismically, the necking domain corresponds to a crustal wedge (e.g. the **taper** geometry, Osmundsen and Redfield (2011)), with the seismic reflection-identified Moho and top basement that converge, defining a crustal thinning from ~30–40 km to less than 10 km (Peron-Pinvidic and Manatschal 2009; Mohn *et al.* 2010; Sutra *et al.* 2013). Both top basement and Moho mark an inflection point defining this drastic crustal thinning, so, stratigraphically, the necking domain is also characterized by a marked basinward increase in the total accommodation space (Sutra and Manatschal 2012). The necking domain is defined between the "**crustal hinge**" and the "**coupling point**", which is the point where the first brittle fault transects the entire crust (Sutra *et al.* 2013). It therefore delimits the decoupled domain (inboard) from the coupled one (outboard) and corresponds to the location where no ductile layers exist anymore in the crust. This very specific region corresponds to the "**crustal embrittlement**" defined by Pérez-Gussinyé *et al.* (2003).

Evaluation

The mid-Norwegian rifted margin segments of Møre, Vøring and Lofoten are among the best-imaged extensional systems in the world (see the case example description in Chapter 3). Thousands of seismic reflection datasets in both 2D and 3D are available, allowing for highly detailed structural mapping. Thanks to this exceptional dataset, it has been shown that the necking domain is characterized by strong structural variabilities along strike, ranging from extremely abrupt necking in the Møre margin segment to more gentle crustal thinning in the adjacent Vøring area (e.g. Brekke 2000). Osmundsen and Peron-Pinvidic (2018) attributed these geometrical variations to lateral changes in the breakaway complex architecture and tectonic role (e.g. necking vs. hyperextension) (see section 2.4.2.5). The fault geometries bordering the necking domain represent the structural response to variations in crustal thickness, rheology and inherited structure. The global shape of the necking domain (e.g. abrupt vs. smooth) reflects the **lateral variability in the efficiency of crustal coupling** (Figure 2.32). Although now better characterized, the overall tectonic, uplift and subsidence history of this particular domain remain largely unconstrained and further detailed mapping is needed.

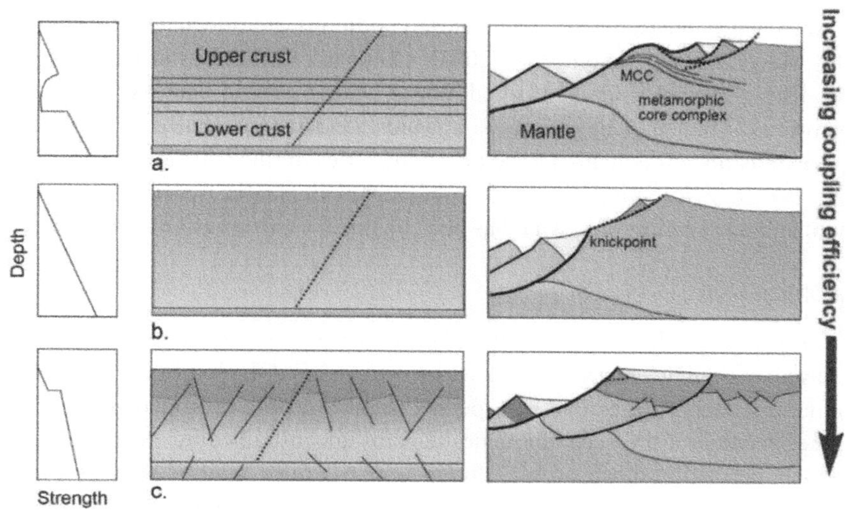

Figure 2.32. *Cartoon showing different simplified coupling scenarios based on observations along the outer necking breakaway complex in the Møre and southern Vøring margin. a) Truncation of a thick crust and remnant ductile layer by a large-magnitude fault. The fault develops a ramp-flat geometry as it flattens in the ductile layer before it incises it and ramps down to couple with structures in the lower crust and upper mantle. Flow in the ductile crust, isostatic footwall uplift and successive fault excision leads to the development of a core complex. Early detachment becomes abandoned, uplifted and eroded as is the front of the core complex. b) Truncation of a thick crust without a ductile layer by a large-magnitude fault. Structural inheritance may be necessary to focus deformation in the thick, cold crust. Isostatic uplift leads to rotation, deactivation and incision of the oldest fault by a younger, steeper splay and formation of a roughly convex detachment with planar segments separated by knickpoints. Long-wavelength footwall uplift and erosion of older, incised and uplifted fault planes over substantial areas. c) Planar to listric faults incising previously extended crust with a high density of inherited faults. Fault geometry is relatively simple. Footwall uplift has a relatively short wavelength (source: Osmundsen and Peron-Pinvidic 2018)*

2.5.1.3. *The distal domain*

Theory

The **distal domain** stands in between the crustal neck and the more magmatic outer domain (Figure 2.30). It is the domain where deformation is coupled on a crustal scale, where large magnitudes of extension lead to the removal of ductile layers within the continental crust, allowing faults to cut from the top of the crust into the mantle, and potentially allowing seawater to promote serpentinization processes when pressure and temperature conditions are adequate (Escartín *et al.* 2001; Mevel 2003; Skelton *et al.* 2005). Outboard of the crustal neck, basement geometries comprise a series of tilted blocks, with a general decrease in block size oceanwards, flanked by concave-upward detachment faults. This is the **hyperextension** sub-domain. Outboard of this series of tilted crustal blocks, if at one stage, the crust is entirely excised, subcontinental mantle material may become exhumed at the seafloor into what has been named the **exhumation sub-domain**.

Evaluation

This last decade has seen an unprecedented improvement in the seismic imaging of the distal margin geometries, thanks to the release of high-resolution long-offset seismic reflection profiles that now permit detailed geometrical mapping. However, the interpretation of the deep lithologies remains unconstrained by other methods and extremely debated in most cases. Unfortunately, basement samples are only available for very few distal margins. These include: the Iberia–Newfoundland conjugates, the Great Australian Bight and the Red Sea at Zabargad Island. In these three cases, the samples validated the presence of exhumed serpentinized mantle (Tucholke and Sibuet 2007; Gillard *et al.* 2015). For the other worldwide cases, observations are generally based on geophysical modeling (seismic, potential field), and various types of interpretation proposed in the literature can be contradictory, the main reason being the **limitations of the geophysical methods** (indirect observations).

In distal domain studies, the standard study protocol usually involves: (1) seismic reflection interpretation, supplemented by (2) refraction and/or (3) potential field modeling. Refraction and potential field modeling are used to counterbalance the absence of reliable seismic reflection observations at depth, a problem that is often encountered in distal margins. Magnetic, density and velocity values and their gradients are regularly used to identify either sedimentary, crystalline, magmatic or mantle material. The issue with this technique is that **different lithologies, with distinct structural, thermal and geochemical histories can be modeled with similar sets of geophysical parameters** (Saltus and Blakely 2011). **Rock velocity values, densities and magnetic parameters are actually not conclusive of any lithology (e.g. the same density value can correspond to different rocks)**. In addition, it is also commonly accepted that distal margins probably undergo major multiphase deformation processes during rifting, including high tectonic activity, magmatism, hydration and metamorphism. This complex tectono-thermal evolution most likely leads to a marked alteration of the physical properties of rocks in the distal and outer margins, and hence of their geophysical signals. Fracturing, shearing, melt percolation/infiltration/intrusion, hydration and serpentinization are rock-altering processes thought to be widespread in distal rifted margins and in terms of geophysical signals, these can drastically affect the velocity, density and magnetic values. The result is that multiple rock types may have overlapping geophysical values (Figure 2.33), which, in the absence of drilled samples, simply precludes any accurate interpretation of these lithologies.

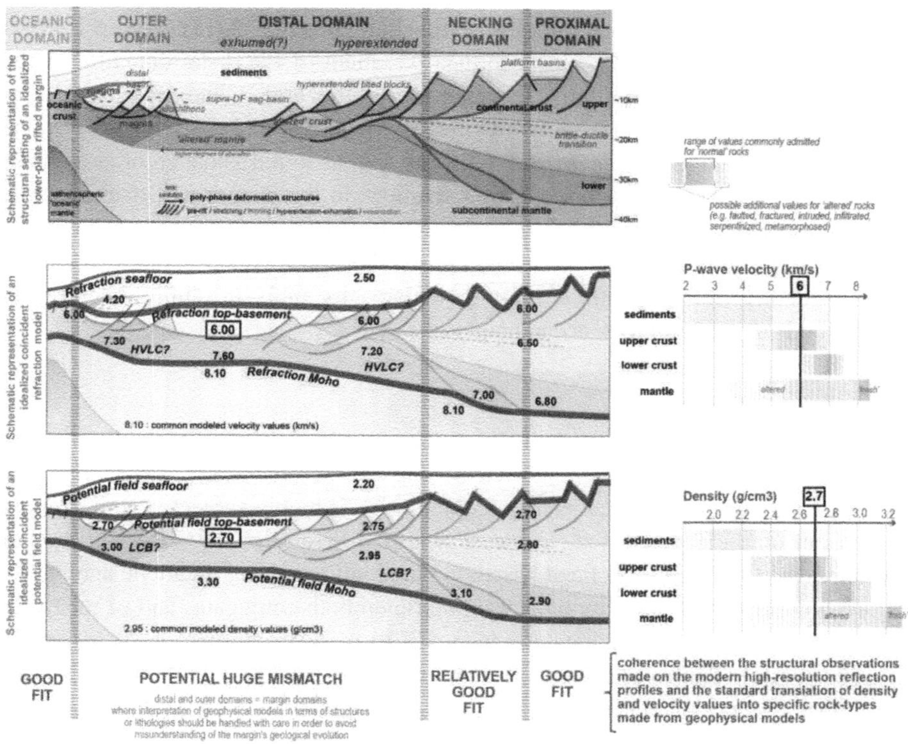

Figure 2.33. *Illustration of the potential fits and misfits between the structural observations made on seismic reflection profiles and the standard translation of density and velocity values into specific rock-types made from geophysical models. Top left: schematic representation of the structural setting of an idealized lower-plate margin (according to Peron-Pinvidic* et al. *2013, 2015). Center left: the same schematic section with the projected commonly modeled main refraction horizons (seafloor, top-basement, Moho). Bottom left: the same schematic section with the projected commonly modeled main potential field horizons (seafloor, top-basement, Moho). Right column: average values of P-wave velocities (in km/s) and of densities (in g/cm3) usually attributed for sedimentary, continental crust and mantle rocks. Note the squared values on the sections and graphs that point to a case example where the interpretation of geophysical models in terms of structures and lithologies fails to relate to the correct structural and stratigraphic setting of the distal margin. DF= detachment fault, HVLC= high-velocity lower crust, LCB= lower crustal body (source: Peron-Pinvidic* et al. *2016)*

2.5.1.4. *The outer domain*

Theory

The **outer domain** is by definition the key domain in the distinction between magma-poor and magma-rich margins (Peron-Pinvidic *et al.* 2013) (Figure 2.30). It is located between the (usually) ill-defined basement of the distal domain and the oceanic crust. Its inboard limit corresponds to the area where the volume of magmatic additions becomes more significant. Magmatic additions can correspond to clearly identifiable sills, lavas and SDRs, as well as to intrusions in or at the base of the basement, or infiltration and melt channels in the deeper part of the mantle that are more difficult to identify with geophysical datasets. Therefore, contrary to the proximal and necking domains, the inboard limit of the outer domain can rarely be confidently defined with a sharp boundary. Its outboard limit theoretically corresponds to the continent–ocean boundary (COB).

Evaluation

The definition of the outer domain is related to the definition of the **continent–ocean boundary (COB)**. Although COB is a well-accepted term, its definition is often ambiguous at best since its mapped definition is highly subjective and strongly dependent on the dataset used (Heine *et al.* 2013; Peron-Pinvidic *et al.* 2015). For instance, the COB can be defined based on the identification of magnetic anomalies, which are interpreted as continuous, symmetric on both conjugates and of relatively high intensity. These anomalies are supposed to mark the formation of magmatic oceanic crust. In that case, several drawbacks may arise as the definition of magnetic anomalies depends on (1) the interpreter's visual judgment and interpretation, (2) the quality of the dataset and (3) if breakup occurred during a period of time when magnetic reversals were regular enough to permit a distinct record of the first accreted oceanic crust, which is not, for instance, the case during the Cretaceous Normal-Polarity Super Chron. Additionally, the magnetic anomaly definition of the COB also makes strong assumptions regarding the magmatic nature of the adjacent oceanic crust adjacent to the rifted margin, when it need not be the case (see the below section on the oceanic domain). The definition of the COB can alternatively be based on seismic reflection mapping, refraction data or potential field modeling. Each of these definitions can be defended and each has advantages and disadvantages. The result is that for the same margin, depending on the preferred method, interpreters can define very different COBs, potentially up to tens of km apart (Figure 2.34), greatly reducing the robustness of the defined limits of the outer domain. The COB is discussed further in section 6.4.

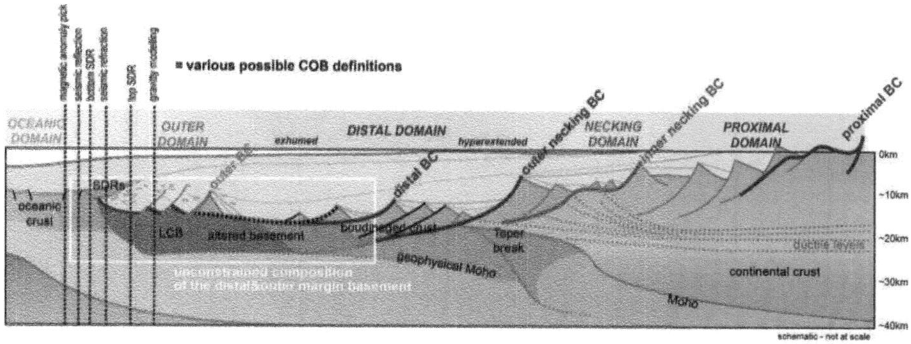

Figure 2.34. *Schematic representation of the structural setting of an idealized rifted margin, indicating the many various possible definitions of continent–ocean boundaries (COB), depending on the dataset and interpreter used (source: Peron-Pinvidic and Osmundsen 2018)*

2.5.1.5. *The oceanic domain*

Theory

Within the framework of rift evolution, when the asthenosphere has risen sufficiently high in the overall extending system, the rift can enter the "**magmatic mode**" due to decompression melting (White and McKenzie 1989). Decompression melting notably depends on the lithospheric extension rate, mantle composition and thermal state. This magmatic mode of rifting consists of the "weakening" of the rift axis by the introduction of a hot magmatic material by means of underplating, intrusion and extrusion. When the thermal state reaches an equilibrium at the oceanic stage, the spreading is interpreted to be settled (Cannat *et al.* 2009). The oceanic crust is typically composed of pillow lavas, dike complexes and gabbroic/ultramafic cumulates (Dick *et al.* 2003, 2006) (Figure 2.35). This is the tripartite description, historically referred to as the **Penrose-type Ophiolites** from the Penrose Conference in 1972. It was recognized early on that the accretion rate is a determining parameter in the overall architecture and composition of the oceanic crust. The triple-layered magmatic stratigraphy of the Penrose ophiolites is encountered in relatively fast-spreading settings. When this spreading rate diminishes, the lithostratigraphy of the oceanic crust changes dramatically. In very slow spreading environments (<1 cm/yr), the produced oceanic crust is thinner (4–5 km) and the accretionary process can lead to the exhumation of wide regions of mantle without significant amounts of magma (e.g. Gakkel Ridge). This differs greatly from the extreme 20 km-thick crustal thickness estimated for the oceanic crust near Iceland.

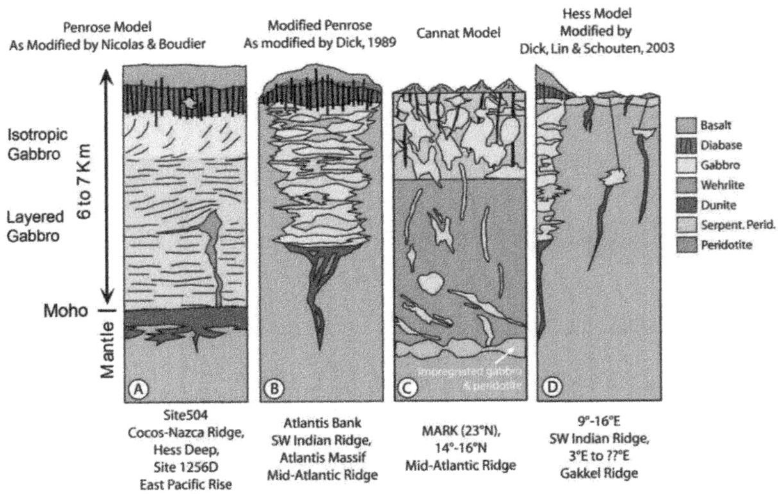

Figure 2.35. *Models for crustal accretion at ocean ridges. a) Classic interpretation of the Penrose model for a fast-spreading ridge based on the Oman Ophiolite. b) Penrose model, as modified for slow-spreading ridges based on the abundance of peridotite and frequent absence of gabbro at transforms following focused melt-flow models. c) Model for the anomalous 14–16°N area of the Mid-Atlantic Ridge. d) Model for magmatic and amagmatic accretionary segments at ultraslow spreading ridges (source: adapted from Dick* et al. *(2006))*

Evaluation

By definition, the oceanic domain includes areas of newly accreted igneous basement that are strictly oceanic in origin and represent steady-state formation of oceanic crust. Although this principal definition sounds straightforward, the distinction from the distal/outer margin may not be that simple.

Rifted margins and oceanic ridges have long been presented as two opposing end-member entities: the formation of the former being tectonically driven, while the latter is typically defined as magmatically driven. From studies of slow-spreading ridges (e.g. Reston Ranero 2011; Sauter *et al.* 2013), it is now recognized that oceanic crust forms not only through magmatic addition (see the above definition), but also through tectonic extensional processes (e.g. Reston 2018; Cannat *et al.* 2019). Fifty percent of the Atlantic ridge system has been proven to be shaped by detachment faults and core complexes, structurally similar to continental extensional settings (Escartín *et al.* 2008; Figure 2.36). The similarities between oceanic and continental core complexes are distinctly noticeable (Whitney *et al.* 2014). These observations beg the question of the intrinsic origin and nature of the first domains of oceanic crust, as opposed to that of distal rifted margins. The existence of "**transition zones**" where processes would gradually switch from rifting to drifting is regularly advocated; these are often called "**proto-oceanic**". However, similarly to the margin domain "**OCT**" ("**ocean–continent transition**"), this "proto-oceanic" domain is very poorly characterized.

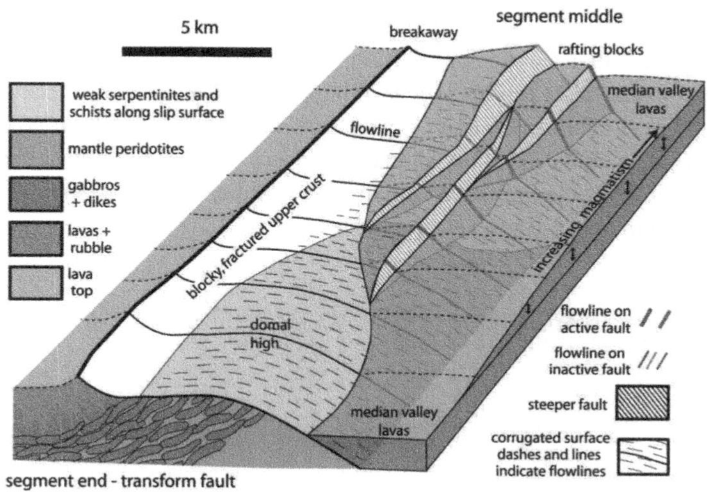

Figure 2.36. *Schematic representation of a slow-spreading ridge segment, showing a lateral transition from the unroofed footwall of a detachment at the segment end to a detachment buried beneath a series of rafted blocks towards the segment middle (source: adapted from Reston and Ranero 2011)*

2.5.1.6. *Continental ribbons and microcontinents*

In addition to the main structural geometries mentioned earlier, rifts and rifted margins may encompass additional crustal blocks such as continental ribbons and microcontinents.

Continental ribbons

Continental ribbons are observed in proximal margin settings and correspond to pieces of crust that have been thinned less than the surrounding marginal domains (Lister *et al.* 1986). They are separated from the non-rifted continents by sedimentary basins that failed to go to complete plate separation, often displaying a "V-shaped" geometry (Figure 2.37). These inboard basins present various amounts of thinning in crustal thickness, ranging from 20 km for the Hatton Basin, 5–7 km for the Rockall trough, <2 km for the Porcupine Basin and 6–8 km for the Galicia

Interior Basin (Shannon *et al.* 1999; Pérez-Gussinyé *et al.* 2003; O'Reilly *et al.* 2006). The close geographic proximity of these basins in the North Atlantic Ocean suggests a complex evolution of the related rift systems from minorly extended, to highly thinned continental crust, to localized exhumation of mantle rocks and formation of embryonic oceanic crust.

Based on the Iberia–Newfoundland transect, Peron-Pinvidic and Manatschal (2010) assumed that continental ribbons are bounded by major top-basement detachment faults, also referred to as thinning or necking faults (see the above section on the structural domains), relayed by detachment and high-angle normal faults in the adjacent basins or distal margins. These families of faults are interpreted to be responsible for the individualization of the continental ribbons and can explain the thinning of the blocks towards their edges.

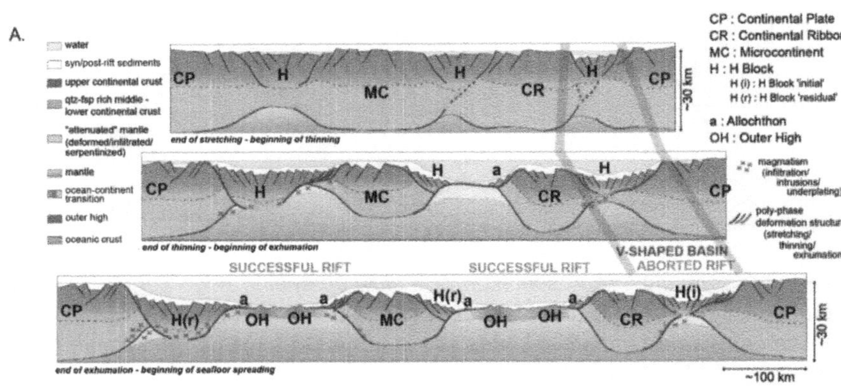

Figure 2.37a. *Schematic model of rifting evolution illustrating the formation of the different categories of crustal blocks. The genesis, evolution, final shape and position in the margin of each block are related to the distinct modes of deformation affecting the margin during rifting (source: adapted from Peron-Pinvidic and Manatschal 2010)*

Microcontinents

Microcontinents are defined as distinct bodies, rooted by continental crust and surrounded by oceanic crust. It has been observed early on that microcontinents are often associated with magmatism, in the form of SDRs, intrusions or underplating (e.g. the type example of the Seychelles; Collier *et al.* 2009). These observations led to the definition of specific magmatic provinces, which were explained by the influence of a mantle thermal anomaly. In that sense, Müller *et al.* (2001) proposed that the formation of microcontinents is intimately linked to **mantle plumes** (see Figure 2.38). In this model, the arrival of a plume results in the lithospheric weakening of areas located further inboard from the active rifting, or away from an existing spreading center. These areas may undergo rifting and eventually force the spreading axis to jump, resulting in the isolation of a portion of continental crust and the successful formation of a microcontinent. The Jan Mayen microcontinent, associated with the Icelandic plume, has been proposed to be one of the key examples of such an evolution. Models of rift/ridge propagation showed later that inherited structures (van Wijk Blackman 2005; Koehn *et al.* 2007; van Wijk *et al.* 2008), stress-dependent damages around the microplate borders (Hieronymus 2004), magmatic heating rate, spreading rate, seafloor ages and the location of the hotspot (Mittelstaedt *et al.* 2008) can also be important parameters to consider.

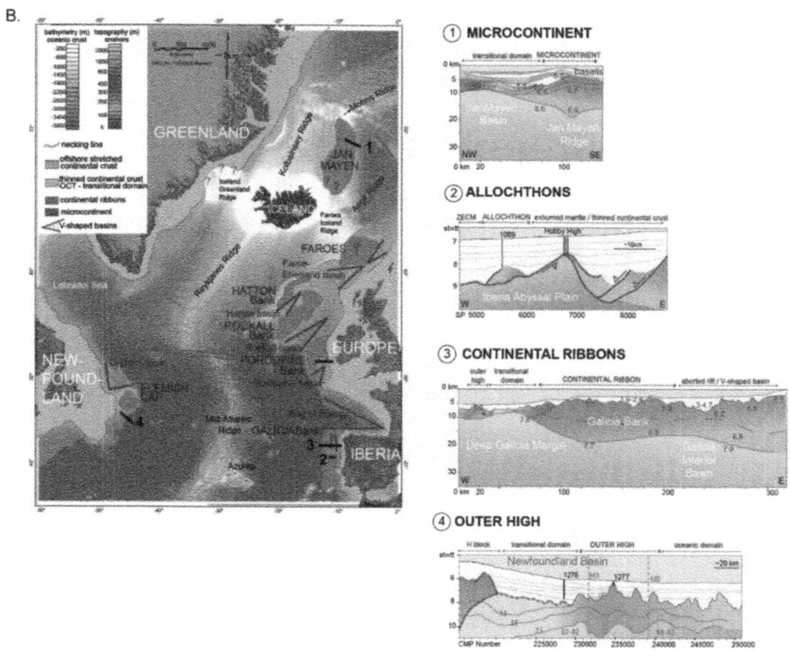

Figure 2.37b. *Left: bathymetric map of the North Atlantic Ocean with mapping (based on bibliography, magnetic, gravimetric and seismic data) of different categories of crustal blocks (continental ribbon and microcontinent) and of crustal settings from un-thinned (onshore continent), stretched (offshore proximal margins; brown areas) to highly thinned continental crust (distal margins, ocean–continent transitions, V-shaped basins; green areas). OCT: Ocean–Continent Transition. ZECM: Zone of Exhumed Continental Mantle. Right: data observations (interpretation of seismic refraction/reflection extracts) of the different categories of crustal blocks (microcontinent, continental ribbon, H-block, allochthon, outer high) in the southern part of the North Atlantic (see the map for profile locations) (source: adapted from Peron-Pinvidic and Manatschal 2010)*

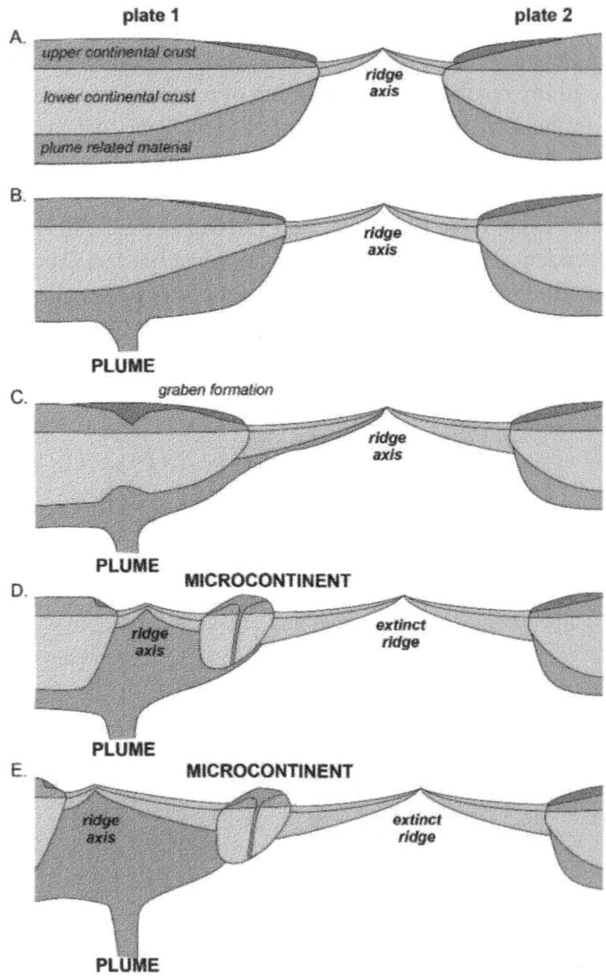

Figure 2.38. *Conceptual model for plume-related microcontinent formation. a) Fragmentation of two large plates has caused a young ocean basin to form. b) Young rifted margin moves into the vicinity of the existing mantle plume, enhancing yield strength minimum landward of rifted margin by conductive heating. c) and d) This process results in renewed rifting, jump of active ridge towards hotspot and isolation of microcontinent. Further ridge jumps towards hotspot cause subsequent asymmetries in oceanic crustal accretion, resulting in excess accretion and series of extinct ridges on plates, including microcontinent, if the hotspot remains under the opposite ridge flank. e) If the ridge crosses the hotspot, the sign of asymmetries in crustal accretion would switch, causing excess accretion on the ridge flank opposite the microcontinent, as in the case of Jan Mayen (source: adapted from Müller* et al. *2001)*

The concept may have to be re-investigated in more detail, as the community now recognize the probable existence of far more **continental-derived microplates** than previously envisaged, like the following provinces that were previously identified as magmatic: the Rio Grande Rise (Mohriak *et al.* 2010; Santos *et al.* 2014), the western Australia Batavia Knoll and Gulden Draak Knoll (Gardner *et al.* 2015), the Elan Bank (Gaina *et al.* 2003), the Mascaregn terranes (Torsvik *et al.* 2013), the East Greenland Ridge, the Hovgård Ridge and associated Boreas Basin (Dossing *et al.* 2008; Døssing Funck 2012), and the Lomonosov Ridge (Poselov *et al.* 2012). Therefore, the isolation of microplates during rifting evolution appears to not be as exceptional as theorized in the past. And as mantle plumes are not found in all extensional settings, other mechanisms must be involved. As summarized in a review of microcontinents, oceanic plateaus and continental fragments by Tetreault and Buiter (2012), it seems that there is not "one recipe" for microplate isolation. Various other scenarios are possible, like multiphase rifting, regional misalignment of rift axes and subsequent linkage that could also be advocated in the Jan Mayen case.

Osmundsen and Peron-Pinvidic (2018) demonstrated that breakaway complexes are structurally important in determining the final architecture of the rift system. These fault complexes are laterally variable, with sets of major boundary faults that may change their tectonic role laterally, from, for example, a "necking-" to "distal-" breakaway complex. Osmundsen and Peron-Pinvidic (2018) interpreted that this lateral structural variability can cause the isolation of some margin units that, depending on the final breakup line location, can ultimately correspond to microcontinents (Figure 2.39).

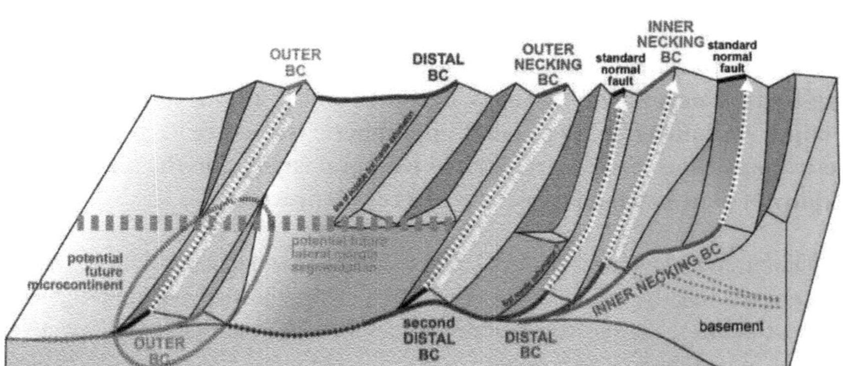

Figure 2.39. *Block diagram illustrating the lateral change of the structural role of the various fault planes encountered in a rifted margin. A necking breakaway complex can, for instance, evolve along strike into a distal breakaway complex (source: Peron-Pinvidic and Osmundsen 2018)*

2.5.2. *The 3D perspective*

It is fundamental to note that studies in large-magnitude continental extension have only been primarily examined in two dimensions in the past. This is historically due to data limitations and computing facilities that were, up until very recently, only available in 2D. In this last decade, publicly available 3D geophysical datasets, as well as dense grids of high-quality 2D data, have enabled the imaging of rifts and rifted margins in a third dimension. And it is only very recently that computing facilities allowed numerical simulations in 3D (e.g. Allken *et al.* 2012; Braun *et al.* 2013; Brune and Autin 2013; Koptev *et al.* 2018; Naliboff *et al.* 2020).

The 2D approach is obviously an important yet oversimplified approach to the mechanics of extension. It is well known that many parameters vary across the margin during rifting, with potentially important effects on rifting evolution. However, for convenience, representativity and ease of communication, 2D profiles, interpretations and models are still preferred.

High-resolution 2D geophysical imaging and numerical investigations suggest that both **rifts and rifted margins have large variations in the architecture in the dip direction** (e.g. margin domain width, tectonic structures, sedimentation, magmatism). Many of these variations are interpreted to reflect the initial rheological structure and the extension velocity (e.g. Lavier and Manatschal 2006; Brune *et al.* 2014; Huismans and Beaumont 2014; Naliboff and Buiter 2015; Tetreault and Buiter 2018). But these models and datasets fail to capture the variable margin-parallel and rift-parallel components. Indeed, many extensional systems display substantial along-strike variability in terms of structures, sedimentary architecture and magmatic products, and this variability often results in the definition of boundaries between distinct margin segments. **The origin of such along-strike heterogeneity is likely derived from multiple competing factors, including variations in the initial lithospheric structure (e.g. inheritance), reactivation of shear zones, spatial and temporal changes in the magnitude and obliquity of plate driving forces, as well as other processes.**

Imaging and modeling, such 3D processes at reasonable resolutions, are only now computationally feasible. However, still, few studies have investigated the origins of along-strike heterogeneities. To date, these investigations have invoked structural inheritance (Balázs *et al.* 2018), rift obliquity (Brune *et al.* 2014), rift propagation (Mondy *et al.* 2017), rift linkage (Le Pourhiet *et al.* 2017) or large-magnitude fault interaction (Osmundsen and Peron-Pinvidic 2018) as primary controlling mechanisms. Naliboff *et al.* (2020) recently suggested that along-strike variations in rift and rifted margin structure may naturally arise from the complex evolution of fault networks, without any large-scale pre-rift anisotropies, with further heterogeneity arising from variable lithospheric structure and plate driving forces.

3D datasets and models will undoubtedly emerge as a major study protocol in the coming years. The next generation of interpretations and concepts should encompass the third dimension, in order to understand how the lateral variability of several parameters steadily observed in all rifted systems worldwide impacts rifting.

2.5.3. *The parameters influencing rift evolution*

As described in the previous sections, rifted margins are shaped by distinct phases of deformation that can be recognized in most extensional systems. However, in detail, rifted margins exhibit different geometries from one another, mainly in terms of the margin domain widths, uplift and subsidence histories, and sedimentary and magma volumes. Numerical modeling has been utilized from early on to investigate the range of parameters that actually control these geometries, and parameters like the rheology, extension rate and pre-existing structural configuration (e.g. thickness of the various lithospheric layers, inherited heterogeneities) have been identified as the key influences on rift evolution and the resulting final architecture.

The margins' geometrical variations can be explained by the interaction of a series of key parameters, such as:

1) the lithospheric and asthenospheric state, with the thermal structure, rheology, compositional variations, layer-thickness variations and the presence of fluids (siliceous and aquatic) (e.g. Goetze and Evans 1979; Kirby and Kronenberg 1987; Buck 1991; Burov and Diament 1995; Hopper and Buck 1996; Burov and Poliakov 2001; Davis and Kusznir 2004; Pérez-Gussinyé *et al.* 2006; Huismans and Beaumont 2008; Huismans and Beaumont 2011; Rupke *et al.* 2013);

2) the inheritance of the previous phases of deformation, which can be structural, compositional and/or thermal (Manatschal *et al.* 2014);

3) the geometry of the rift opening with the angle of rift obliquity (Heine *et al.* 2013);

4) temporal factors like the extension rate and their spatial variations, temporal variations and the length of the rifting time period (e.g. Allemand and Brun 1991; Bassi *et al.* 1993; Newman and White 1997; Newman and White 1999; Davis and Kusznir 2004; Pérez-Gussinyé *et al.* 2006);

5) the sedimentary and magmatic input during rifting (e.g. Ebinger Casey 2001, Rooney 2019; Geoffroy 2005; Bialas *et al.* 2010).

Temperature plays a key role in rifting processes. The thermal structure of the lithosphere and asthenosphere directly controls and impacts many crucial mechanical and kinematic parameters, like the lithospheric rheology, subsidence and uplift evolution, metamorphism, magmas genesis, deformation modes, and so on (e.g. Chapman and Rybach 1985; Louden *et al.* 1991).

Rheology directly controls the lithospheric response to stress and thus plays a major role in determining the rift mode. At the margin-scale, it has been promptly identified that a "weak" lithospheric rheology leads to a different rift mode than a "strong" rheology (e.g. Buck 1991). At a more local scale, rheology directly affects the morphology, seismicity and distribution of faults and fault families in rift zones.

The continental lithosphere consists of a series of brittle and ductile layers, some of which are inherited from previous tectonic phases. During continental extension, the brittle layers accommodate shear stress by faulting, while the ductile layers deform by viscous flow – so the spatial repartition of brittle versus ductile layers is of primary importance for the rift evolution and geometry. In order to explain unexpected observations and/or test some hypotheses, more complex strength profiles have been investigated throughout the last decades of research, notably by numerical modeling. Huismans and Beaumont (2008), for instance, have tested configurations, including multiple successions of brittle and ductile layers, and highlighted the crucial role of the ductile layers in governing the rifting evolution and final geometries. Lithospheric rheologies are more complex than the monomineralic diffusion-creep flow laws that are commonly used in numerical studies; therefore, different constitutive relationships (i.e. dislocation-creep, Pierls-creep) and polymineralic flow laws should be incorporated (Kohlstedt 2007).

Time is also a determining parameter in the study of rifting processes: the absolute age of a rift places its development in a regional or global context. Unfortunately, it is often difficult to robustly assign time constraints on the evolution of a rift. At the margin scale, where most observations are made offshore, the necessary constraints are simply not accessible. Constraining timing would require systematic rock drilling and sampling of all stratigraphic units, ideally together with fault plane dating, what is unrealistic. Therefore, the age of formation of the various marginal domains is usually extrapolated based on scarce data points, which is unfortunate and may lead to misinterpretations.

The role of **inheritance** in controlling the evolution of rifted margins has received increasing interest in these last years. Obviously, all present-day rifted margins were developed in a continental lithosphere that had undergone earlier tectonic phases. The idealized homogeneous lithospheric layering envisioned for the pre-rift stage in most geodynamic models is perhaps too simple an approximation. Lithospheric inheritance is clearly present, but its role during rifting has been poorly investigated. Progress has been made in defining different types of inheritance, including structural, thermal and compositional aspects (Manatschal *et al.* 2014). However, it remains very difficult to characterize their importance on the scale of rifted margins and to understand the way they are related to rift-induced processes, such as serpentinization and magmatism. A better understanding of the pre-rift settings would enable a better characterization of the initial thickness of the various individual lithospheric layers in play during rifting, along with their initial composition and rheology (see section 6.1 for an illustration).

Within rift studies, the structure of the mantle lithosphere has received relatively little attention. The **mantle's** thermal state and composition has been recently revealed to be of crucial importance: its depleted or enriched status will determine its ability to produce magma during the extensional phases, directly impacting the rift evolution and margin geometries. A fertile hot mantle is less dense than a cold depleted mantle, which will result in distinct buoyancy behaviors and uplift/subsidence curves for the overall margin (e.g. Hirth and Kohlstedt 1996). It is also now recognized that the subcontinental mantle can contain important lateral compositional variations (notably due to earlier tectonic phases), resulting in similar lateral density variations, and hence variable buoyancy forces that will directly impact rift development (e.g. Müntener *et al.* 2010).

The influence of **melt production** and **magmatic products** in the overall rifting evolution is actually quite poorly constrained in rifted margin studies. It is clearly observed that some margins contain more magmatic features than others. However, the actual volumes are still unconstrained and thus very much debated, together with the timing and spatial distribution of magma production. The interactions between tectonic and magmatic processes need further study: both have long been known to be deeply associated (e.g. Ebinger and Casey 2001; Ebinger 2005). However, the tendency to categorize rifts within end-member models (e.g. the volcanic vs. non-volcanic margins) ignores the need for more detailed research in these relationships.

It can be mentioned that the presence of magmatic features and the assumed large volumes are often interpreted to be related to anomalous mantle temperatures (e.g. Skogseid *et al.* 1992). The elevated temperatures are regularly proposed to be related to a **hot-spot** or **plume** (e.g. White and McKenzie 1989). Because of the thermal anomalies plumes produce, their presence would drastically influence the rift evolution and breakup stages. However, debates remain regarding the actual existence of such mantle features. Even though plume-rift models are attractive, some researchers argue that (1) the actual volumes of magma in rifts and rifted margins are unknown, so there may not be a need for particular mantle thermal conditions (e.g. Peron-Pinvidic *et al.* 2010; Wangen *et al.* 2011), (2) increased mantle temperatures are not always associated with plumes and can be generated by other processes (e.g. van Wijk *et al.* 2001) and (3) some magma-rich regions do not correspond with any known mantle plumes. These debates extend beyond the scope of this contribution. For further information, the reader is referred to publications specifically dedicated to such topics.

In conclusion: **all of the above-mentioned parameters are determining factors in creating a rift's geometry, its spatial and temporal evolution and thus its final architecture of the margins**. These key parameters are today becoming much better identified and understood. The advanced numerical techniques now available allow researchers to systematically explore the impact of such parameters on rift evolution and geometries (e.g. Huismans Beaumont 2007; Regenauer-Lieb *et al.* 2008; Brune *et al.* 2014). Additionally, a new generation of models have more recently provided more realistic descriptions of 3D strain distribution and of the tectonic, magmatic, isostatic and thermal evolution of rifted margins (e.g. Brune *et al.* 2014; Naliboff *et al.* 2020). However, many questions remain unanswered and the possible feedback effects between these parameters are still poorly understood.

2.6. Future research

As tentatively illustrated in this book, notably with the "theory" and "evaluation" sections used in the previous sections, rifts and rifted margins are geological objects that are not yet fully understood. Research is very active, and a series of unsolved and challenging questions are being debated. The knowledge evolution achieved within the last three to four decades is tremendous, partly driven by the contribution from modern high-resolution data that permit unprecedented imaging. However, **with new data came new observations, which focused attention on key questions that are still unanswered**.

The stratigraphic architecture of rifted margins is a good example for how new data and observations have changed our understanding. Sedimentary geometries in rifts and rifted margins have long been explained within the framework of classical models, with syn-tectonic packages containing growth structures in fault-bounded half-graben basins. However, it is now well accepted that margins do not evolve solely through high-angle normal faulting and half-graben environments; only a small fraction of the entire extensional system may apply to this mode of deformation. Modern concepts now include polyphase, in-and-out of sequence fault-systems, concave-downward detachments, zones of newly accreted basement with low-angle top-basement fault surface, among others. These configurations ultimately led to the creation of new depositional surfaces in tectonic contexts that are very different from standard half-grabens. Additionally, the overall tectonic evolution is now recognized to be polyphase, with a succession of distinct phases of tectonic activity that further propagate in the rift domain (see section 2.5). Within that framework, terms such as syn-rift, post-rift and breakup unconformity need to be revisited. For example, a sedimentary sequence that is considered as syn-rift in a certain marginal domain may correspond to a pre-rift stage in another marginal domain (see Chapter 1). At the local scale, the complex structural evolution predicted by the new rifted margin models has important controls on the depositional settings, but this is still poorly understood.

The rifted margin community agrees on the fact that our strategy for future research should focus on developing **multi- and inter-disciplinary projects** (Peron-Pinvidic *et al.* 2019): the idea is to not only work together from our own disciplines (multi-disciplinary), but to also work with other communities (inter-disciplinary), for example, developing more active collaborations with colleagues from the mid-ocean ridge, orogen and subduction communities. Mid-ocean ridges, orogens and subduction zones are geological subjects of primary importance in the overall evolution of the Earth and have direct relationships with rifts and rifted margins. This would help refine our scientific interrogations and better target key questions in more efficient ways. For instance, it is still not understood how an oceanic spreading ridge initiates at the end of the rifting evolution. As no strict boundary is observed between the distal margin and the oceanic crust (see section 6.3), the evolution from rift to drift is likely rather gradual and the related processes probably share numerous similarities. A more developed collaboration between the two related communities would therefore be extremely beneficial. Similarly, it is recognized that the architecture of rifted margins can have a huge impact on orogenic evolution and geometry. A better integration of the new models developed by the rift margin community into orogenic concepts may help further constrain different aspects of orogenesis.

Data is of course a crucial component in every research task. The evolution of knowledge achieved by our community in these last decades clearly illustrates the huge impact that advanced datasets can have on our understanding of geological processes. Higher-resolution data, additional data and better access are crucial:

1) A higher spatial and temporal resolution of geological and geophysical datasets would permit more detailed analyses and the filling of data gaps that prevented full characterization of some geometries and/or their lateral variations.

2) Additional rock sampling is necessary in many settings to strengthen the existing datasets and improve the interpretations. In particular, this would be extremely beneficial in the distal margins: deep stratigraphy and basement samples from distal settings are not available to determine the non-uniqueness of geophysical interpretations, except for a limited number of cases (Red Sea, Southern Australia, Iberia–Newfoundland; see section 5.1.3). Deep see drilling constraints would finally settle the intense debates regarding the composition and tectonic evolution of the distal margin basement and deep stratigraphy. Research drilling campaigns are essential for rifted margin research, like through IODP (International Ocean Discovery Program, see: http://www.iodp.org/). Based on group discussions, the "Imagining Rifting" workshop came to the conclusion that the community should target already well-constrained systems (e.g. Iberia, South China Sea) in the near-future, in order to achieve a balance between scientific challenges, accessibility and representativity (Peron-Pinvidic *et al.* 2019).

3) Based on the previous two points, the **access to the data** is of primary importance, but is still a regular problem for many researchers. Regarding rifted margins, numerous high-quality seismic reflection datasets are not accessible to the majority of researchers because of industry confidentiality. Making these data, or at least a part of these data, public would surely lead to a breakthrough in our understanding of offshore extensional geometries. Many colleagues in academia and industry are advocating **open access** for subsurface databases. And indeed, this is crucial for both of the communities: the use of similar datasets, formats and software would probably facilitate scientific exchange and promote collaborations, which is always important in knowledge evolution.

Below, we discuss five research questions that are still actively debated in the community, to illustrate the huge potential of research in rifts and rifted margins. This list includes margin segmentation, final rifting evolution, distal margin basement composition and lithospheric breakup. Many other questions are equally unanswered and debated, like the influence of inheritance on the overall rift evolution, salt history and influence on rift evolution, spatial and temporal evolution of fault rupture geometries, the volume, dynamics, timing and geometries of emplacement of magma, and the variation of rheology during extension.

2.6.1. *Question 1: What is early rifting?*

As summarized in the previous sections, the last few decades of research in rifted margin architecture have greatly benefited from the high-resolution long-offset seismic reflection dataset, which now covers most of the rifted systems worldwide. Based on the unexpected observations made in distal settings (see section 5.1.3),

the main research efforts have been focused on the deep, or distal, margins in the last decade. Various detailed studies have led to a profound revision of our understanding of the related structures. Early models have been greatly revised and although numerous questions remain, we consider that our understanding of the distal geometries and related geological processes has made a significant step forward.

Within the framework of today's accepted models (see section 2.5), the proximal domain is defined as the margin domain that records the early phases of rift-related deformation. Since deformation progressively migrates and localizes from the continent towards the ocean, the basement and sedimentary geometries related to the first increments of extension should be preserved in the proximal domain. They are not overprinted by later phases of deformation, as is the case for the other margin domains, where various deformational phases superposed in time and space give rise to very complex geometries.

Conceptually, the proximal domain corresponds to a continental crust that is only moderately stretched. The relatively simple structural geometries are dominated by Andersonian-type high-angle planar or listric concave-upwards normal faults (e.g. Lavier and Manatschal 2006; Peron-Pinvidic *et al.* 2013; Tugend *et al.* 2015) that flank half-graben basins and sole out at shallow to mid-crustal levels. Associated syn-tectonic sedimentary successions are often wedge-shaped, thickening towards the footwall block (Gawthorpe and Leeder 2000).

Thus, the interpretation of the margin proximal domain still follows standard concepts. **Based on our experience with the profound revision of distal domains' geometries, it is a possibility that the genetic evolution of the proximal domain is also misunderstood.** A careful re-investigation of the geometries and evolution of this domain is therefore opportune and favorable.

The mid-Norwegian margin is a relevant laboratory for this re-investigation. The rift, like many others, settled on the remnants of a pre-existing orogen, the Caledonides. The first extensional episodes that affected what today corresponds to the mid-Norwegian margin began in Devonian time with the "**collapse**" of the Caledonian orogen (Andersen Jamtveit 1990; Séranne 1992). When and how rifting exactly began is actually a matter of definition (e.g. Dewey 1988; Platt and England 1994; Osmundsen *et al.* 2002). The region contains several extensional geometries, but these can be related to either the collapse of the mountain range or to the North-East Atlantic rifting. The distinction in many cases is not straightforward, and some researchers believe that orogenic collapse initiated rifting.

Detailed onshore field studies have shown that the collapse of the orogen was accommodated by major corrugated shear zones, displaying core-complex geometries involving significant >50 km displacements (e.g. Andersen *et al.* 1994). On the other hand, the first phase of rifting that is assumed to shape the proximal domain of the margin is traditionally interpreted in terms of the standard normal faulting, leading to the formation of (half-)grabens flanked by relatively high-angle faults (see above). **The deformation modes envisioned for the orogenic collapse and the early stages of rifting are thus considered radically different**. Key questions relate to how, when and where the extending system evolves from one deformation mode to the other.

Based on observations from seismic reflection data and evidence for reactivation of detachments in East Greenland and Norway (e.g. Rotevatn *et al.* 2018), it has been proposed that the transition between orogenic collapse and early rifting corresponds to a specific, previously unspecified tectonic phase (Peron-Pinvidic *et al.* 2020). Termed the "**reactivation phase**" in Figures 2.40 and 2.41, this stage constitutes a critical phase in the overall evolution of rifted margins that develop from former orogens. It corresponds to a period of extensive reactivation of some of the collapse-related structures with major detachment faulting, leading to further crustal thinning and basement exhumation (e.g. the Trøndelag Platform). This specific deformational phase precedes the standard rift evolution, giving rise to previously unrecognized geometries in the margin's proximal domain. In the Norwegian case, crustal thinning appears to have significantly varied along the margin, with the largest magnitudes of the reactivation phase extension recorded inboard of the future Vøring margin.

The transition from the detachment zones to the high-angle normal faults of the stretching phase is interpreted to occur after a period of intense sedimentary influx into the basins. This influx is associated with the drastic crustal thinning of the reactivation phase. The accumulated sedimentary thicknesses, together with a probable tectonic quiescence, led to a thermal and rheological readjustment that may have induced new decoupled levels in the lithosphere, which enhanced the development of a subsequently more standard fault network.

Thermal re-equilibration typically includes diagenesis and other post-depositional processes which impact the physical properties of the rocks at the base of the existing sedimentary basins and at the base of the crust. Various studies have reported the imprint of a pervasive low-grade metamorphism in sub-greenschist facies in onshore south-western Norway, which is interpreted to reflect the burial of the Devonian sediments to significant depths (Norcliffe *et al.* 2018), although shear heating and the transfer of remnant heat from the exhumed footwall may have also played a role in these processes (Souche *et al.* 2013). Offshore, similar metamorphism in the Devonian rocks would explain the difficulty in identifying the top of the basement in potential field or refraction datasets (Breivik *et al.* 2011). Because of densification processes related to large basinal thicknesses, the upper basement rocks may become undistinguishable from the densified sedimentary layers above. Although they are lithologically different, these rocks may become geophysically indistinguishable (Peron-Pinvidic *et al.* 2016).

A similar uncertainty can be hypothesized for the position of the Moho in the proximal domain. The basement is locally thinned enough to allow faults to cut through the crust and down into the mantle, providing potential conduits for water to circulate downwards. This process can then lead to the serpentinization of some local mantle bodies, a metamorphic process known to affect the physical properties and geophysical signals of the affected rocks. In these cases, the Moho determined from seismic refraction (e.g. mantle rocks are assumed to have 8 km/s P-wave velocities) may not correspond to the lithological base of the crust.

Based on this, the post-reactivation phase in offshore mid-Norway is interpreted to have led to a local redefinition of the geophysical top basement and Moho. It is assumed that the lithosphere reached a new equilibrium at that time, which promoted the development of a different deformational mode in the next tectonic phase: subsequent rifting focused on the less-thinned adjacent areas (Figure 2.40).

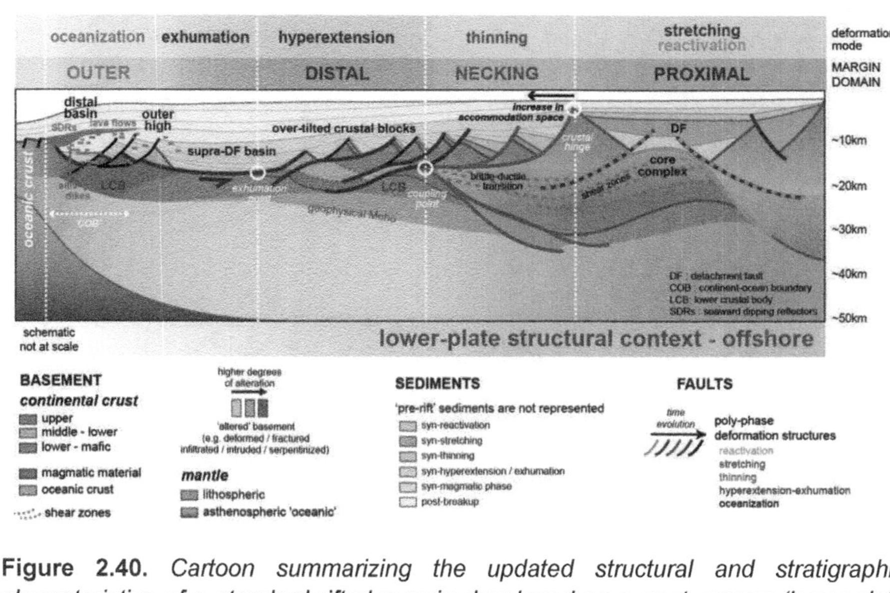

Figure 2.40. *Cartoon summarizing the updated structural and stratigraphic characteristics of a standard rifted margin developed on a past orogen (lower-plate settings) (source: Peron Pinvidic and Osmundsen 2020, adapted from Peron-Pinvidic* et al. *2013)*

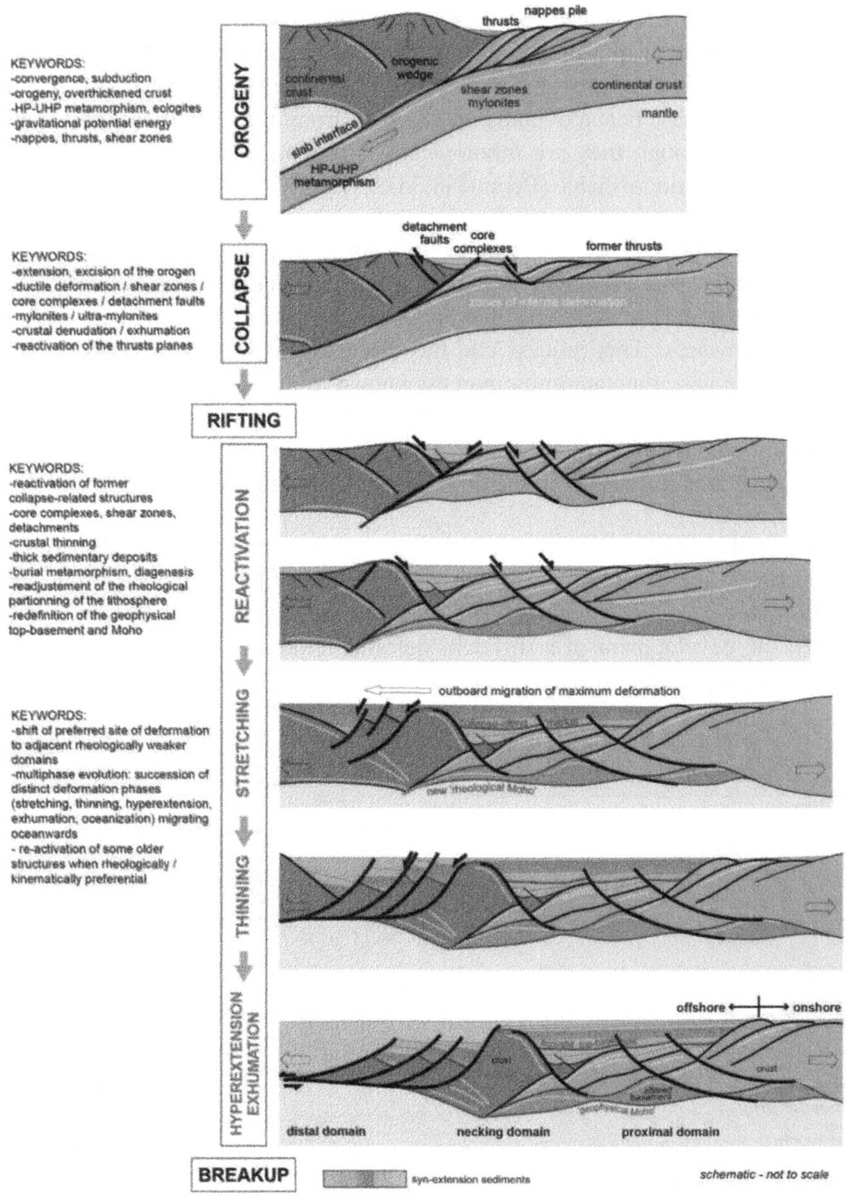

Figure 2.41. *Schematic representation of the structural crustal evolution from an orogenic context to a rifted margin, taking into account the recent observations made on the proximal domain of the mid-Norwegian margin (source: Peron-Pinvidic* et al. *2020)*

2.6.2. *Question 2: What is final rifting?*

Based on modern definitions of rifted margins, it is now well accepted that the distal domain conceptually encompasses a hyperextended sub-domain with a possible exhumation sub-domain, and corresponds to high crustal thinning factors and a significant increase in accommodation space (see section 5.1.3). Based on geophysical modeling (seismic and potential field), various interpretations have been proposed to account for the nature and origin of the basement in these distal settings. Given the complexity of the tectonic processes this region has endured, complex lithological assemblages, such as extremely deformed mid- to lower crustal material and/or altered mantle can be expected. Furthermore, the distal and outer domains are often associated with materials which can have strong masking effects on geophysical signals, such as intrusive and/or extrusive magmatic rocks (e.g. the Møre and Vøring margins; Skogseid *et al.* 1992) or salt (e.g. the Angola margin; Peron-Pinvidic *et al.* 2015). These features commonly prevent clear imaging, characterization, modeling and mapping of structural geometries at depth. **The result is that the distal and outer domains of rifted margins are still poorly constrained**. This is unfortunate as these domains are unique and constitute the transition between two distinct regions formed by two distinct geological processes: the rifted margin inboard, primarily formed by tectonic extension of the continental lithosphere, and the oceanic crust outboard, primarily formed by asthenospheric magmatic additions. The dominant driving processes responsible for the formation of these domains are thus questionable: which process, tectonic or magmatic, lithospheric or asthenospheric, is the most important factor for their formation?

In detail, the composition and architecture of distal rifted margins remain a matter of intense debate. Since proposed margin geometries are usually based on the study of seismic reflection profiles, a large range of interpretations will be possible from a single dataset. Geophysical observations are commonly not sufficient to provide a unique solution in areas of poor data coverage, or at depths where the data resolution typically deteriorates. In these cases, a standard method is to supplement the seismic reflection-based interpretation with models derived from seismic refraction or potential field datasets. The advantage of this combined approach lies in data integration and in the exploitation of the strengths of the different data types. The composite section arising from these combined datasets is thus viewed as the optimal representation.

However, throughout the last decade, a new generation of long-offset seismic reflection profiles has permitted an unprecedented improvement in imaging the stratigraphy, top-basement and Moho, and considerably clarifying structural geometries at depths exceeding 10 km. Consequently, margin-scale architecture can now be mapped and studied with more confidence. Key datasets, such as those acquired and interpreted in the South China Sea (e.g. Savva *et al.* 2013) and offshore Brazil (e.g. Stica *et al.* 2014), have ignited a profound revision of our understanding of rifted margin geometries. With the improved imagery also came an understanding that bodies constructed from geophysical modeling do not always correspond to the geological structures observed on the seismic reflection profiles. Significant mismatches between the interpretations of different types of datasets are now recognized (Peron-Pinvidic *et al.* 2016).

The ambiguity of potential field and refraction methods exists because different lithologies with distinct structural, thermal and geochemical histories can correspond to similar sets of geophysical parameters (e.g. Saltus and Blakely 2011) (see section 5). The obvious result is that a standardized translation of geophysical values into lithologies may ignore alternative scenarios of interpretation. This problem has been commonly identified in specific marginal areas, such as the distal domain, and in the transition to the oceanic crust, where the structural entities are affected by complex multi-phase processes with composite tectonic, geochemical and/or metamorphic histories that can lead to profound modifications of their geophysical characteristics. In the absence of deep drilling, calibration of the lithological composition of rocks in distal margin settings is impossible. **Thus, the architecture and evolution of distal margins and their transition to oceanic crust are poorly constrained in most regions.**

2.6.3. *Question 3: What is breakup?*

"Breakup" is a widely used term commonly used to designate the end of rifting and the beginning of oceanic drifting. It is defined as a unique, instantaneous and localized process, after which the formation of new basement is no longer dominated by lithospheric-scale tectonic extension, but rather by magmatic processes in the asthenosphere. The "COB" (continent–ocean boundary) is the identifying feature used to track this fundamental lithospheric-scale process. In theory, once mapped, the COB corresponds to both a line on a map and to a specific instant in time. Apart from this straightforward definition (which likely led to the term's widespread use), no robust official definition has been given in the literature. This is most likely because no consensus has been reached on the structural, stratigraphic and magmatic geometries of the distal rifted margins and oceanic domains that record this geological process.

Various structural models have been proposed to explain basement accretion in the distal margin and towards the proto-oceanic domain. The majority of the proposed models heavily depend on large-scale normal faults, usually termed "detachment faults" (Wernicke 1985; Lister *et al.* 1986; Davis and Lister 1988). These faults are special because of their displacement (usually >20 km). The geometry, either concave-upward or concave-downward, typically includes multiple parts – the most active being the ones at higher angles, whereas the lower-angle parts are interpreted as segments that have been flexurally rotated during unroofing after their main phase of activity (Buck 1988; Tucholke *et al.* 1998). Within that context, rolling-hinge-type detachments are often advocated for as the driving mechanism for basement exhumation (Lavier *et al.* 1999). Models that invoke "flipping detachments" provide a robust mechanism for how extensive surfaces of mantle can be exhumed during the final rifting evolution and accretion at ultraslow spreading centers (Reston and McDermott 2011; Gillard *et al.* 2015; Reston 2018). However, even though these detachments are widely accepted and interpreted as primary structures, little information exists on their initiation at the rift axis. How, where and when these detachments initiate remain topics to be tackled.

Based on numerical simulations of continental rifting, Peron-Pinvidic and Naliboff (2020) proposed some additional theories on this exhumation process (see section 2.4.2.5). In this study, a scenario is proposed where mantle exhumation at the final rifting stage operates with multiple concave-downward detachment faults that successively cut the seafloor at the rift axis, later migrating off-axis and deactivating when the hanging wall moves away. The resulting accreted basement is an amalgamation of different lithologies, exhumed from various depths at various times. However, these numerical simulations do not include magma, fluid circulation and serpentinization processes, which may influence the deformation history of the fault system. For instance, in oceanic accretion, Bickert *et al.* (2020) have shown that the geometries of the detachment faults at ultraslow-spreading ridges are actually influenced by the amount and location of serpentinite, together with the grain size reduction related to the previous stages of faulting. **Thus, further studies must be conducted to constrain the actual role of melts, fluids and serpentinization on the basement exhumation process, in order to capture the overall rift-to-drift process.**

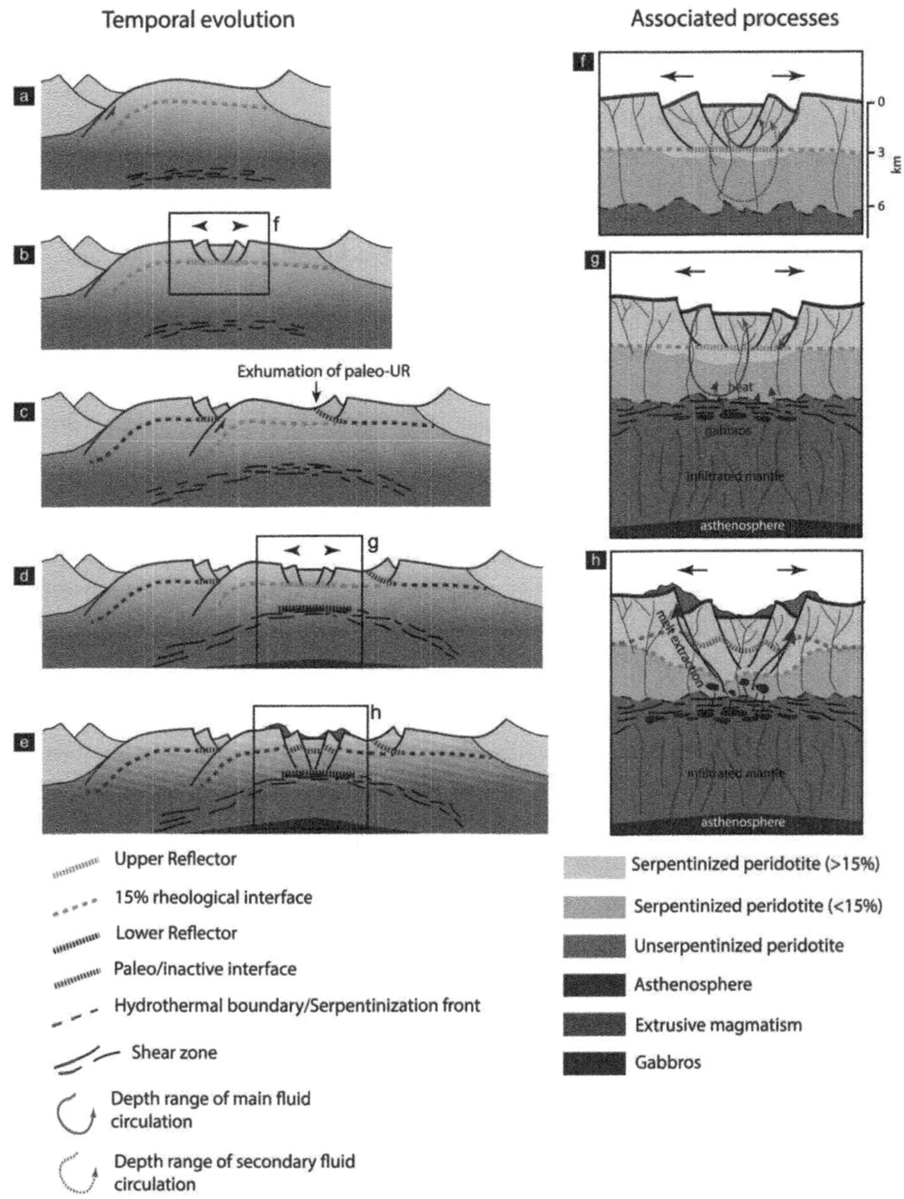

Figure 2.42. *Geologic model showing the temporal evolution of the distal margin, with specific focus on the formation and evolution of the different decoupling interfaces in distal domains of magma-poor rifted margins. Panels (f–h) highlight the processes (hydration, shearing, magmatism) and their interactions associated with the formation/evolution of the decoupling interfaces (source: Gillard* et al. *2019)*

Within that framework, the rheological evolution of the distal margin and oceanic/proto-oceanic basement appears to be a key factor, since tectonic activity and serpentinization are directly dependent on the lithology. The existence of specific rheological interfaces in the distal margin has been recently proposed by Gillard *et al.* (2019). Based on the interpretation of seismic profiles from the Antarctica and Newfoundland-Labrador margins, Gillard *et al.* (2019) propose that the distal margin's structural geometries are governed by a combination of concave-downward detachment faults and high-angle normal faults (Figure 2.42). The latter would develop in the footwall of the larger-scale detachments and interact with hydration and magmatic processes, leading to the formation of decoupling interfaces, at ~3 and 6 km depths. These interfaces are interpreted as rheological boundaries governing the fault population geometry and evolution, and it is the coupling of the deformation between the upper and lower layers defined by these interfaces that is interpreted to mark the breakup of the continental lithosphere. In this magma-poor, slow-spreading setting, the process is gradual and no strict COB seems to be definable. The existence of these rheological boundaries would have a direct impact on the fault activity, fluid circulation, serpentinization and magmatism of the domain, which are fundamental processes. **Thus, the role and feedback effects of the basement rheological evolution, together with magmatic processes and fluid circulation, need to be further described and constrained**.

2.6.4. *Question 4: What is the COB?*

The "COB" (continent–ocean boundary) is defined as the spatial and temporal line where basement with (sub)continental affinity is juxtaposed against oceanic crust. Like "breakup", "COB" is a widely used term in the geosciences, but its definition is often ambiguous, as its mapping is highly subjective and strongly dependent on the dataset used (Heine *et al.* 2013; Peron-Pinvidic *et al.* 2015).

The identification of the COB can alternatively be based on seismic reflection or refraction data and is defined by the first appearance of the characteristic three-layer geometry attributed to "Penrose-type" oceanic crust. In these cases, the identification very often suffers from the presence of salt or volcanics, which can mask most of the underlying seismic signals. Additionally, this mapping methodology assumes a classical oceanic layered geometry which may not necessarily be the case, depending on the magmatic supply and accretion processes. Some of the different ways to define the COB, with their related problems, are listed below:

1) Based on seismic reflection profiles, the COB definition can be associated with the identification of SDRs. But then, the question becomes if the COB should

be mapped at the top or at the bottom of the magmatic wedge. Depending on the dimension of the magmatic wedge, the difference in location can be significant. In that context, the origin and composition of the basement underneath SDRs is of primary importance, but very little is known about it. Additionally, the dynamics of SDR formation remain widely discussed (e.g. Pindell *et al.* 2014).

2) The COB can also be defined at the first seismic trace where oceanic crust is "unambiguously" observed (i.e. clear tri-partite basement geometries, Penrose oceanic crust). The problem here is that interpreters are often left with an inboard wide area of so-called "transitional" basement that may very well contain oceanic crust, but is not wholly oceanic crust. If accreted in an "embryonic" state or magma-poor/ultra-slow process, the oceanic crust may not contain enough magmatic material to generate a clearly layered standard basement and can lead to a misidentification of the nature of the basement.

3) The COB can also be defined based on seismic refraction data. This definition is theoretically supposed to correspond to the previous definition: the mapped COB will correspond to the first oceanwards occurrence of a layered basement displaying the characteristic oceanic magmatic layer velocities (e.g. Christensen and Mooney 1995). This actually leads to the same difficulties as in the previous definition; this mapping method may miss a domain where the basement would have been accreted in a non-standard process, although it can be considered as oceanic (e.g. by an oceanic core complex leading to the exposure of mantle rocks with few magmatic occurrences). The additional issue with refraction data resides in the spatial distribution of the sonobuoys, which is often generally substantial (>10 km spacing), and can hamper any robust imaging and identification because of low resolution.

4) Very often and usually for regional investigation purposes (e.g. plate/palinspastic reconstructions), the COB is defined by magnetic anomalies. The first anomaly with clear(er) and high(er) amplitude, often bordering an outboard domain with more organized and linear anomalies than in the rifted margin domain, is interpreted as marking the first blocks of accreted oceanic crust. In that case, the definition depends on the interpreter's visual judgment, on the quality and resolution of the dataset, and if breakup occurred during a period of time when magnetic reversals were regular enough to permit a distinct record of the first accreted oceanic crust, which is not the case during the Cretaceous Normal-Polarity Super Chron. Key studies have shown that this method, if taken as standalone justification for the COB definition, may result in maps that are difficult to reconcile with other datasets. In the Southern Iberia Abyssal Plain, Srivastava *et al.* (2000) interpreted M-series magnetic anomalies up to M17 within the distal margin, whereas ODP drillings revealed the presence of sub-continental mantle more than 150 km further oceanwards. And in the Pelotas margin, Stica *et al.* (2014) demonstrated that the

mapped magnetic anomalies cross-cut the margin structures, mapped SDRs and COB (as defined by interpretation of seismic reflection data) at a high angle. Therefore, there can be a mismatch between the magnetic signal and the structural geometries of the rifted margin. Various types of rocks (dikes, sills, volcanics, serpentinites) can have strong magnetic properties (Pariso and Johnson 1993; Oufi *et al.* 2002). Depending on the structural context, the genesis of symmetric contemporaneous anomalies is possible. While the mapping of these magnetic anomalies is recognized as robust in steady-state oceanic accretion settings (Harrison 1987), it is not the case for every type of ridge (Cannat *et al.* 2019) or in rift-drift transition areas (Bronner *et al.* 2014). Peron-Pinvidic and Naliboff (2020) proposed an exhumation detachment model with a scenario where mantle is exhumed in a complex structural pattern, giving rise to wide zones of exposed mantle with composite geology. Depending on where the leading detachment cut the footwall of the previous fault, the rocks exhumed by the deactivated detachment may be isolated on one plate or the other. A consequence of this process is the complicated unroofing histories of seafloor rocks. While symmetric patterns may prevail at the margin-wide scale (although probably of weak intensity; Bronner *et al.* 2014), magnetic anomalies may not be created at the same time and do not mark the formation of the same basement block. Based on these issues, magnetic anomalies in slow-accretion environments or rift-drift transition areas may not be isochrons and thus cannot be used to map COBs and rift opening kinematics, unless systematic sampling for dating is possible.

5) Another method for defining the COB is to map it based on potential field modeling results. In this approach, the COB is defined as the line where the outer high magmatic underplating is the thickest, based on the *a priori* definition that breakup corresponds to a pulse in the magmatic input into the rifted margin (e.g. Corti *et al.* 2003). The limitation for this definition is similar to many of the previous methods: since the definition is modeling-derived, it also tends to be very interpreter/modeler-dependent.

6) Finally, the COB can be mapped by attempting to combine all, or some, sets of the above definitions, which can be very challenging.

Each of these definitions has advantages and disadvantages. The result is that for the same margin, interpreters can define very different COBs depending on the definition that is picked, potentially up to hundreds of km apart (Figure 2.43).

Figure 2.43 shows an example from the North-East Atlantic, with the white lines representing some of the various COBs proposed in the literature (Mosar *et al.* 2002; Sigmond 2002; Hamann *et al.* 2005; Scott *et al.* 2005; Gaina *et al.* 2009; Voss *et al.* 2009; Faleide *et al.* 2010; Peron-Pinvidic *et al.* 2012; Tsikalas *et al.* 2012). The significant discrepancy between the various COBs is striking, in particular, for East Greenland. The basic exercise of closing the oceanic domain to juxtapose the COBs, which is one of the basic steps in every plate tectonic reconstruction, is obviously not a straightforward task.

The resulting uncertainties have huge impacts on various other margin-related studies like 2D section reconstructions, plate kinematics exercises and basin modeling.

Plate tectonic reconstruction models are fundamental constraints used in every rifted margin study. Regarding rift basins, these models can provide (1) the identification of conjugate structures, which reduces the risk of structural misinterpretation, and (2) a basic estimate of the extension accumulated during rifting. These reconstructions can either be global or local, but they always require the definition of COBs as they depend on the fundamental distinction between domains created solely by oceanic accretion and domains derived from continental material. **Today's non-unique definition of the COB renders the exercise rather difficult at a worldwide scale, as the input COB line directly depends on the dataset availability and quality, and the interpreter's judgment.**

Based on the recent advances in our understanding of rifting processes, the fact that the COBs mapped on different datasets (e.g. reflection, refraction or magnetic) give rise to very different locations on the map may not be surprising. Each line (e.g. reflection-based, refraction-based or magnetic-based) may actually correspond to the geometrical or geophysical characteristic of an internal process of the breakup phase. In short: the reflection-based definition will tend to point to the structural evolution (e.g. identification of specific fault systems), while the refraction-based definition will tend to point to the basement evolution (e.g. basement thickening due to magma input), and the magnetic-based definition to the magmatic content evolution (e.g. extrusion of basalts that can record magnetic information). In most cases, the related COBs will not match, simply because they image different features.

The methodology and dataset used should always be described in detail when a COB is proposed in the literature, in order to avoid misunderstanding and allow for global comparisons. When dealing with plate reconstructions, it is fundamental to select a dataset and a definition that can be uniformly applied to the entire studied area, in order to not confuse the various COB lines and end with plates whose defined boundaries differ from one site to the other.

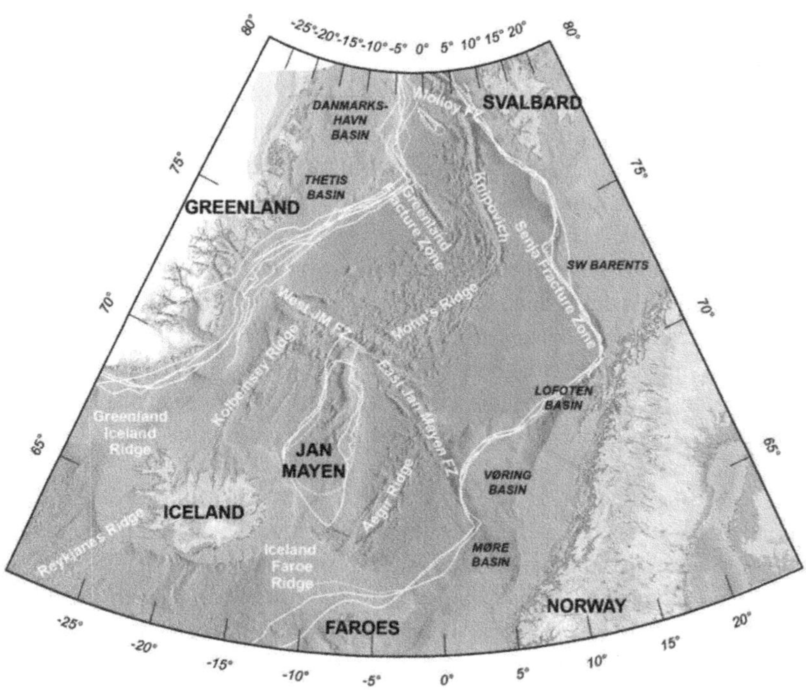

Figure 2.43. *Bathymetric map of the North-East Atlantic, with the white lines representing some of the various COBs published in the literature (after Mosar* et al. *2002; Sigmond 2002; Hamann* et al. *2005; Scott* et al. *2005; Gaina* et al. *2009; Voss* et al. *2009; Faleide* et al. *2010; Peron-Pinvidic* et al. *2012; Tsikalas* et al. *2012)*

2.6.5. *Question 5: Why are rifted margins segmented?*

The terminology dealing with the segmentation and sub-system limits is vast and often confusing, usually reflecting the uncertainties regarding their formation.

In the oceanic domain, the term **transform fault** is used when the connection from one sub-system to the other is governed by one distinct fault (Gibbs 1984), and the term **transfer zone** is used when this connection corresponds to a network of faults. Lister *et al.* (1986) proposed that comparable structural features can cause the along-strike segmentation of rifted margins and introduced the term **transfers** for continental domains, following the concept of the **upper and lower plates**. The term **accommodation zone** is also used heavily in the literature (Rosendahl 1987; Acocella *et al.* 2005), often when the connection is structurally less well defined, such as when the transfer of strain is more diffused, operating over a wider area than a sole fault plane.

In oceanic domains, the identification of transfers from one major sub-system to the other is relatively straightforward. In continental domains, such as rifted margins, the identification is much more difficult. The issue is to understand and correctly constrain what these transfers are in terms of tectonic structures, magmatic contents, stratigraphic records and other parameters. A precise unified definition would greatly aid in the accurate mapping of margin architecture. However, determining a unified definition is found to be highly problematic. Depending on the selected characteristics being investigated (e.g. fault geometry, sills extent, stratigraphic architecture, magnetic signal, bathymetry), the transfer may be (1) complicated to identify (in the case of diffuse and diverse magmatic activity without clear termination) and (2) substantially different from one set of characteristics to the other.

One example of inconsistencies in defining the transfer line is in the case of the Møre to Vøring segment transfer in the mid-Norwegian margin. Studies with a focus on the oceanic domain tend to extend the outline of the Jan Mayen Fracture Zone south-eastward along the SE flank of the Vøring margin segment, which does not match the outline of tectonic structures or stratigraphic and magmatic features mapped on seismic reflection data. Faults, sills and sedimentary packages rarely end along the same line. Interpreters must often deal with diffuse transitions rather than clear boundaries; therefore, definitions of limits are regularly a matter of debate.

In addition to the uncertainties in the definition and identification of transfers, a key question that remains is the timing of the transfer activity relative to the evolution of the margin. Specifically, when does margin compartmentalization initiate? The birth and location of offsetting structures between sub-basins are not straightforward. Are these transfer features inherited pre-rift features (e.g. structural, thermal), or do they develop during rifting? In the latter case, are the influencing parameters extension rate, obliquity, sedimentary/magmatic supply, source location/efficiency or something else? Various key factors can be responsible for the structural variability and segmentation of rift systems, making it difficult to understand which one is the controlling factor. Based on the example of the Gulf of California, which has significant along-strike variations in rifting style and magmatism, Lizarralde *et al.* (2007) proposed that mantle depletion or fertility could play a significant role. However, the authors also highlight the fact that sedimentation rate and volume could play a major role. In that case, it means that either pre-rift or syn-rift parameters can be the controlling factors. In the South Atlantic, Heine *et al.* (2013) showed that the Pangea rifting and breakup have been influenced by changes in the opening obliquity and rate of extension, which are syn-rift parameters. In a recent contribution, Osmundsen and Peron-Pinvidic (2018) proposed instead that the breakaway complexes play a fundamental role in the overall architecture and evolution of rifted margins (see section 2.4). Based on the observations from the mid-Norwegian margin, they showed that breakaway complexes can accommodate the successive role of stretching and necking at some margin segments. For instance, the deformation phases stack on the same fault systems in the Møre segment, while at other margin segments, each deformation phase is accommodated by a different set of faults (e.g. the nice distinction of breakaway complexes for parts of the Vøring segment). Peron-Pinvidic and Osmundsen (2018) then proposed that the lateral evolution of these breakaway complexes may be the key parameter defining margin segmentation. The along-strike change in the structural role of the major faults creates distinct margin compartments that have specific tectonic settings and present distinct sedimentary and magmatic records, and therefore different geophysical characteristics. The underlying question remaining is why does this lateral shift in the role of a fault system occur and when does it begin?

2.7. References

Acocella, V., Morvillo, P., Funiciello, R. (2005). What controls relay ramps and transfer faults within rift zones? Insights from analogue models. *Journal of Structural Geology*, 27(3), 397–408.

D'Acremont, E., Leroy, S., Beslier, M.O., Bellahsen, N., Fournier, M., Robin, C., Maia, M., Gente, P. (2005). Structure and evolution of the eastern Gulf of Aden conjugate margins from seismic reflection data. *Geophysical Journal International*, 160(3), 869–890.

Allemand, P. and Brun, J.P. (1991). Width of continental rifts and rheological layering of the lithosphere. *Tectonophysics*, 188(1–2), 63–69.

Allken, V., Huismans, R.S., Thieulot, C. (2012). Factors controlling the mode of rift interaction in brittle-ductile coupled systems: A 3D numerical study. *Geochemistry, Geophysics, Geosystems*, 13(5).

Amante, C. and Eakins, B.W. (2009). ETOPO1 1 arc-minute global relief model: Procedures, data sources and analysis. Report, NOAA, NGDC.

Andersen, T.B. and Jamtveit, B. (1990). Uplift of deep crust during orogenic extensional collapse: A model based on field studies in the Sogn–Sunnfjord area of Western Norway. *Tectonics*, 9, 1097–1111.

Andersen, T.B., Osmundsen, P.T., Jolivet, L. (1994). Deep crustal fabrics and a model for the extensional collapse of the southwest Norwegian Caledonides. *Journal of Structural Geology*, 16, 1191–1203.

Andersen, T.B., Corfu, F., Labrousse, L., Osmundsen, P.-T. (2012). Evidence for hyperextension along the pre-Caledonian margin of Baltica. *Journal of the Geological Society*, 169(5), 601–612.

Andrés-Martínez, M., Pérez-Gussinyé, M., Armitage, J., Morgan, J.P. (2019). Thermomechanical implications of sediment transport for the architecture and evolution of continental rifts and margins. *Tectonics*, 38(2), 641–665.

Aragón-Arreola, M., Morandi, M., Martín-Barajas, A., Delgado-Argote, L., González-Fernández, A. (2005). Structure of the rift basins in the central Gulf of California: Kinematic implications for oblique rifting. *Tectonophysics*, 409(1), 19–38.

Autin, J., Leroy, S., Beslier, M.O., d'Acremont, E., Razin, P., Ribodetti, A., Bellahsen, N., Robin, C., Al Toubi, K. (2010). Continental break-up history of a deep magma-poor margin based on seismic reflection data (northeastern Gulf of Aden margin, offshore Oman). *Geophysical Journal International*, 180(2), 501–519.

Balázs, A., Matenco, L., Vogt, K., Cloetingh, S., Gerya, T. (2018). Extensional polarity change in continental rifts: Inferences from 3-D numerical modeling and observations. *Journal of Geophysical Research: Solid Earth*, 123(9), 8073–8094.

Ball, P., Eagles, G., Ebinger, C., McClay, K., Totterdell, J. (2013). The spatial and temporal evolution of strain during the separation of Australia and Antarctica. *Geochemistry, Geophysics, Geosystems*, 14(8), 2771–2799.

Basile, C. (2015). Transform continental margins – Part 1: Concepts and models. *Tectonophysics*, 661, 1–10.

Basile, C. and Braun, J. (2016). The initiation of pull-apart basins and transform continental margins: Results from numerical experiments of kinematic partitioning in divergent settings. *Terra Nova*, 28(2), 120–127.

Bassi, G., Keen, C.E., Potter, P. (1993). Contrasting styles of rifting: Models and examples from the Eastern Canadian Margin. *Tectonics*, 12(3), 639–655.

Becker, K., Franke, D., Schnabel, M., Schreckenberger, B., Heyde, I., Krawczyk, C.M. (2012). The crustal structure of the southern Argentine margin. *Geophysical Journal International*, 189(3), 1483–1504.

Berndt, C., Planke, S., Alvestad, E., Tsikalas, F., Rasmussen, T. (2001). Seismic volcanostratigraphy of the Norwegian Margin: Constraints on tectonomagmatic break-up processes. *Journal of the Geological Society*, 158, 413–426.

Bialas, R.W., Buck, W.R., Qin, R. (2010). How much magma is required to rift a continent? *Earth and Planetary Science Letters*, 292(1–2), 68–78.

Bickert, M., Lavier, L., Cannat, M. (2020). How do detachment faults form at ultraslow mid-ocean ridges in a thick axial lithosphere? *Earth and Planetary Science Letters*, 533, 116048.

Bird, D. (2001). Shear margins: Continent-ocean transform and fracture zone boundaries. *The Leading Edge*, 20(2), 150–159.

Blaich, O.A., Faleide, J.I., Tsikalas, F., Franke, D., Leon, E. (2009). Crustal-scale architecture and segmentation of the Argentine margin and its conjugate off South Africa. *Geophysical Journal International*, 178(1), 85–105.

Blystad, P., Brekke, H., Færseth, R.B., Larsen, B.T., Skogseid, J., Tørudbakken, B. (1995). Structural elements of the Norwegian continental shelf, Part II. The Norwegian Sea Region. *Norwegian Petroleum Directorate Bulletin*, 8, 0–45.

Boillot, G., Grimaud, S., Mauffret, A., Mougenot, D., Kornprobst, J., Mergoil-Daniel, J., Torrent, G. (1980). Ocean-continent boundary off the Iberia margin: A serpentinite diapir west of the Galicia Bank. *Earth and Planetary Science Letters*, 48, 23–34.

Bondevik, S., Løvholt, F., Harbitz, C., Mangerud, J., Dawson, A., Svendsen, J.I. (2005). The Storegga Slide tsunami – Comparing field observations with numerical simulations. In *Ormen Lange – An Integrated Study for Safe Field Development in the Storegga Submarine Area*, Solheim, A., Bryn, P., Berg, K., Sejrup, H.P. Mienert, J. (eds). Elsevier, Oxford.

Braun, J., Deschamps, F., Rouby, D., Dauteuil, O. (2013). Flexure of the lithosphere and the geodynamical evolution of non-cylindrical rifted passive margins: Results from a numerical model incorporating variable elastic thickness, surface processes and 3D thermal subsidence. *Tectonophysics*, 604, 72–82.

Breivik, A.J., Mjelde, R., Raum, T., Faleide, J.I., Murai, Y., Flueh, E.R. (2011). Crustal structure beneath the Trøndelag Platform and adjacent areas of the mid-Norwegian margin, as derived from wide-angle seismic and potential field data. *Norwegian Journal of Geology*, 90(4), 141–161.

Brekke, H. (2000). The tectonic evolution of the Norwegian Sea Continental Margin with emphasis on the Vøring and Møre Basins. In *Dynamics of the Norwegian Margin: Geological Society Special Publication*, Nøttvedt, A. (ed.). Geological Society of London, UK.

Bronner, A., Sauter, D., Manatschal, G., Peron-Pinvidic, G., Munschy, M. (2011). Magmatic breakup as an explanation for magnetic anomalies at magma-poor rifted margins. *Nature Geoscience*, 4(8), 549–553.

Bronner, A., Sauter, D., Munschy, M., Carlut, J., Searle, R., Cannat, M., Manatschal, G. (2014). Magnetic signature of large exhumed mantle domains of the Southwest Indian Ridge – Results from a deep-tow geophysical survey over 0 to 11 Ma old seafloor. *Solid Earth*, 5(1), 339–354.

Brun, J.P. and Beslier, M.O. (1996). Mantle exhumation at passive margins. *Earth and Planetary Science Letters*, 142(1), 161–173.

Brune, S. and Autin, J. (2013). The rift to break-up evolution of the Gulf of Aden: Insights from 3D numerical lithospheric-scale modelling. *Tectonophysics*, 607, 65–79.

Brune, S., Heine, C., Pérez-Gussinyé, M., Sobolev, S.V. (2014). Rift migration explains continental margin asymmetry and crustal hyper-extension. *Nature Communications*, 5.

Brune, S., Williams, S.E., Müller, R.D. (2018). Oblique rifting: The rule, not the exception. *Solid Earth*, 9(5), 1187–1206.

Bryn, P., Berg, K., Forsberg, C.F., Solheim, A., Kvalstad, T.J. (2005). Explaining the Storegga Slide. *Marine and Petroleum Geology*, 22(1), 11–19.

Buck, W.R. (1988). Flexural rotation of normal faults. *Tectonics*, 7, 959–973.

Buck, W.R. (1991). Modes of continental lithospheric extension. *Journal of Geophysical Research*, 96(B12), 20161–20178.

Buck, W.R., Lavier, L.L., Poliakov, A.N.B. (2005). Modes of faulting at mid-ocean ridges. *Nature*, 434(7034), 719–723.

Buiter, S. and Ellis S. (2012). SULEC: Benchmarking a new ALE finite-element code. EGU General Assembly Conference Abstracts. April.

Burov, E.B. and Diament, M. (1995). The effective elastic thickness (Te) of continental lithosphere: What does it really mean? *Journal of Geophysical Research: Solid Earth*, 100(B3), 3905–3927.

Burov, E. and Poliakov, A. (2001). Erosion and rheology controls on synrift and postrift evolution: Verifying old and new ideas using a fully coupled numerical model. *Journal of Geophysical Research: Solid Earth*, 106(B8), 16461–16481.

Cannat, M., Manatschal, G., Sauter, D., Péron-Pinvidic, G. (2009). Assessing the conditions of continental breakup at magma-poor rifted margins: What can we learn from slow spreading mid-ocean ridges? *Comptes rendus geosciences*, 341(5), 406–427.

Cannat, M., Sauter, D., Lavier, L., Bickert, M., Momoh, E., Leroy, S. (2019). On spreading modes and magma supply at slow and ultraslow mid-ocean ridges. *Earth and Planetary Science Letters*, 519, 223–233.

Chapman, D.S. and Rybach, L. (1985). Heat flow anomalies and their interpretation. *Journal of Geodynamics*, 4(1), 3–37.

Christensen, N.I. and Mooney, W.D. (1995). Seismic velocity structure and composition of the continental crust: A global view. *Journal of Geophysical Research*, 100(B6), 9761–9788.

Clerc, C. and Lagabrielle, Y. (2014). Thermal control on the modes of crustal thinning leading to mantle exhumation: Insights from the Cretaceous Pyrenean hot paleomargins. *Tectonics*, 33(7), 1340–1359.

Clerc, C., Jolivet, L., Ringenbach, J.-C. (2015). Ductile extensional shear zones in the lower crust of a passive margin. *Earth and Planetary Science Letters*, 431, 1–7.

Clerc, C., Lagabrielle, Y., Labaume, P., Ringenbach, J.-C., Vauchez, A., Nalpas, T., Bousquet, R., Ballard, J.-F., Lahfid, A., Fourcade, S. (2016). Basement – Cover decoupling and progressive exhumation of metamorphic sediments at hot rifted margin. Insights from the Northeastern Pyrenean analog. *Tectonophysics*, 686, 82–97.

Clerc, C., Ringenbach, J.-C., Jolivet, L., Ballard, J.-F. (2018). Rifted margins: Ductile deformation, boudinage, continentward-dipping normal faults and the role of the weak lower crust. *Gondwana Research*, 53, 20–40.

Coffin, M.F. and Eldholm, O. (1994). Large igneous provinces – Crustal structure, dimensions, and external consequences. *Reviews of Geophysics*, 32(1), 1–36.

Collier, J.S., Minshull, T.A., Hammond, J.O.S., Whitmarsh, R.B., Kendall, J.M., Sansom, V., Lane, C.I., Rumpker, G. (2009). Factors influencing magmatism during continental breakup: New insights from a wide-angle seismic experiment across the conjugate Seychelles-Indian margins. *Journal of Geophysical Research-Solid Earth*, 114. B03101, doi:10.1029/2008JB005898.

Corti, G., Bonini, M., Conticelli, S., Innocenti, F., Manetti, P., Sokoutis, D. (2003). Analogue modelling of continental extension: A review focused on the relations between the patterns of deformation and the presence of magma. *Earth-Science Reviews*, 63(3–4), 169–247.

Crosby, A.G., Fishwick, S., White, N. (2010). Structure and evolution of the intracratonic Congo Basin. *Geochemistry, Geophysics, Geosystems*, 11(6).

Davis, M. and Kusznir, N.J. (2004). Depth-dependent lithospheric stretching at rifted continental margins. In *Rheology and Deformation of the Lithosphere at Continental Margins*, Karner, G.D., Taylor, B., Driscoll, N.W., Kohlsted, D.L. (eds). Columbia University Press, New York.

Davis, G.A. and Lister, G.A. (1988). Detachment faulting in continental extension; Perspective from the southwestern U.S. Cordillera. In *Processes in Continental Lithospheric Deformation*, Clark S.P., Burchfiel B.C., Suppe J. (eds). Geological Society of America, Boulder.

Dewey, J.F. (1988). Extensional collapse of orogens. *Tectonics*, 7, 1123–1139.

Dick, H.J.B., Lin, J., Schouten, H. (2003). An ultraslow-spreading class of ocean ridge. *Nature*, 426(6965), 405–412.

Dick, H.J.B., Natland, J.H., Ildefonse, B. (2006). Past and future impact of deep drilling in the ocanic crust and mantle. *Oceanography*, 19, 74–82.

Døssing, A. and Funck, T. (2012). Greenland fracture zone–east greenland ridge(s) revisited: Indications of a C22-change in plate motion? *Journal of Geophysical Research: Solid Earth*, 117(B1).

Døssing, A., Dahl-Jensen, T., Thybo, H., Mjelde, R., Nishimura, Y. (2008). East Greenland Ridge in the North Atlantic Ocean: An integrated geophysical study of a continental silver in a boundary transform fault setting. *Journal of Geophysical Research-Solid Earth*, 113(B10).

Driscoll, N.W. and Karner, G.D. (1998). Lower crustal extension across the Northern Carnarvon basin, Australia: Evidence for an eastward dipping detachment. *Journal of Geophysical Research: Solid Earth*, 103(B3), 4975–4991.

Ebinger, C. (2005). Continental break-up: The East African perspective. *Astronomy & Geophysics*, 46(2), 16–21.

Ebinger, C. and Casey, M. (2001). Continental breakup in magmatic provinces: An Ethiopian example. *Geology*, 29, 527–530.

Ebinger, C., Ayele, A., Keir, D., Rowland, J., Yirgu, G., Wright, T., Belachew, M., Hamling, I. (2010). Length and timescales of rift faulting and magma intrusion: The afar rifting cycle from 2005 to present. *Annual Review of Earth and Planetary Sciences*, 38(1), 439–466.

Eldholm, O., Thiede, J., Taylor, B. (1989). Evolution of the Vøring volcanic margin. *Proceedings of the Ocean Drilling Program, Scientific Results*, 104, 1033–1065.

Escartín, J., Hirth, G., Evans, B. (2001). Strength of slightly serpentinized peridotites: Implications for the tectonics of oceanic lithosphere. *Geology*, 29(11), 1023–1026.

Escartín, J., Smith, D.K., Cann, J., Schouten, H., Langmuir, C.H., Escrig, S. (2008). Central role of detachment faults in accretion of slow-spreading oceanic lithosphere. *Nature*, 455, 790.

Faleide, J.I., Bjørlykke, K., Gabrielsen, R.H. (2010). Geology of the Norwegian Continental Shelf. In *Petroleum Geoscience: From Sedimentary Environments to Rock Physics*, Bjørlykke, K. (ed.). Springer-Verlag, Berlin, Heidelberg.

Fossen, H., Gabrielsen, R.H., Faleide, J.I., Hurich, C.A. (2014). Crustal stretching in the Scandinavian Caledonides as revealed by deep seismic data. *Geology*, 42(9), 791–794.

Foulger, G. (2002). Plumes, or plate tectonic processes? *Astronomy and Geophysics*, 43(6), 19–23.

Franke, D. (2013). Rifting, lithosphere breakup and volcanism: Comparison of magma-poor and volcanic rifted margins. *Marine and Petroleum Geology*, 43(0), 63–87.

Franke, D., Barckhausen, U., Baristeas, N., Engels, M., Ladage, S., Lutz, R., Montano, J., Pellejera, N., Ramos, E.G., Schnabel, M. (2011). The continent-ocean transition at the southeastern margin of the South China Sea. *Marine and Petroleum Geology*, 28(6), 1187–1204.

Funck, T., Jackson, H.R., Louden, K.E., Dehler, S.A., Wu, Y. (2004). Crustal structure of the northern Nova Scotia rifted continental margin (eastern Canada). *Journal of Geophysical Research–Solid Earth*, 109(B9).

Gaina, C., Müller, R.D., Brown, B., Ishihara, T. (2003). Microcontinent formation around Australia. *Geological Society of Australia Special Publication*, 22, 399–410.

Gaina, C., Gernigon, L., Ball, P. (2009). Palaeocene-Recent plate boundaries in the NE Atlantic and the formation of the Jan Mayen microcontinent. *Journal of the Geological Society*, 166, 601–616.

Gardner, R.L., Daczko, N.R., Halpin, J.A., Whittaker, J.M. (2015). Discovery of a microcontinent (Gulden Draak Knoll) offshore Western Australia: Implications for East Gondwana reconstructions. *Gondwana Research*, 28(3), 1019–1031.

Gawthorpe, R.L. and Leeder, M.R. (2000). Tectono-sedimentary evolution of active extensional basins. *Basin Research*, 12(3–4), 195–218.

Geoffroy, L. (2005). Volcanic passive margins. *Comptes rendus geoscience*, 337(16), 1395–1408.

Gibbs, A. (1984). Structural evolution of extensional basin margins. *Journal of the Geological Society*, 141(4), 609–620.

Gillard, M., Autin, J., Manatschal, G., Sauter, D., Munschy, M., Schaming, M. (2015). Tectonomagmatic evolution of the final stages of rifting along the deep conjugate Australian-Antarctic magma-poor rifted margins: Constraints from seismic observations. *Tectonics*, 34(4), 753–783.

Gillard, M., Tugend, J., Müntener, O., Manatschal, G., Karner, G.D., Autin, J., Sauter, D., Figueredo, P.H., Ulrich, M. (2019). The role of serpentinization and magmatism in the formation of decoupling interfaces at magma-poor rifted margins. *Earth-Science Reviews*, 196, 102882.

Goetze, C. and Evans, B. (1979). Stress and temperature in the bending lithosphere as constrained by experimental rock mechanics. *Geophysical Journal International*, 59(3), 463–478.

Haflidason, H., Sejrup, H.P., Nygård, A., Mienert, J., Bryn, P., Lien, R., Forsberg, C.F., Berg, K., Masson, D. (2004). The Storegga Slide: Architecture, geometry and slide development. *Marine Geology*, 213(1), 201–234.

Hamann, N.E., Whittaker, R.C., Stemmerik, L. (2005). Geological development of the Northeast Greenland Shelf. In *Petroleum Geology: North-West Europe and Global Perspectives – Proceedings of the 6th Petroleum Geology Conference*, Doré, A.G. Vinning, B.A. (eds). Geological Society, London.

Harrison, C. (1987). Marine magnetic anomalies – The origin of the stripes. *Annual Review of Earth and Planetary Sciences*, 15, 505–543.

Heine, C. and Brune, S. (2014). Oblique rifting of the Equatorial Atlantic: Why there is no Saharan Atlantic Ocean. *Geology*, 42(3), 211–214.

Heine, C., Zoethout, J., Müller, R.D. (2013). Kinematics of the South Atlantic rift. *Solid Earth*, 4, 215–253.

Hieronymus, C.F. (2004). Control on seafloor spreading geometries by stress- and strain-induced lithospheric weakening. *Earth and Planetary Science Letters*, 222(1), 177–189.

Hirth, G. and Kohlstedt, D.L. (1996). Water in the oceanic upper mantle: Implications for rheology, melt extraction and the evolution of the lithosphere. *Earth and Planetary Science Letters*, 144(1), 93–108.

Hopper, J.R. and Buck, W.R. (1996). The effect of lower crustal flow on continental extension and passive margin formation. *Journal of Geophysical Research-Solid Earth*, 101(B9), 20175–20194.

Huismans, R.S. and Beaumont, C. (2007). *Roles of Lithospheric Strain Softening and Heterogeneity in Determining the Geometry of Rifts and Continental Margins*. Geological Society, London.

Huismans, R.S. and Beaumont, C. (2008). Complex rifted continental margins explained by dynamical models of depth-dependent lithospheric extension. *Geology*, 36(2), 163–166.

Huismans, R.S. and Beaumont, C. (2011). Depth-dependent extension, two-stage breakup and cratonic underplating at rifted margins. *Nature*, 473(7345), 74–78.

Huismans, R.S. and Beaumont, C. (2014). Rifted continental margins: The case for depth-dependent extension. *Earth and Planetary Science Letters*, 407, 148–162.

Jarvis, G.T. and McKenzie, D.P. (1980). Sedimentary basin formation with finite extension rates. *Earth and Planetary Science Letters*, 48(1), 42–52.

Keen, C.E., Stockmal, G.S., Welsink, H., Quinlan, G., Mudford, B. (1987). Deep crustal structure and evolution of the rifted margin northeast of Newfoundland – Results from lithoprobe. *Canadian Journal of Earth Sciences*, 24(8), 1537–1549.

Kirby, S.H. and Kronenberg, A.K. (1987). Rheology of the lithosphere: Selected topics. *Reviews of Geophysics*, 25(6), 1219–1244.

Koehn, D., Aanyu, K., Haines, S., Sachau, T. (2007). Rift nucleation, rift propagation and the creation of basement micro-plates within active rifts. *Tectonophysics*, doi:10.1016/j.tecto.2007.1010.1003.

Kohlstedt, D.L. (2007). Properties of rocks and minerals – Constitutive equations, rheological behavior, and viscosity of rocks. *Treatise on Geophysics*, 2, 389–417.

Koptev, A., Burov, E., Gerya, T., Le Pourhiet, L., Leroy, S., Calais, E., Jolivet, L. (2018). Plume-induced continental rifting and break-up in ultra-slow extension context: Insights from 3D numerical modeling. *Tectonophysics*, 746, 121–137.

Kusznir, N.J. (1987). The strength of the continental lithosphere. *Geophysical Journal of the Royal Astronomical Society*, 89(1), 451–451.

Kusznir, N.J., Karner, G.D., Egan, S.S. (1987). Geometric, thermal and isostatic consequences of detachments in continental lithosphere extension and basin formation. In *Sedimentary Basins and Basin Forming Mechanisms*, Beaumont, C. and Tankard, A.J. (eds). Canadian Society of Petroleum Geologists, Memoir 12.

Laberg, J.S., Vorren, T.O., Dowdeswell, J.A., Kenyon, N.H., Taylor, J. (2000). The Andøya Slide and the Andøya Canyon, north-eastern Norwegian–Greenland Sea. *Marine Geology*, 162(2), 259–275.

Laberg, J., Vorren, T.O., Mienert, J., Bryn, P., Lien, R. (2002). The Trænadjupet Slide: A large slope failure affecting the continental margin of Norway 4,000 years ago. *Geo-Marine Letters*, 22(1), 19–24.

Lavier, L.L. and Manatschal, G. (2006). A mechanism to thin the continental lithosphere at magma-poor margins. *Nature*, 440, 324–328.

Lavier, L.L., Buck, W.R., Poliakov, A.N.B. (1999). Selfconsistent rolling-hinge model for the evolution of large-offset, low-angle normal faults. *Geology*, 27, 1127–1130.

Le Pichon, X. and Sibuet, J.C. (1981). Passive margins: A model of formation. *Journal of Geophysical Research*, 86(B5), 3708–3720.

Le Pourhiet, L., May, D.A., Huille, L., Watremez, L., Leroy, S. (2017). A genetic link between transform and hyper-extended margins. *Earth and Planetary Science Letters*, 465, 184–192.

Lemoine, M., Tricart, P., Boillot, G. (1987). Ultramafic and gabbroic ocean-floor of the Ligurian Tethys (Alps, Corsica, Apennines) – In search of a genetic model. *Geology*, 15(7), 622–625.

Leroy, S., Gente, P., Fournier, M., d'Acremont, E., Patriat, P., Beslier, M.O., Bellahsen, N., Maia, M., Blais, A., Perrot, J., Al-Kathiri, A., Merkouriev, S., Fleury, J.M., Ruellan, P.Y., Lepvrier, C., Huchon, P. (2004). From rifting to spreading in the eastern Gulf of Aden: A geophysical survey of a young oceanic basin from margin to margin. *Terra Nova*, 16(4), 185–192.

Leynaud, D., Mienert, J., Vanneste, M. (2009). Submarine mass movements on glaciated and non-glaciated European continental margins: A review of triggering mechanisms and preconditions to failure. *Marine and Petroleum Geology*, 26(5), 618–632.

Lister, G.S., Etheridge, M.A., Symonds, P.A. (1986). Detachment faulting and the evolution of passive continental margins. *Geology*, 14, 246–250.

Lister, G.S., Etheridge, M.A., Symonds, P.A. (1991). Detachment models for the formation of passive continental margins. *Tectonics*, 10(5), 1038–1064.

Lizarralde, D., Axen, G.J., Brown, H.E., Fletcher, J.M., Gonzalez-Fernandez, A., Harding, A.J., Holbrook, W.S., Kent, G.M., Paramo, P., Sutherland, F., Umhoefer, P.J. (2007). Variation in styles of rifting in the Gulf of California. *Nature*, 448(7152), 466–469.

Louden, K.E., Sibuet, J.C., Foucher, J.P. (1991). Variations in heat-flow across the Goban Spur and Galicia Bank continental margins. *Journal of Geophysical Research-Solid Earth*, 96(B10), 16131–16150.

Manatschal, G. (2004). New models for evolution of magma-poor rifted margins based on a review of data and concepts from West Iberia and the Alps. *International Journal of Earth Sciences*, 93(3), 432–466.

Manatschal, G. and Bernoulli, D. (1998). Rifting and early evolution of ancient ocean basins: The record of the Mesozoic Tethys and of the Galicia-Newfoundland margins. *Marine Geophysical Researches*, 20(4), 371–381.

Manatschal, G., Lavier, L., Chenin, P. (2014). The role of inheritance in structuring hyperextended rift systems: Some considerations based on observations and numerical modeling. *Gondwana Research*, 27(1), 140–164.

Mann, P., Gahagan, L., Gordon, M. (2001). Tectonic setting of the world's giant oil fields. *World Oil*, 222(9), 42–50.

Marrett, R. and Allmendinger, R.W. (1992). Amount of extension on "small" faults: An example from the Viking graben. *Geology*, 20(1), 47–50.

Martinsen, O.J., Sømme, T.O., Thurmond, J.B., Helland-Hansen, W., Lunt, I. (2010). Source-to-sink systems on passive margins: Theory and practice with an example from the Norwegian continental margin. *Geological Society, London,* Petroleum Geology Conference Series, 7(1), 913–920.

Masini, E., Manatschal, G., Tugend, J., Mohn, G., Flament, J.M. (2014). The tectono-sedimentary evolution of a hyper-extended rift basin: The example of the Arzacq–Mauléon rift system (Western Pyrenees, SW France). *International Journal of Earth Sciences (Geol Rundsch)*, 103, 1569–1596.

McDermott, K.G. and Reston, T. (2015). To see, or not to see? Rifted margin extension. *Geology*, 43(11), 967–970.

McKenzie, D.P. (1978). Some remarks on the development of sedimentary basins. *Earth and Planetary Science Letters*, 40, 20–32.

Mercier De Lépinay, M., Loncke, L., Basile, C., Roest, W.R., Patriat, M., Maillard, A., De Clarens, P. (2016). Transform continental margins – Part 2: A worldwide review. *Tectonophysics*, 693, 96–115.

Mevel, C. (2003). Serpentinization of abyssal peridotites at mid-ocean ridges. *Comptes rendus geoscience*, 335(10–11), 825–852.

Mittelstaedt, E., Ito, G., Behn, M.D. (2008). Mid-ocean ridge jumps associated with hotspot magmatism. *Earth and Planetary Science Letters*, 266(3–4), 256–270.

Mjelde, R., Shimamura, H., Kanazawa, T., Kodaira, S., Raum, T., Shiobara, H. (2003). Crustal lineaments, distribution of lower crustal intrusives and structural evolution of the Vøring Margin, NE Atlantic; new insight from wide-angle seismic models. *Tectonophysics*, 369(3–4), 199–218.

Mjelde, R., Raum, T., Murai, Y., Takanami, T. (2007). Continent-ocean-transitions: Review, and a new tectono-magmatic model of the Vøring Plateau, NE Atlantic. *Journal of Geodynamics*, 43(3), 374–392.

Mjelde, R., Raum, T., Kandilarov, A., Murai, Y., Takanami, T. (2009). Crustal structure and evolution of the outer More Margin, NE Atlantic. *Tectonophysics*, 468(1–4), 224–243.

Mjelde, R., Goncharov, A., Müller, R.D. (2012). The Moho: Boundary above upper mantle peridotites or lower crustal eclogites? A global review and new interpretations for passive margins. *Tectonophysics*, 609, 636–650.

Mohn, G., Manatschal, G., Müntener, O., Beltrando, M., Masini, E. (2010). Unravelling the interaction between tectonic and sedimentary processes during lithospheric thinning in the Alpine Tethys margins. *International Journal of Earth Sciences*, 99, 75–101.

Mohriak, W.U., Nóbrega, M., Odegard, M.E., Gomes, B.S., Dickson, W.G. (2010). Geological and geophysical interpretation of the Rio Grande Rise, south-eastern Brazilian margin: Extensional tectonics and rifting of continental and oceanic crusts. *Petroleum Geoscience*, 16(3), 231–245.

Mondy, L.S., Rey, P.F., Duclaux, G., Moresi, L. (2017). The role of asthenospheric flow during rift propagation and breakup. *Geology*, 46(2), 103–106.

Mosar, J., Lewis, G., Torsvik, T.H. (2002). North Atlantic sea-floor spreading rates: Implications for the Tertiary development of inversion structures of the Norwegian-Greenland Sea. *Journal of the Geological Society*, 159, 503–515.

Moulin, M., Aslanian, D., Unternehr, P. (2010). A new starting point for the South and Equatorial Atlantic Ocean. *Earth-Science Reviews*, 98(1–2), 1–37.

Müller, R.D., Gaina, C., Roest, W.R., Hansen, D.L. (2001). A recipe for microcontinent formation. *Geology*, 29(3), 203–206.

Müller, R.D., Cannon, J., Qin, X., Watson, R.J., Gurnis, M., Williams, S., Pfaffelmoser, T., Seton, M., Russell, S.H.J., Zahirovic, S. (2018). GPlates: Building a virtual Earth through deep time. *Geochemistry, Geophysics, Geosystems*, 19(7), 2243–2261.

Müntener, O. and Manatschal, G. (2006). High degrees of melt extraction recorded by spinel harzburgite of the Newfoundland margin: The role of inheritance and consequences for the evolution of the southern North Atlantic. *Earth and Planetary Science Letters*, 252(3–4), 437–452.

Müntener, O., Manatschal, G., Desmurs, L., Pettke, T. (2010). Plagioclase peridotites in ocean-continent transitions: Refertilized mantle domains generated by melt stagnation in the shallow mantle lithosphere. *Journal of Petrology*, 51(1–2), 255–294.

Naliboff, J.B. and Buiter, S.J.H. (2015). Rift reactivation and migration during multiphase extension. *Earth and Planetary Science Letters*, 421, 58–67.

Naliboff, J.B., Buiter, S.J.H., Péron-Pinvidic, G., Osmundsen, P.T., Tetreault, J. (2017). Complex fault interaction controls continental rifting. *Nature Communications*, 8(1), 1179.

Naliboff, J.B., Glerum, A., Brune, S., Péron-Pinvidic, G., Wrona, T. (2020). Development of 3-D rift heterogeneity through fault network evolution. *Geophysical Research Letters*, 47(13), e2019GL086611.

Newman, R. and White, N. (1997). Rheology of the continental lithosphere inferred from sedimentary basins. *Nature*, 385, 621–624.

Newman, R. and White, N. (1999). The dynamics of extensional sedimentary basins: Constraints from subsidence inversion. In *Response of the Earth's Lithosphere to Extension*, White, R.S., Hardman, R.F.P., Watts, A.B., Whitmarsh, R.B. (eds). The Royal Society, London.

Norcliffe, J.R., Paton, D.A., Mortimer, E.J., McCaig, A.M., Nicholls, H., Rodriguez, K., Hodgson, N., Van Der Spuy, D. (2018). Laterally confined volcanic successions (LCVS); recording rift-jumps during the formation of magma-rich margins. *Earth and Planetary Science Letters*, 504, 53–63.

O'Reilly, B.M., Hauser, F., Ravaut, C., Shannon, P.M., Readman, P.W. (2006). Crustal thinning, mantle exhumation and serpentinization in the Porcupine Basin, offshore Ireland: Evidence from wide-angle seismic data. *Journal of the Geological Society*, 163, 775–787.

Osmundsen, P.T. and Peron-Pinvidic, G. (2018). Crustal-scale fault interaction at rifted margins and the formation of domain-bounding breakaway complexes: Insights from offshore Norway. *Tectonics*, 37(3), 935–964.

Osmundsen, P.T. and Redfield, T. (2011). Crustal taper and topography at passive continental margins. *Terra Nova*, 23(6), 349–361.

Osmundsen, P.T., Sommaruga, A., Skilbrei, J.R., Olesen, O. (2002). Deep structure of the Mid Norway rifted margin. *Norwegian Journal of Geology*, 82(4), 205–224.

Osmundsen, P.T., Peron-Pinvidic, G., Ebbing, J., Erratt, D., Fjellanger, E., Bergslien, D., Syvertsen, S.E. (2016). Extension, hyperextension and mantle exhumation offshore Norway: A discussion based on 6 crustal transects. *Norwegian Journal of Geology*, 96(4), 343–372.

Oufi, O., Cannat, M., Horen, H. (2002). Magnetic properties of variably serpentinized abyssal peridotites. *Journal of Geophysical Research*, 107(B5).

Pariso, J.E. and Johnson, H.P. (1993). Do layer 3 rocks make a significant contribution to marine magnetic anomalies? *In situ* magnetization of gabbros at ocean drilling program hole 735 b. *Journal of Geophysical Research*, 98, 16033–16052.

Pérez-Gussinyé, M., Ranero, C.R., Reston, T.J., Sawyer, D. (2003). Mechanisms of extension at nonvolcanic margins: Evidence from the Galicia interior basin, west of Iberia. *Journal of Geophysical Research-Solid Earth*, 108(B5), 2245–2264.

Pérez-Gussinyé, M., Morgan, J.P., Reston, T.J., Ranero, C.R. (2006). The rift to drift transition at non-volcanic margins: Insights from numerical modelling. *Earth and Planetary Science Letters*, 244(1–2), 458–473.

Peron-Pinvidic, G. and Manatschal, G. (2009). The final rifting evolution at deep magma-poor passive margins from Iberia-Newfoundland: A new point of view. *International Journal of Earth Sciences*, 98(7), 1581–1597.

Peron-Pinvidic, G. and Manatschal, G. (2010). From microcontinents to extensional allochthons: Witnesses of how continents break apart? *Petroleum Geosciences, Rift Thematic Volume*, 16, 189–197.

Peron-Pinvidic, G. and Manatschal, G. (2019). Rifted margins: State of the art and future challenges. *Frontiers in Earth Science*, 7(218).

Peron-Pinvidic, G. and Naliboff, J. (2020). The exhumation detachment factory. *Geology*, 635–639.

Peron-Pinvidic, G. and Osmundsen, P.T. (2016). Architecture of the distal and outer domains of the mid-Norwegian Vøring rifted margin: Insights from the Rån Ridge system. *Marine and Petroleum Geology*, 77, 280–299.

Peron-Pinvidic, G. and Osmundsen, P.T. (2018). The Mid Norwegian – NE Greenland conjugate margins: Rifting evolution, margin segmentation, and breakup. *Marine and Petroleum Geology*, 98, 162–184.

Peron-Pinvidic, G. and Osmundsen, P.T. (2020). From orogeny to rifting: Insights from the Norwegian "reactivation phase". *Scientific Reports*, 10(1), 14860.

Peron-Pinvidic, G., Manatschal, G., Minshull, T.A., Sawyer, D.S. (2007). Tectonosedimentary evolution of the deep Iberia-Newfoundland margins: Evidence for a complex breakup history. *Tectonics*, 26(2), 1–119.

Peron-Pinvidic, G., Shillington, D.J., Tucholke, B.E. (2010). Characterization of sills associated with the U reflection on the Newfoundland margin: Evidence for widespread early post-rift magmatism on a magma-poor rifted margin. *Geophysical Journal International*, 182(1), 113–136.

Peron-Pinvidic, G., Gernigon, L., Gaina, C., Ball, P. (2012). Insights from the Jan Mayen system in the Norwegian–Greenland sea I. Mapping of a microcontinent. *Geophysical Journal International*, 191(2), 385–412.

Peron-Pinvidic, G., Manatschal, G., Osmundsen, P.T. (2013). Structural comparison of archetypal Atlantic rifted margins: A review of observations and concepts. *Marine and Petroleum Geology*, 43, 21–47.

Peron-Pinvidic, G., Manatschal, G., Masini, E., Sutra, E., Flament, J.M., Haupert, I., Unternehr, P. (2015). Unravelling the along-strike variability of the Angola–Gabon rifted margin: A mapping approach. *Geological Society, London, Special Publications*, 438, 49–76.

Peron-Pinvidic, G., Osmundsen, P.T., Ebbing, J. (2016). Mismatch of geophysical datasets in distal rifted margins studies. *Terra Nova*, 28(5), 340–347.

Peron-Pinvidic, G., Osmundsen, P.T., Bunkholt, H. (2020). The proximal domain of the Mid-Norwegian rifted margin: The Trøndelag Platform revisited. *Tectonophysics*, 790, 228551.

Pindell, J., Graham, R., Horn, B. (2014). Rapid outer marginal collapse at the rift to drift transition of passive margin evolution, with a Gulf of Mexico case study. *Basin Research*, 26(6), 701–725.

Planke, S., Symonds, P.A., Alvestad, E., Skogseid, J. (2000). Seismic volcanostratigraphy of large-volume basaltic extrusive complexes on rifted margins. *Journal of Geophysical Research-Solid Earth*, 105(B8), 19335–19351.

Planke, S., Rasmussen, T., Rey, S.S., Myklebust, R. (2005). Seismic characteristics and distribution of volcanic intrusions and hydrothermal vent complexes in the Vøring and Møre basins. In *Petroleum Geology: North-West Europe and Global Perspectives – Proceedings of the 6th Petroleum Geology Conference*, Dore, A.G., Vining, B.A. (eds). Geological Society, London.

Platt, J.P. and England, P.C. (1994). Convective removal of lithosphere beneath mountain belts; thermal and mechanical consequences. *American Journal of Science*, 294(March), 307–336.

Poliakov, A.N.B., Podladchikov, Y., Talbot, C. (1993). Initiation of salt diapirs with frictional overburdens: Numerical experiments. *Tectonophysics*, 228(3), 199–210.

Poselov, V.A., Avetisov, G.P., Butsenko, V.V., Zholondz, S.M., Kaminsky, V.D., Pavlov, S.P. (2012). The Lomonosov Ridge as a natural extension of the Eurasian continental margin into the Arctic Basin. *Russian Geology and Geophysics*, 53(12), 1276–1290.

Proffett Jr., J.M. (1977). Cenozoic geology of the Yerington district, Nevada, and implications for the nature and origin of Basin and Range faulting. *GSA Bulletin*, 88(2), 247–266.

Ranero, C.R. and Pérez-Gussinyé, M. (2010). Sequential faulting explains the asymmetry and extension discrepancy of conjugate margins. *Nature*, 468, 294–297.

Regenauer-Lieb, K., Rosenbaum, G., Weinberg, R.F. (2008). Strain localisation and weakening of the lithosphere during extension. *Tectonophysics*, 458(1–4), 96–104.

Reston, T.J. (1988). Evidence for shear zones in the lower crust offshore Britain. *Tectonics*, 7(5), 929–945.

Reston, T.J. (2005). Polyphase faulting during the development of the west Galicia rifted margin. *Earth and Planetary Science Letters*, 237(3–4), 561–576.

Reston, T.J. (2007). Extension discrepancy at North Atlantic nonvolcanic rifted margins: Depth-dependent stretching or unrecognized faulting? *Geology*, 35(4), 367–370.

Reston, T.J. (2009). The structure, evolution and symmetry of the magma-poor rifted margins of the North and Central Atlantic: A synthesis. *Tectonophysics*, 468(1–4), 6–27.

Reston, T.J. (2018). Flipping detachments: The kinematics of ultraslow spreading ridges. *Earth and Planetary Science Letters*, 503, 144–157.

Reston, T.J. and McDermott, K.G. (2011). Successive detachment faults and mantle unroofing at magma-poor rifted margins. *Geology*, 39(11), 1071–1074.

Reston, T.J. and McDermott, K.G. (2014). An assessment of the cause of the "extension discrepancy" with reference to the west Galicia margin. *Basin Research*, 26(1), 135–153.

Reston, T.J. and Ranero, C.R. (2011). The 3-D geometry of detachment faulting at mid-ocean ridges. *Geochemistry, Geophysics, Geosystems*, 12(7).

Roberts, A.M., Lundin, E.R., Kusznir, N.J. (1997). Subsidence of the Vøring Basin and the influence of the Atlantic continental margin. *Journal of the Geological Society*, 154(3), 551–557.

Rooney, T.O. (2019). The Cenozoic Magmatism of East Africa: Part II – Rifting of the mobile belt. *Lithos*, 105291.

Rosendahl, B.R. (1987). Architecture of continental rifts with special reference to East Africa. *Annual Review of Earth and Planetary Sciences*, 15(1), 445–503.

Rotevatn, A., Kristensen, T.B., Ksienzyk, A.K., Wemmer, K., Henstra, G.A., Midtkandal, I., Grundvåg, S.-A., Andresen, A. (2018). Structural inheritance and rapid rift-length establishment in a multiphase rift: The East Greenland Rift System and its Caledonian orogenic ancestry. *Tectonics*, 37(6), 1858–1875.

Rouméjon, S., Cannat, M., Agrinier, P., Godard, M., Andreani, M. (2015). Serpentinization and fluid pathways in tectonically exhumed peridotites from the Southwest Indian Ridge (62–65°E). *Journal of Petrology*, 56(4), 703–734.

Rouzo, S., Klingelhöfer, F., Jonquet-Kolstø, H., Karpuz, R., Kravik, K., Mjelde, R., Murai, Y., Raum, T., Shimamura, H., Williamson, P., Géli, L. (2006). 2-D and 3-D modelling of wide-angle seismic data: An example from the Vøring volcanic passive margin. *Marine Geophysical Research*, 27(3), 181–199.

Royden, L. and Keen, C.E. (1980). Rifting process and thermal evolution of the continental margin of Eastern Canada determined from subsidence curves. *Earth and Planetary Science Letters*, 51(2), 343–361.

Rupke, L.H., Schmid, D.W., Pérez-Gussinyé, M., Hartz, E. (2013). Interrelation between rifting, faulting, sedimentation, and mantle serpentinization during continental margin formation – Including examples from the Norwegian Sea. *Gcube*, 14(10), 4351–4368.

Saltus, R.W. and Blakely, R.J. (2011). Unique geologic insights from "non-unique" gravity and magnetic interpretation. *GSA Today*, 21(12), 4–11.

Santos, R.V., Glasmacher, U.A., Geldmacher, J. (2014). Scientific drilling in the South Atlantic. *Eos, Transactions American Geophysical Union*, 95(27), 249–249.

Sauter, D., Cannat, M., Roumejon, S., Andreani, M., Birot, D., Bronner, A., Brunelli, D., Carlut, J., Delacour, A., Guyader, V., Macleod, C.J., Manatschal, G., Mendel, V., Menez, B., Pasini, V., Ruellan, E., Searle, R. (2013). Continuous exhumation of mantle-derived rocks at the Southwest Indian Ridge for 11 million years. *Nature Geoscience*, 6(4), 314–320.

Savva, D., Meresse, F., Pubellier, M., Chamot-Rooke, N., Lavier, L., Po, K.W., Franke, D., Steuer, S., Sapin, F., Auxietre, J.L., Lamy, G. (2013). Seismic evidence of hyper-stretched crust and mantle exhumation offshore Vietnam. *Tectonophysics*, 608(0), 72–83.

Scott, R.A., Ramsey, L.A., Jones, S.M., Sinclair, S., Pickles, C.S. (2005). Development of the Jan Mayen microcontinent by linked propgation and retreat of spreading ridges. In *Onshore-Offshore Relationships on the North Atlantic Margin*, Wandås, B.T.G., Nystuen, J.P., Eide, E. Gradstein, F. (eds). Norwegian Petroleum Society (NPF), Special Publication, Oslo.

Séranne, M. (1992). Late Paleozoic kinematics of the Møre-Trøndelag Fault Zone and adjacent areas, central Norway. *Norsk Geologisk Tidsskrift*, 72, 141–158.

Seton, M., Müller, R.D., Zahirovic, S., Gaina, C., Torsvik, T., Shephard, G., Talsma, A., Gurnis, M., Turner, M., Maus, S., Chandler, M. (2012). Global continental and ocean basin reconstructions since 200 Ma. *Earth-Science Reviews*, 113(3–4), 212–270.

Shannon, P.M., Jacob, A.W.B., O'Reilly, B., Hauser, F., Readman, P.W., Makris, J. (1999). Structural setting, geological development and basin modelling in the Rockall Trough. In *Petroleum Geology of Northwest Europe: Proceedings of the 5th Conference*, Fleet, A.J. Boldy, S.A.R. (eds). Geological Society, London.

Sibuet, J.C. (1992). Formation of nonvolcanic passive margins – A composite model applies to the conjugate Galicia and Southeastern Flemish Cap Margins. *Geophysical Research Letters*, 19(8), 769–772.

Sigmond, E.M.O. (2002). Geological map of land and sea area of Northern Europe, scale 1:4,000,000. Gelogical Survey of Norway, Trondheim.

Skelton, A., Whitmarsh, R.B., Arghe, F., Crill, P., Koyi, H. (2005). Constraining the rate and extent of mantle serpentinization from seismic and petrological data: Implications for chemosynthesis and tectonic processes. *Geofluids*, 5(3), 153–164.

Skogseid, J., Pedersen, T., Eldholm, O., Larsen, B.T. (1992). Tectonism and magmatism during NE Atlantic continental break-up: The Vøring basin. In *Magmatism and the Causes of Continental Break-up*, Storey, B.C., Alabaster, T. Plankhurst, R.J. (eds). *Geological Society, London, Special Publications*, 305–320.

Souche, A., Medvedev, S., Andersen, T.B., Dabrowski, M. (2013). Shear heating in extensional detachments: Implications for the thermal history of the Devonian basins of W Norway. *Tectonophysics*, 608, 1073–1085.

Srivastava, S.P., Sibuet, J.C., Cande, S., Roest, W.R., Reid, I.D. (2000). Magnetic evidence for slow seafloor spreading during the formation of the Newfoundland and Iberian margins. *Earth and Planetary Science Letters*, 182(1), 61–76.

Stica, J.M., Zalán, P.V., Ferrari, A.L. (2014). The evolution of rifting on the volcanic margin of the Pelotas Basin and the contextualization of the Paraná–Etendeka LIP in the separation of Gondwana in the South Atlantic. *Marine and Petroleum Geology*, 50, 1–21.

Straume, E.O., Gaina, C., Medvedev, S., Hochmuth, K., Gohl, K., Whittaker, J.M., Abdul Fattah, R., Doornenbal, J.C., Hopper, J.R. (2019). GlobSed: Updated total sediment thickness in the world's oceans. *Geochemistry, Geophysics, Geosystems*, 20(4), 1756–1772.

Sutra, E. and Manatschal, G. (2012). How does the continental crust thin in a hyperextended rifted margin? Insights from the Iberia margin. *Geology*, 40, 139–142.

Sutra, E., Manatschal, G., Mohn, G., Unternehr, P. (2013). Quantification and restoration of extensional deformation along the Western Iberia and Newfoundland rifted margins. *Geochemistry, Geophysics, Geosystems*, 14(8), 2575–2597.

Svensen, H., Planke, S., Malthe-Sørenssen, A., Jamtveit, B., Myklebust, R., Rasmussen Eidem, T., Rey, S.S. (2004). Release of methane from a volcanic basin as a mechanism for initial Eocene global warming. *Nature*, 429(6991), 542–545.

Tetreault, J.L. and Buiter, S.J.H. (2012). Geodynamic models of terrane accretion: Testing the fate of island arcs, oceanic plateaus, and continental fragments in subduction zones. *Journal of Geophysical Research*, 117(B08403).

Tetreault, J.L. and Buiter, S.J.H. (2018). The influence of extension rate and crustal rheology on the evolution of passive margins from rifting to break-up. *Tectonophysics*, 746, 155–172.

Thinon, I., Matias, L., Rehault, J.P., Hirn, A., Fidalgo-Gonzalez, L., Avedik, F. (2003). Deep structure of the Armorican Basin (Bay of Biscay): A review of Norgasis seismic reflection and refraction data. *Journal of the Geological Society*, 160, 99–116.

Torsvik, T.H., Amundsen, H., Hartz, E.H., Corfu, F., Kusznir, N., Gaina, C., Doubrovine, P.V., Steinberger, B., Ashwal, L.D., Jamtveit, B. (2013). A Precambrian microcontinent in the Indian Ocean. *Nature Geoscience*, 6(3), 223–227.

Tsikalas, F., Faleide, J.I., Eldholm, O., Antonio Blaich, O. (2012). The NE Atlantic conjugate margins. In *Regional Geology and Tectonics: Phanerozoic Passive Margins, Cratonic Basins and Global Tectonic Maps*, Bally, D.G.R.W. (ed.). Elsevier, Boston.

Tucholke, B.E. and Sibuet, J.C. (2007). Leg 210 synthesis: Tectonic, magmatic, and sedimentary evolution of the Newfoundland-Iberia rift. In *Proc. ODP, Sci. Results*, Tucholke, B.E., Sibuet, J.C., Klaus, A. (eds). College Station.

Tucholke, B.E., Lin, J., Kleinrock, M.C. (1998). Megamullions and mullion structure defining oceanic metamorphic core complexes on the mid-Atlantic ridge. *Journal of Geophysical Research-Solid Earth*, 103(B5), 9857–9866.

Tucholke, B.E., Sawyer, D.S., Sibuet, J.C. (2007). Breakup of the Newfoundland – Iberia rift. In *Imaging, Mapping and Modelling Continental Lithosphere Extension and Breakup, Geological Society of London Special Publication 282*, Karner, G.D., Manatschal, G., Pinhiero, L.M. (eds). Geological Society, London.

Tugend, J., Manatschal, G., Kusznir, N.J., Masini, E. (2015). Characterizing and identifying structural domains at rifted continental margins: Application to the Bay of Biscay margins and its Western Pyrenean fossil remnants. *Geological Society, London, Special Publications*, 413, 23–47.

Tugend, J., Gillard, M., Manatschal, G., Nirrengarten, M., Harkin, C., Epin, M.-E., Sauter, D., Autin, J., Kusznir, N., McDermott, K.G. (2018). Reappraisal of the magma-rich versus magma-poor rifted margin archetypes. *Geological Society, London, Special Publications*, 476, SP476.479.

Van Wijk, J.W. and Blackman, D.K. (2005). Dynamics of continental rift propagation: The end-member modes. *Earth and Planetary Science Letters*, 229(3–4), 247–258.

Van Wijk, J.W., Huismans, R.S., Ter Voorde, M., Cloetingh, S. (2001). Melt generation at volcanic continental margins: No need for a mantle plume? *Geophysical Research Letters*, 28(20), 3995–3998.

Van Wijk, J.W., Lawrence, J.F., Driscoll, N.W. (2008). Formation of the Transantarctic Mountains related to extension of the West Antarctic Rift system. *Tectonophysics*, 458(1–4), 117–126.

Voss, M., Schmidt-Aursch, M.C., Jokat, W. (2009). Variations in magmatic processes along the East Greenland volcanic margin. *Geophysical Journal International*, 177(2), 755–782.

Wangen, M., Mjelde, R., Faleide, J.I. (2011). The extension of the Voringmargin (NE Atlantic) in case of different degrees of magmatic underplating. *Basin Research*, 23, 83–100.

Watts, A.B., Karner, G., Steckler, M.S., Kent, P., Bott, M.H.P., Mckenzie, D.P., Williams, C.A. (1982). Lithospheric flexure and the evolution of sedimentary basins. *Philosophical Transactions of the Royal Society of London. Series A, Mathematical and Physical Sciences*, 305(1489), 249–281.

Welsink, H. and Tankard, A. (2012). Extensional tectonics and stratigraphy of the Mesozoic Jeanne d'Arc basin, Grand Banks of Newfoundland. In *Regional Geology and Tectonics: Phanerozoic Rift Systems and Sedimentary Basins*, Roberts, D.G., Bally, A.W. (eds). Elsevier, Boston.

Wernicke, B. (1985). Uniform-sense normal simple shear of the continental lithosphere. *Canadian Journal of Earth Sciences*, 22, 108–125.

White, R.S. and McKenzie, D. (1989). Magmatism at rift zones: The generation of volcanic continental margins and flood basalts. *Journal of Geophysical Research*, 94, 7685–7729.

Whitmarsh, R.B. and Sawyer, D.S. (1996). The ocean-continent transition beneath the Iberia Abyssal Plain and continental rifting to seafloor spreading processes. In *Proceedings of the Ocean Drilling Program – Scientific Results*, Whitmarsh, R.B., Sawyer, D.S., Klaus, A., Masson, D.G. (eds). College Station.

Whitmarsh, R.B., Manatschal, G., Minshull, T.A. (2001). Evolution of magma-poor continental margins from rifting to seafloor spreading. *Nature*, 413(6852), 150–154.

Whitney, D.L., Teyssier, C., Rey, P., Buck, R. (2014). Continental and oceanic core complexes. *GSA Bulletin*, 125(3/4), 273–298.

Wilson, J.T. (1966). Did the Atlantic close and then re-open? *Nature*, 211(5050), 676–681.

Withjack, M.O. and Jamison, W.R. (1986). Deformation produced by oblique rifting. *Tectonophysics*, 126(2), 99–124.

Zalán, P.V., Severino, M.G., Rigoti, C.A., Magnavita, L.P., Oliveira, J.a.B., Viana, A.R. (eds) (2011). An entirely new 3-D view of the crustal and mantle structure of a South Atlantic Passive Margin – Santos, Campos and Espírito Santo Basins, Brazil. *AAPG Annual Conference and Exhibition*, Houston.

Ziegler, P.A. (1983). Crustal thinning and subsidence in the North Sea. *Nature*, 304(5926), 561–561.

Zielke, O. and Strecker, M.R. (2009). Recurrence of large earthquakes in magmatic continental rifts: Insights from a Paleoseismic Study along the Laikipia–Marmanet Fault, Subukia Valley, Kenya Rift. *Bulletin of the Seismological Society of America*, 99(1), 61–70.

Zwaan, F., Schreurs, G., Adam, J. (2018). Effects of sedimentation on rift segment evolution and rift interaction in orthogonal and oblique extensional settings: Insights from analogue models analysed with 4D X-ray computed tomography and digital volume correlation techniques. *Global and Planetary Change*, 171, 110–133.

PART 2

How Do We Study Rifted Margins?

Summary of Contributions

Part 2 summarizes how we, as researchers, study rifted margins. Various disciplines of Earth Science are involved in such studies (e.g. geology, geophysics and geochemistry). We have selected the four main methods of investigations: (1) onshore geological observations, (2) offshore geophysical methods, (3) analogue modeling and (4) numerical modeling.

Chapter 3, by Mohn and co-authors, explains the methodological approach used to recognize, analyze and interpret rifted margins in orogens. Processes developed during continental collision may lead to the isolation of pieces of rifted margins into the orogen; these may be obducted or subducted and subsequently exhumed by diverse mechanisms. The result is that direct access, mapping and sampling of portions of ancient rifted margins are possible on some outcrops, providing unique datasets to study the architecture and evolution of rifted margins. After some reminders on the orogenic processes and standard orogen geometries, the chapter introduces the fundamental methods and clarifies how to identify fossil-rifted margins in orogenic systems. The authors include useful case examples that are often referred to in the scientific literature (e.g. the Alps, Pyrenees and Rif belt).

Present-day rifted margins are mostly located offshore, under water. Direct access to rocks is therefore extremely limited, as opposed to the onshore cases. Drillings and dredgings are possible but logistically expensive, and therefore remain exclusive options of investigation. Consequently, remote sensing methods are favored for these offshore contexts. **Chapter 4, by Autin and Watremez, describes the main geophysical methods used when studying modern rifted margins.** These allow for high-resolution imaging and parametrization of most geometries characterizing the various parts of rifted margins. Autin and Watremez review the seismic reflection, seismic refraction and potential field methods, before focusing on specific features characteristic of rifted margins and presenting their common geophysical signals with illustrative data examples.

The combined onshore and offshore studies allow access to a robust dataset that gives insight into the present-day architecture of the rifted margins. Age and duration of geological events are usually extremely difficult to constrain, unfortunately. Only a few rock samples are available, and thus very little age dating has been performed relative to the dimension of the structures. In addition, robust temporal constraints would require the systematic drilling, sampling and analysis of kilometers of rocks, which is unrealistic for submarine rocks. Geodynamic modeling is an alternative strategy developed by the scientific community to assess the temporal and spatial evolution of geological features in rifted margins. Geodynamic experiments can be performed with analogue models, reproducing geological environments with specific mediums in laboratory settings, or with numerical experiments based on rock physics and continuum mechanics.

M. Perez-Gussinye provides in Chapter 5 an historical overview of our knowledge evolution regarding numerical experiments of rifting. The author then reviews the fundamentals of the numerical methods by summarizing the various numerical techniques and listing the governing equations. Modern 2D and 3D geodynamic modeling strategies are illustrated through a selection of model applications.

Chapter 6, by Zwaan and Schreurs, explains how analogue modeling can be insightful by simulating extensional processes. The authors show the importance of choosing the correct materials and experimental set-up to simulate rift-related processes within short time periods in the laboratory, allowing scientists to test and investigate hypotheses in an efficient manner. They review the techniques used today to model and analyze analogue model results and present some case examples that allow the study of rift processes at both crustal and lithospheric scales, in both 2D and 3D settings.

3

How Can We Identify and Study Remnants of Rifted Margins in Orogens?

Geoffroy MOHN[1], Julie TUGEND[1,2], Benoît PETRI[3], Anders MCCARTHY[1,4], Dominique FRIZON DE LAMOTTE[1]

[1] *GEC, CY Cergy Paris Université, Cergy-Pontoise, France*

[2] *Sorbonne Université, CNRS-INSU, Institut des Sciences de la Terre Paris, ISTeP UMR 7193, Paris, France*

[3] *EOST – Institut Terre et Environnement de Strasbourg, CNRS UMR 7063, Université de Strasbourg, France*

[4] *Department of Earth Sciences, ETH-Zürich, Switzerland*

This chapter aims to provide an overview of the methodology developed over the past few decades to recognize the remnants of rifted margins in orogens. Direct access to present-day rifted margins is usually hampered by deep-water conditions and thick post-rift sedimentary cover. As illustrated by the so-called Wilson cycle, former oceanic domains and rifted margins eventually became incorporated in orogenic systems (Figure 3.1), which enables the investigation of remnant margins exposed in the field. Consequently, the study of remnant margins cropping out in orogens worldwide represents an appropriate complement to the study of present-day rifted margins. Historically, an iterative approach between present-day margins (offshore) and fossil analogues (onshore) has led to significant improvements in our understanding of the formation of rifted margins. In addition, deciphering the relative contribution of rift-related and orogeny-related processes is fundamental to

---

For a color version of all the figures in this chapter, see www.iste.co.uk/peronpinvidic/margins1.zip.

Continental Rifted Margins 1,
coordinated by Gwenn PERON-PINVIDIC. © ISTE Ltd 2022

understand orogenic dynamics and the importance of rift inheritance. Taking into account the initial complexities of rifted margins prior to convergence may therefore have significant implications on the understanding of mountain building mechanisms. By combining examples from the Pyrenees, Western and Central Alps as well as from the Rif belt (Figure 3.2), key first-order elements, which might help to identify remnants of rifted margins in orogens, are described including (1) the identification of coherent tectonic units, (2) deciphering the pre-, syn- and post-rift stratigraphic record, (3) determining the pre- and syn-rift history of the basement and (4) constraining syn-rift deformation style and structures.

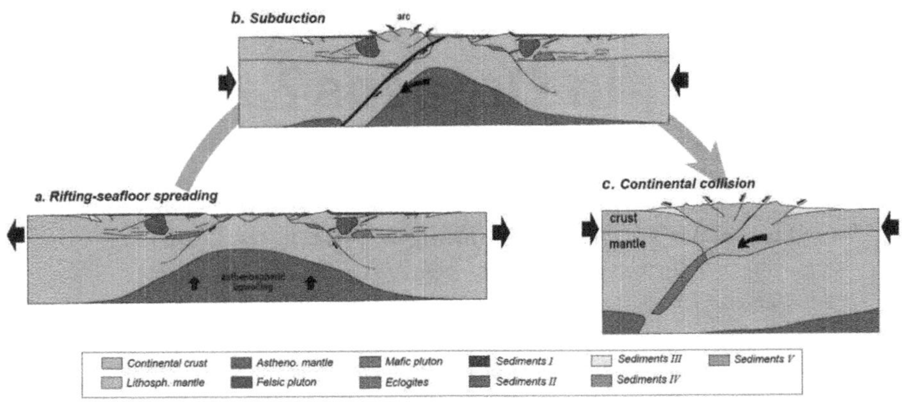

Figure 3.1. *Diagrams highlighting three major stages of the Wilson Cycle: a) rifting/rifted margin formation, b) subduction of an ocean basin, c) continental collision. Stage c) shows the reactivation and imbrication of former rifted margins in orogenic systems (source: modified from Petri 2014)*

3.1. Introduction

It has long been recognized that orogenic belts result from the reactivation and closure of former sedimentary basins [geosynclines (Dana 1873; Haug 1925; Shardt 1893; Lugeon and Argand 1905)]. At the beginning of the 20th century in the Alps, Gustav Steinmann (Steinmann 1905) described in the axial part of "geosynclines", the quasi-systematic association of magmatic rocks and deep-sea marine sediments, including serpentinite, diabase and radiolarite. Later named "the Steinmann trinity", its identification has been used to recognize "sub-oceanic" domains (for further details see Bernoulli *et al.* (2003) and Desmurs *et al.* (2001)). Since the 1960s, extensive studies in the Alpine and Apennine systems reported evidence of rift-related basins (e.g. Monte Generoso Basin in the Southern Alps; Bernoulli

1964), and ophiolites, which were associated with the extensive exposure of serpentinized mantle. Such ophiolitic series are now interpreted as continent-ocean transitions and oceanic crust formed at ultra-slow to slow-spreading oceanic ridges (e.g. Elter 1972; Trumpy 1975, Lemoine *et al.* 1987; Lagabrielle and Cannat 1990).

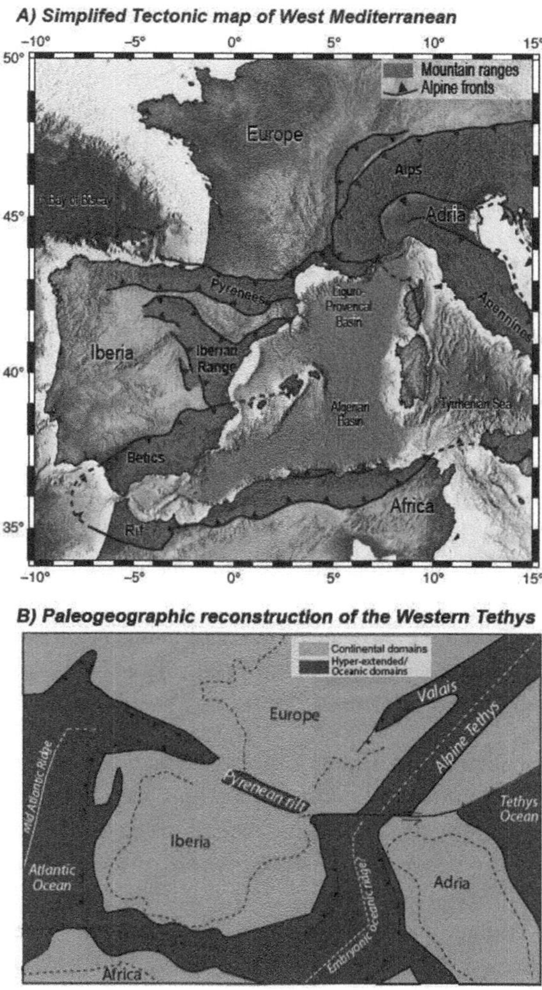

Figure 3.2. *a) Tectonic map of the West Mediterranean region showing the location of the Alps, Pyrenees and Rif belt, b) paleogeographic reconstruction of Alpine and Maghrebian Tethys during the Lower Cretaceous (source: modified from McCarthy* et al. *2021)*

An important breakthrough in our understanding of continental breakup and formation of rifted margins occurred by studying offshore systems through seismic imaging combined with dredging, deep diving submarine explorations and drilling expeditions such as the Deep Sea Drilling Project (DSDP), Ocean Drilling Program (ODP) and International Ocean Discovery Program (IODP) expeditions (Sibuet and Ryan 1979; Boillot *et al.* 1987; Whitmarsh *et al.* 2001; Manatschal 2004; Tuchloke and Sibuet 2007; Larsen *et al.* 2018). Offshore studies unraveled the complexities of rifted margin crustal architecture and the diversity of continental breakup mechanisms, which was simplified into two end-members now referred to as magma-poor or magma-rich rifted margins (Franke 2013; Péron-Pinvidic *et al.* 2013; Doré and Lundin 2015; Tugend *et al.* 2020).

Several studies have highlighted the similarities between the conjugate offshore Iberia-Newfoundland magma-poor margins and the onshore fossil remnants of the Alpine Tethys preserved in the Western and Central Alps and the Apennines (Manatschal and Bernoulli 1999; Wilson *et al.* 2001). Similar comparisons have recently been made through seismic and petrological studies of the Australian–Antarctic conjugate magma-poor margins as well (Gillard *et al.* 2015, 2017; McCarthy *et al.* 2020a). Over the years, evidence for analogues of magma-poor margins has been identified in several orogenic belts, including the Pyrenees (Jammes *et al.* 2009, Lagabrielle *et al.* 2010), the Caledonian orogen (Andersen *et al.* 2012; Jakob *et al.* 2017, 2019) and the Rif belt (Gimeno-Vives *et al.* 2019). By contrast, although magma-rich rifted margins represent an estimated 50% of the present-day passive margins (Coffin and Eldhom 1994; Skogseid, 2001), remnants of this rifted margin type remained unrecognized in orogens until the characterization of remnants of the magma-rich rifted margin of Baltica related to the opening of the Iapetus ocean and subsequently preserved in the Caledonian orogen (e.g. Kjoll *et al.* 2019) (see Chapter 8 of Volume 2).

In the following section, we first briefly provide key notions about orogens and rifted margins before describing a multi-method approach to recognize different rift domains in orogenic systems. The European Alps and the Pyrenees represent pioneering natural laboratories for the development of these methodologies. These orogens can be qualified in some respects as a particular type of orogenic system (referred to as the Alpine-type orogen, for example, Sengör (1990)) being the final result of the inversion and closure of narrow oceanic basins or even only hyper-extended rift basins (Manatschal *et al.* 2015; Mohn *et al.* 2015; Chenin *et al.* 2017; McCarthy *et al.* 2020b, 2021). The methodology developed hereafter can be applied to other orogenic systems (e.g. the Caledonides), although careful consideration of certain characteristics, such as the origins of ophiolites, is required. We will highlight this methodology using several case studies in distinct systems belonging to Alpine type orogens (the Alps and the Pyrenees) and also to other types such as the Rif (Northern Morocco) (Figure 3.2).

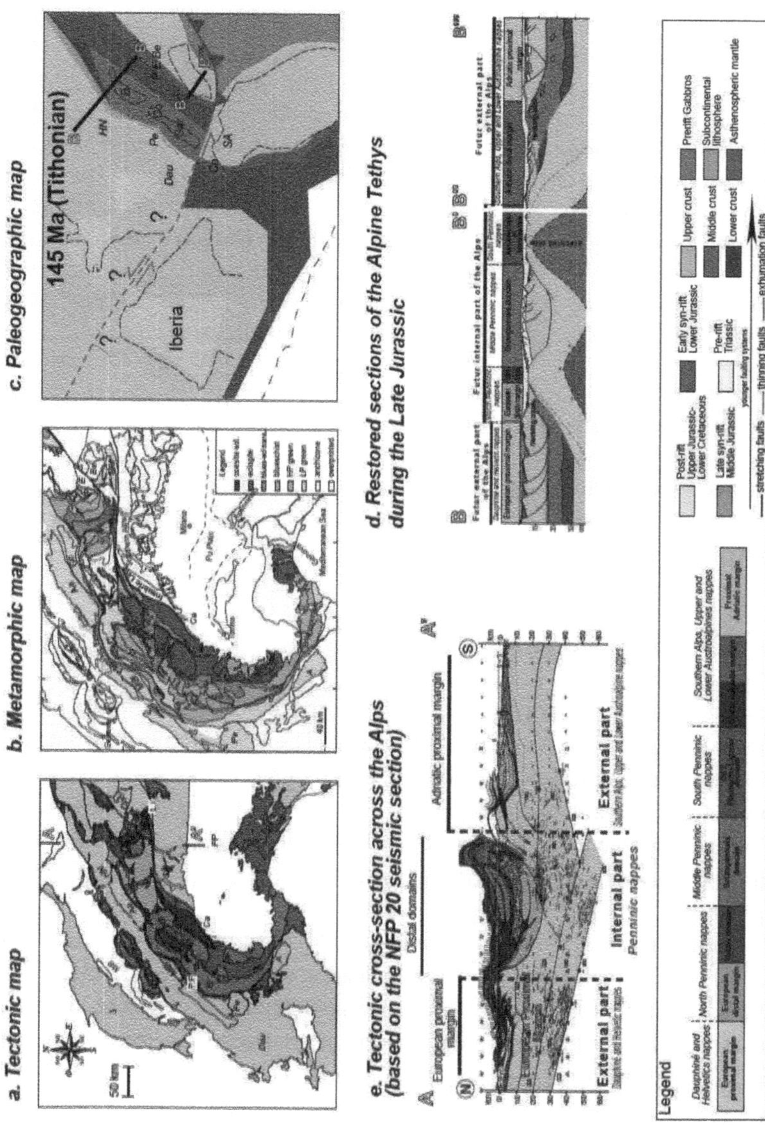

Figure 3.3. *Maps and cross-sections across the Alpine orogen showing the relationship between tectonic units, orogeny-related metamorphic facies and former paleogeographic domains*

CONTINUATION OF CAPTION FOR FIGURE 3.3.– *Proximal margins are generally sampled in the external zones and were only weakly inverted during orogenic-related deformation and metamorphism. Distal margins are often sampled in the internal zones and record significant orogenic deformation and metamorphism. a) Tectonic map (source: Mohn* et al. *2010, modified from Schmid* et al. *2004). b) Metamorphic map showing the highest metamorphic facies and age recorded by the distinct Tectonic units of the orogen (based on Beltrando* et al. *2010 and Bousquet* et al. *2008). c) Paleogeographic map of the Alpine Tethys during the Late Jurassic (source: Mohn* et al. *2010). d) Restored cross-section across the Alpine Tethys showing the Late Jurassic structure of the rifted margin (source: from Mohn* et al. *2010). e) Interpreted NFP20 seismic section across the Central Alps showing the integration of paleogeographic units in the orogen (source: modified from Schmid et al. 2016). A: Argentera massif, Aa: Aar massif, B: Briançonnais nappes, Be: Bernina nappe, E: Err nappe, MBl: Mont Blanc massif, MM: Malenco-Margna nappes, P: Platta nappe, Pe: Pelvoux massif.*

3.2. Brief considerations on orogens

The following section aims to give a brief overview of the main characteristics of orogenic systems that should be taken into account when investigating fossil remnants of rifted margins.

3.2.1. *Structure of an orogen*

The characterization in space and time of orogeny-related deformation is a prerequisite to investigate the preceding rift history. The general understanding of an orogenic belt relies among others on knowledge of its present-day structure, and also on its tectonic, magmatic and metamorphic evolution, often intimately linked with the rift-related paleogeography (Figure 3.3). Several tectonic units are usually recognized in an orogenic belt based on their paleogeographic affinity and orogenic evolution.

Often, the orogenic systems are divided into external and internal zones (Figure 3.3). The external zones are usually characterized by weak to moderate orogeny-related deformation and low-grade or no metamorphism with a tectonic style associating folds and thrust-faults. Geophysical surveys across these zones generally image a limited thickening of the continental crust (e.g. from 30 to 40 km thick in the External European Zone of the Alps (Thouvenot *et al.* 2007)).

By contrast, the internal zones (or axial zone) have been affected by strong and pervasive deformation with a high-grade metamorphic overprint related to subduction, collision and crustal thickening processes. This results in a thickened

(> 40 km) continental crust and the development of large allochthonous tectonic nappes detached from their original substratum and transported over long distances. The occurrence of syn-orogenic deposits (i.e. Flysch) is common, associated with post- to pre-rift sedimentary sequences and ophiolites, being the remnants of rift basins prior to their incorporation into the orogen. The term Flysch can be misleading and needs to be used with caution as its significance can change from one orogenic system to another: Flysch was originally defined as syn-orogenic deposits in the Alpine belt (Caron *et al.* 1989), yet they correspond to post-rift sequence in the Rif and Betic belts, and syn- to post-rift sequences in the Pyrenees.

3.2.2. *The significance of Ophiolitic assemblages*

Ophiolites are sections of former oceanic lithosphere and/or continent-ocean transitions typically incorporated in orogenic domains or obducted along former continental margins. The occurrence of one or several ophiolitic units is commonly interpreted as suture zones, delimiting two distinct paleogeographic domains, often continental in origin.

The classical 7-km thick Penrose-type oceanic sequence (Anonymous 1972) is formed of depleted ultramafic rocks overlaid by isotropic and sheeted gabbros, a sheeted dyke complex and thick lava flows and pillow lavas. However, ophiolites, in many cases, differ from the classical Penrose model, and their structure and composition depend on their geodynamic origin. Many ophiolites representing remnants of the Neo-Tethys (e.g. Semail Ophiolite, Ricou 1971) or found along the southwest Pacific (e.g. New Caledonia Ophiolite, Cluzel *et al.* 2001; Ulrich *et al.* 2010) represent supra-subduction zone oceanic lithosphere, and therefore record oceanic crust forming above an evolving subduction zone (e.g. Stern *et al.* 2012; Maffione *et al.* 2017). These ophiolites are typically dominated by highly depleted mantle rocks (harzburgite to dunite) with fertile mantle domains (lherzolite) being less prominent. Gabbro and sheeted dyke complex can be laterally discontinuous, or absent (or not preserved). Magmatic rocks associated with these ophiolites vary from mid-ocean ridge basalts to magmas recording the influence of the dehydrating and partial melting of the subducting oceanic lithosphere (e.g. boninites and calc-alkaline magmas).

Ophiolites found in the Western and Central Alps (Manatschal and Müntener 2009, Lagabrielle and Cannat 1990), Pyrenees (Lagabrielle and Bodinier 2008, Jammes *et al.* 2009) and Rif system (e.g. Michard *et al.* 1992, 2007; Gimeno-Vives *et al.* 2019) are however fundamentally distinct (McCarthy *et al.* 2020b and references therein). They are dominated by heterogeneous, yet generally fertile mantle rocks (lherzolites) (e.g. Müntener *et al.* 2010; Picazo *et al.* 2016), sparse mid-ocean ridge magmatism and/or alkaline magmatism and are overlain by thick sedimentary sequences. Moreover, field-relations show that a portion of these mantle rocks are associated with hyper-thinned continental crust and are interpreted to represent subcontinental lithosphere (Hermann *et al.* 1997). Other parts of these ophiolites show gabbroic core complexes and exhumed mantle rocks (Lagabrielle *et al.* 2015; Decrausaz *et al.* 2021) that are similar to ultra-slow spreading along the Southwest Indian Ridge and Mid-Atlantic Ridge (e.g. Lagabrielle and Cannat 1990; Cannat *et al.* 2019).

Therefore, ophiolites likely derive from a variety of geodynamic contexts (see Stern *et al.* 2012) with two main origins being: (1) ophiolites formed above nascent subduction zones (e.g. fore-arc settings) and rapidly obducted and (2) narrow domains of oceanic lithosphere and ocean-continent transition zones obducted along passive margins. The mechanisms and tectonic settings leading to the emplacement of ophiolites together with the nature of substratum on which they lie has been the subject of several studies and classifications (e.g. Dilek 2003; Dilek and Furnes 2014). These works emphasized the complex correlation between the age of magmatic rock formation, the age of emplacement, and the nature and geological contexts of the ophiolite. In conclusion, the different natures and modes of emplacement of ophiolitic sequences over continental domains need to be taken into account when aiming to reconstruct the rifted margin evolution and their subsequent incorporation in orogens.

3.2.3. *Classification and types of orogens*

Orogens are classified using various nomenclatures and subclasses depending on the parameters considered, such as: thermal and mechanical evolution, process of convergence, occurrence of magmatic arc as well as nature and size of lost (subducted) oceanic/rifted domains (for further details see Jamieson and Beaumont 2013 and references therein). A detailed review of the distinct type of orogenic belts is beyond the scope of this chapter. We refer the reader to Sengör (1990, 1991). Nevertheless, as emphasized by Sengör (1991) and more recently by Chenin *et al.* (2017) and McCarthy *et al.* (2020b, 2021), the initial size of the oceanic domain and/or rifted basins plays a significant control on the evolution of the orogenic system. This point is of particular interest for this study. Indeed, as mentioned previously, Alpine-type orogens are related to the closure of narrow oceanic domains and the imbrication of the former rifted margins. By contrast, orogenic belts derived from the Neo-Tethys system and now exposed from Eastern Europe to Southeast Asia are linked to the closure of wide oceans dominated by the steady-state subduction of oceanic lithosphere. Such belts are often associated with the obduction of portions of oceanic lithosphere, long-lived subduction zones that formed magmatic arcs (e.g. Agard *et al.* 2011) and arc-continent collisions.

3.3. Identifying fossil-rifted margins in orogenic systems: principles and criteria

Rifted margins are commonly divided into distinct rift domains, recording different steps of crustal and lithospheric thinning (Figure 3.4). In spite of the large variability of the styles of rifted margins, these domains share common characteristics from one margin to another. In the following section, we will refer to the classification developed in Chapter 2, which recognized the following domains: the proximal domain, the necking domain and the hyperextension/ exhumation domain.

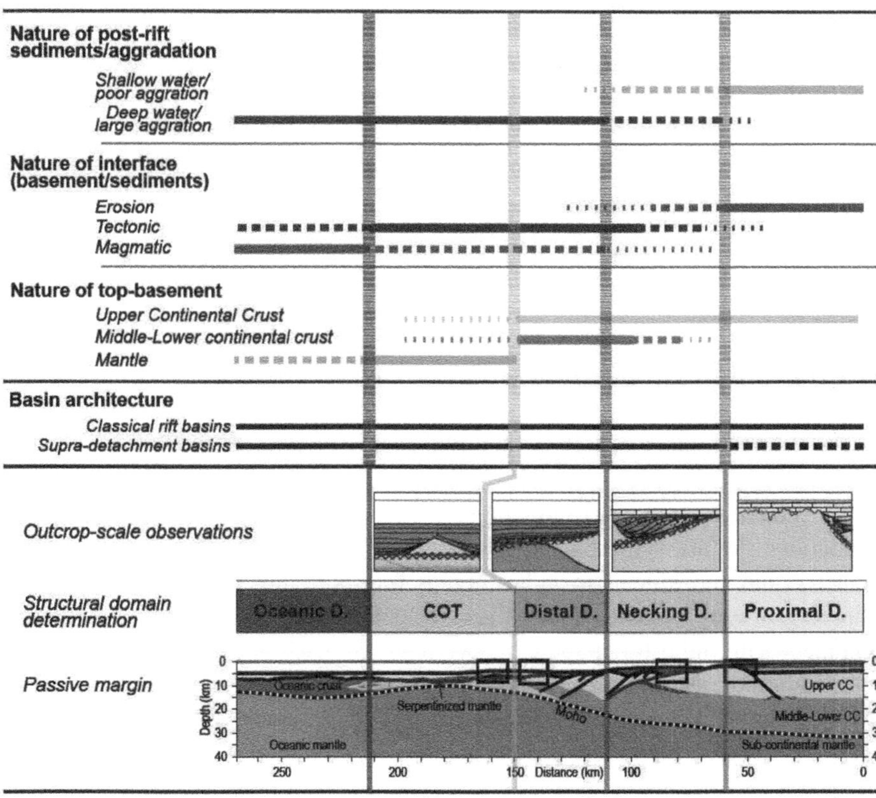

Figure 3.4. *Diagram summarizing the geological characteristics of structural domains of rifted margins. At the bottom, a section across a rifted margin is shown as well as idealized outcrop-scale geological observations illustrating the rift basin architecture, the nature of basement and sediment contacts and depositional environment evolution (source: modified after Tugend* et al. *2015a)*

In orogenic systems, some elements of these rift domains had already been recognized decades ago. Since then, generic approaches and methodologies have been developed in order to identify rift domains preserved in various orogenic systems. These methodologies have been primarily developed in the Pyrenees and in the Alps, both benefiting from an extensive knowledge on the stratigraphic record, basement history, paleogeographic evolution and so on.

In this section, we will highlight the main criteria used for the identification of different rift domains. For further details, we refer the reader to Mohn *et al.* (2010, 2011); Masini *et al.* (2013); Beltrando *et al.* (2014) and Tugend *et al.* (2014, 2015a). In present-day systems, the distinction between different rift domains relies on different relationships between accommodation space and crustal thinning. In fossil analogues, rift-related accommodation space and crustal thinning are no longer preserved and can only be assessed through the study of different lithostratigraphic associations and their rift-related deformation (Figures 3.4–3.6). The way to identify remnants of rifted margins in orogens combines multi-disciplinary observations (Figure 3.4) and relies on: (1) identifying coherent tectonic units, (2) decoding the pre-, syn- and post-rift stratigraphic record, (3) determining pre, syn- and post-rift history of the basement and (4) constraining syn-rift deformation style and structures.

3.3.1. *Identification of coherent tectonic units*

The tectonic units sampling the remnants of rifted margins are generally delimited by first-order orogenic structures (i.e. thrust or strike-slip faults, compressional shear zones), and preserve coherent rift-related lithostratigraphic associations despite syn-orogenic deformation (Figure 3.5). The mapping of the geometry and the kinematic indicators of transport directions of the structures delimiting these tectonic units allow a palinspastic restoration. Restorations combined with lithostratigraphic associations of each tectonic unit enable the identification of distinct paleogeographic domains, which can often be interpreted in terms of former rift domains (Figure 3.5). This approach is generally valid in external zones and, for a more general point of view, in weakly deformed domains.

However, the recognition of former rift domains is often more challenging in internal orogenic zones. There, rift-related structures and associated rift-related lithostratigraphic associations are often reactivated, overprinted or even erased by orogenic deformation and metamorphism. This results in an apparent lack of coherency of the lithostratigraphic association of tectonic units forming the internal zones. The significance of these units has therefore been the subject of long-standing debates, and these units have also been interpreted as a subduction *mélange* in accretionary channels at the subducting plate interface (e.g. Engi *et al.* 2001; Angiboust *et al.* 2014). To address this issue, Beltrando *et al.* (2014) defined three principles, which enable the detection of coherent tectonic units overprinted by polyphase pervasive deformation and metamorphism and representing former magma-poor rifted margins (Figure 3.6):

1) *The coherency of the lithostratigraphic association within a tectonic unit.* Identification of preserved key pre-orogenic markers (e.g. top-basement, base of the first post-rift sediments) and mapping them continuously across the different tectonic zones (Figure 3.6).

2) *The original proximity of the different rock types within a tectonic unit.* The occurrence of basement clasts within adjacent meta-sedimentary units is a good argument to indicate their proximity before orogeny-related deformation and metamorphism (Figure 3.6).

3) *The orogenic pressure–temperature–time (P–T–t) evolution is similar for the different rocks of a given tectonic unit.* Although the entire *P–T–t* path could be difficult to reconstruct, at least the oldest identified metamorphic fabric should be correlated between the different lithologies (Figure 3.6).

3.3.2. *Sedimentary/stratigraphic record and depositional environments*

Variations in the sedimentary/stratigraphic record across distinct tectonic units are robust observations, which have been used for decades to distinguish between different paleogeographic domains (Figures 3.4 and 3.5). The analysis of the sedimentary record is easier if a more general understanding of the tectonic history of the studied orogen is available. This information will help to focus the observations on key sequences relative to the rifting event (pre-, syn- and post-rift). The sedimentary history (e.g. lithological and thickness changes, and depositional environment) of syn- to post-rift sequences provides a first-order proxy for crustal thickness estimations based on analogies with present-day rifted margins. This assumption is only valid for magma-poor settings, where onset of decompression melting and melt extraction occurs late relative to extreme crustal thinning, thereby enabling mantle exhumation (Tugend *et al.* 2020). Rift basins can be filled by variable sediment influx influencing the depositional environments, thickness and geometry of syn- to post-rift sequences, whose end-members have been qualified as sediment-starved and sediment-rich rifted margins (Peron-Pinvidic *et al.* 2013). In the case of sediment-rich systems, syn- to post-rift successions can be several kilometers thick in relation to extreme crustal thinning (Pyrenees, Tugend *et al.* 2015a). For a sediment-starved system, the main argument suggesting significant crustal thinning relies on the deep marine conditions, while syn- to post-rift sedimentary successions may remain quite thin (Western Alps, Masini *et al.* 2013).

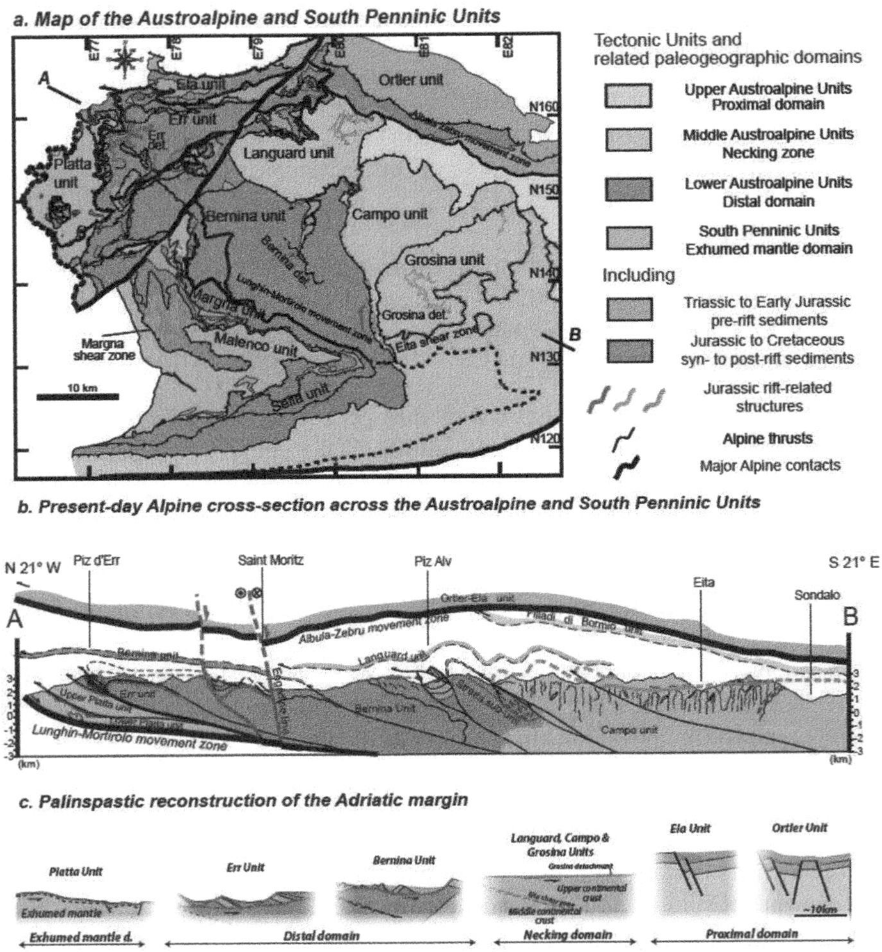

Figure 3.5. *a) Tectonic map of the Austroalpine and South Penninic Units sampling distinct paleogeographic domains derived from the Adriatic margin (southeast Switzerland and northern Italy), b) present-day cross-sections across these units (source: modified after Mohn* et al. *2011), c) reconstruction of the Adriatic margin showing the different paleogeographic domains subsequently sampled in the distinct units separated by Alpine contacts (source: modified after Ribes* et al. *2019)*

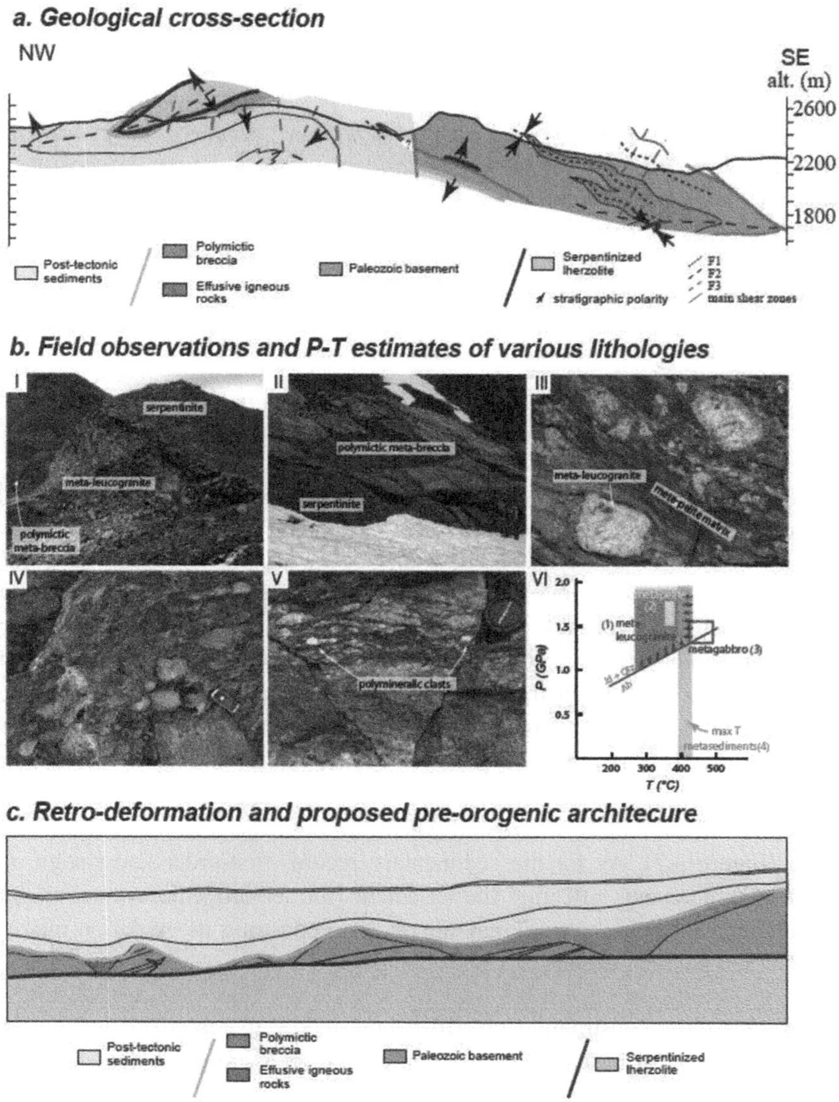

Figure 3.6. *Criteria used in the Punta Rosa Unit to recognize remnants of magma-poor rifted margins in the internal zone of the Alpine orogen (source: modified from Beltrando* et al. *2012, 2014)*

CONTINUATION OF CAPTION FOR FIGURE 3.6.– *a) Geological cross-sections showing the different rock types. b) Field observations and estimates of the P–T evolution for the different lithostratigraphic associations of the Punta Rosa Unit, (I) section of serpentinized mantle peridotites in contact with the Paleozoic basement. The latter is overlain by polymictic breccia, (II) serpentinized mantle directly overlain by polymictic breccia, (III) clast of meta-leucogranites found in post-tectonic sediments, (IV) monomictic breccia reworking serpentinized peridotites, (V) polymineralic clasts of leucogranites interpreted as pre-Alpine fault rocks, (VI) P–T evolution recorded by the main lithologies of the Punta-Rossa Unit. All the studied lithologies reached metamorphic re-equilibration at P > 1.0 GPa, c) retro-deformation and proposed pre-orogenic architecture of the Punta Rossa Unit.*

3.3.3. *Basement lithology and evolution*

The characterization of the basement evolution (being either continental or oceanic in nature) is critical in order to recognize the different rift domains.

Orogenic systems commonly show large exposures of continental basement, which recorded a polyphase evolution, including pre-, syn-, post-rift and orogenic events. The evolution of the continental basement is often difficult to unravel as it requires detailed petrological, structural, thermodynamic and geochronological studies including the determination of pressure–temperature–time–deformation paths (*P–T–t–d*) of basement rocks. The robust understanding of the basement history enables the identification of key events for which the *P–T–t(–d)* conditions can be determined, that is, where and when did these rocks form, re-equilibrate or deform (Figure 3.7). As for the sedimentary record, first-order knowledge of the distinct tectonic events affecting the basement (former orogenic events, magmatic intrusions, metamorphic events) may also help to constrain its evolution relative to pre-, syn- and post-rift events.

Pre-rift events: constraining pre-rift events is key to reconstructing the initial depth and position of basement units (Figure 3.7). They are generally controlled by the last trans-crustal event, such as the last geodynamic event during which the continental crust was pervasively deformed, or during which a trans-crustal magmatic system was emplaced. One caution should be noted: the occurrence of pre-orogenic high-grade metamorphic rocks at the surface (e.g. granulite) is often interpreted as a proof of lower crustal exhumation during rifting. Although possible, additional detailed studies are required to exclude that these high-grade metamorphic rocks were not already exhumed at upper crustal depth prior to rifting by more ancient, and unrelated, tectonic events. Hence, a key control is to also verify that these rocks stayed at the depth of their formation/re-equilibration until the beginning of rifting stages, with petrological, structural and/or geochronological arguments. Once pre-rift events are constrained, one can determine the pre-rift depth of the tectonic units, and also assess the syn-rift basement deformation (see Petri *et al.* 2019).

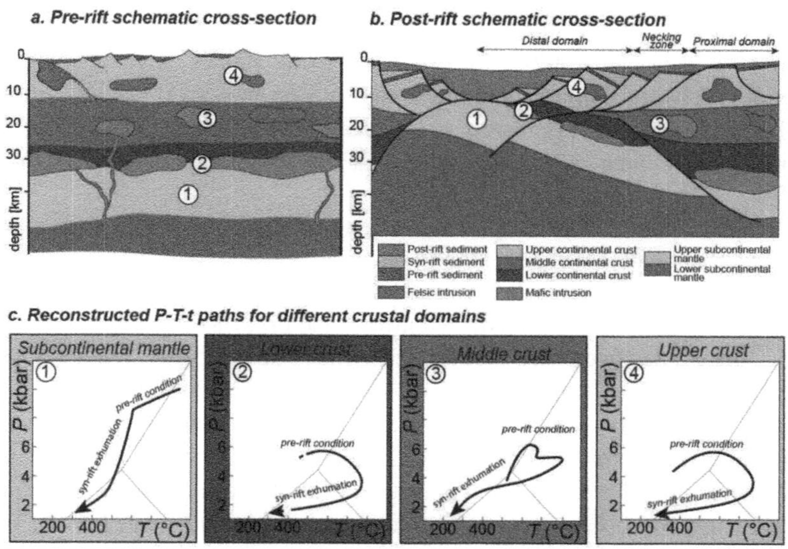

Figure 3.7. *Schematic cross-sections showing the evolution of distinct basement levels during the a) pre-rift and b) post-rift setting (source: modified from Petri* et al. *2019). c) Reconstructed P–T–t paths for the different crustal domains. Pre-rift magmatic intrusions were used as markers based on their P–T–t paths from pre-rift to syn-rift settings documenting the exhumation of the distinct parts of the continental lithosphere across the margin*

Syn-rift events: syn-rift basement metamorphism, deformation and exhumation, attested among others by geochronological data, have long been underestimated and ascribed either to orogenic events or simply disregarded (e.g. Beltrando *et al.* 2015). In the parts of the orogens, which have been less affected by syn-orogenic deformation and metamorphism, the recognition of rift-related events within the basement has been locally described (e.g. Manatschal and Müntener 2009; Mohn *et al.* 2011). Recognizing such syn-rift events is more difficult in internal parts of the orogen where high-grade orogeny-related metamorphic conditions often overprint prior events (e.g. Beltrando *et al.* 2014; Pantet *et al.* 2020). Overall, our understanding of the basement evolution during rifting is important in order to understand the deformation and thermal history of the continental crust, a history which is almost impossible to assess in present-day offshore-rifted margins.

Orogenic systems also expose ophiolites, which correspond to pieces of oceanic lithosphere or fragments of Continent Ocean Transition (COT). As previously discussed, Alpine ophiolites are characterized by a relatively low proportion of mafic magmatic rocks but a large proportion of fertile and serpentinised mantle, contrasting with more classical Penrose-type oceanic crust. As a result, these ophiolites have been considered for a long time as "abnormal" oceanic crust (see review of Lagabrielle 2009). However, numerous studies have recognized the similarities between Alpine ophiolites and present-day COT and/or ultra-slow spreading oceanic crust (Lombardo and Pognante 1982; Tricart and Lemoine 1983; Lagabrielle and Cannat 1990; Manatschal *et al.* 2011; Lagabrielle *et al.* 2015).

The recognition of Alpine ophiolites as remnants of a COT (for further details, see Manatschal and Müntener 2009; McCarthy *et al.* 2020b) is based on the occurrence of serpentinized subcontinental mantle directly overlain by discontinuous continental blocks and/or pre-rift sediments and/or syn-rift tectono-sedimentary breccias, the whole covered by post-rift sedimentary sequences (e.g. Manatschal and Müntener; 2009). The top of the subcontinental mantle is characterized by a brittle deformation overprint interpreted as evidence of a low-angle detachment fault leading to its exhumation. Other Alpine ophiolites are characterized by exhumed mantle including gabbroic bodies that are overlain by various volcanic sequences and post-rift covers. The top of the mantle is also characterized by important shear-zones related to major low-angle detachments formed during extension. Such structural pattern and lithostratigraphic associations are similar to oceanic core complexes in ultra-slow spreading system (e.g. Sauter *et al.* 2013; Cannat *et al.* 2019). The two likely represent a continuum in space and time, passing from exhumation of mantle in COT to ultra-slow spreading accretions with mid-ocean ridge magmatism remaining sparse and short lived (e.g. Li *et al.* 2013; Tribuzio *et al.* 2016).

3.3.4. *Rift-related deformation style: classical rift geometries versus supra-detachment basins*

Different deformation styles and basin architectures are documented across rifted margins from proximal to distal domains. The identification of the initial rift architecture may therefore help to characterize rift domains. At first-order, two different extensional systems are identified based on the amount of extension accommodated along the bounding fault(s) (Figure 3.8) and referred to as low-beta (i.e. stretching factor) and high-beta factors/settings/basins (Wilson *et al.* 2001; Tugend *et al.* 2015a; Ribes *et al.* 2019). Another key feature is the nature of the contact between the basement and the cover, which could be either a primary stratigraphic contact, or a tectonic contact in relation to high-angle or low-angle extensional faults.

Low-beta settings are classically characterized by high-angle normal faults forming typical half-graben geometries or horst and graben alternations (Figure 3.8, e.g. Eberli 1988; Cowie *et al.* 2005). These rift basins are characterized by the tilting of pre-rift sequences and are filled by syn-rift successions showing a general thickening towards the footwall (i.e. growth strata). The syn-rift sedimentary bedding lies at high angle with respect to the fault planes. Exhumation of basement rocks remains limited to the high-angle fault offset. Such rift basins are common in the proximal domains of rifted margins (Alves *et al.* 2006) or in intra-continental basins as in the North Sea (Cowie *et al.* 2005; Claringbould *et al.* 2020), but may occur all along the rifted margin. Special attention is required if salt deposits occur in the stratigraphic record. The occurrence of pre-kinematic salt may alter the previously described geometry. Depending on the degree of coupling between sub- and supra-salt levels, various types of deformation can be observed: normal fault cutting the basement can be associated with forced folding in the supra-salt cover or may just be restricted to sub-salt levels (Figure 3.8) (Jackson and Lewis 2016).

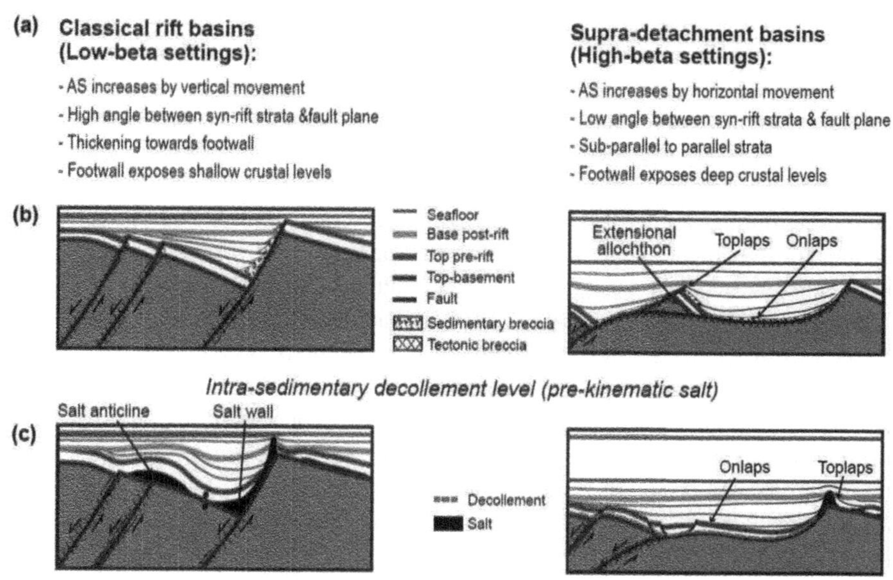

Figure 3.8. *Characteristics and architecture of classical rift and supra-detachment basins, also referred to as low- and high-beta extensional systems. The geometries of syn-rift sediments depend on the extensional setting and presence or absence of a pre-kinematic decoupling layer like salt (bottom cartoons). AS: Accommodation Space*

Several contributions discussed the mechanism of inversion of half graben rift basins in low-beta extensional settings (Buiter and Pfiffner 2003; Dumont *et al.* 2008; Bauville and Schmalholz 2017; Bellanger *et al.* 2015). They emphasize the potential role as a local buttress of the basement blocks delimiting the syn-rift basinal successions. The reactivation of the existing normal fault or the formation of new compressive structures depends among other factors on the initial dip of the inherited normal faults and the orientation of the pre-existing fabric in the basement.

High-beta settings are associated with large offset normal faults (also often referred as extensional/low-angle/top-basement detachment fault or downward-concave faults) in relation to major crustal and lithospheric thinning (De Charpal *et al.* 1978; Lemoine *et al.* 1987; Reston *et al.* 1996; Manatschal *et al.* 2001). These large offset normal faults can accommodate the exhumation of continental basement or mantle in their footwalls over large distances (Boillot *et al.* 1987; Masini *et al.* 2012). The faults often show a flat geometry at the surface and are interpreted to steepen at depth following the rolling hinge model (Buck 1993; Lavier *et al.* 1999). These faults delimit large sedimentary basins, referred to as supra-detachment basins (Figure 3.8). They may be associated with exhumation of deeper levels of the crust and their juxtaposition with the syn-rift sedimentary succession along tectonic contacts. Such settings are generally restricted to necking zones or distal margin domains.

The overlying sedimentary succession in the hanging-wall is characterized by a complex rift-related architecture: pre-rift sediments form rider blocks to extensional allochthones (hanging-wall blocks) onlapped by syn-rift sediments. In contrast to high-angle normal fault rift basins, the syn-rift sedimentary bedding is sub-parallel to the top-basement, that is, to the outcropping fault. Interactions between pre-kinematic salt and large-offset normal faults have been reported in several studies (Pyrenees, Lagabrielle *et al.* 2010; Labaume and Teixell 2020, Columbrets Basin, Etheve *et al.* 2018). The rheological behavior of salt creates an efficient decoupling layer, partitioning the deformation between the supra-salt cover and underlying basement. This may result in a dismembering of the pre-rift cover forming raft geometries. These complex rift-related geometries need to be integrated to avoid erroneous interpretations and overestimations of the orogeny-related deformation. In conclusion, a detailed assessment of the basement–cover relationship is necessary. Although the basement cover contact is often reactivated during the convergence, the above-mentioned methodology (Beltrando *et al.* 2014) enables the recognition of the partial to complete preservation of rift-related lithostratigraphic associations.

3.4. Application to case studies

This section is an application of the principles and criteria described above on key examples selected across the Alps and the Pyrenees (Western Europe) as well as the Rif belt (Northern Morocco) (Figure 3.2), where distinct rift domains can be identified. All three are Cenozoic orogenic belts that were formed in the context of the convergence between Eurasia, Iberia, Adria and Africa.

The Alps and the Rif preserved the remnants of narrow oceanic basins belonging to the Western Tethys that resulted from polyphased rift events during the Triassic to Middle Jurassic (Lemoine *et al.* 1986; Rosenbaum *et al.* 2002; Handy *et al.* 2010, Van Hinsbergen *et al.* 2020; for the Alps see Chapter 7 of Volume 2). From the Late Cretaceous onward, subduction of these basins was followed by continental collision of the two passive margins during the Cenozoic.

In contrast, the Pyrenees represent former Early Cretaceous hyper-extended rift basins that did not reach continental breakup stages and were inverted and reactivated from the Late Cretaceous to Cenozoic (see Chapter 3.9, Tugend *et al.* 2015b). For more details regarding the evolution of these orogenic belts, the reader is referred to Munoz (2002); Schmid *et al.* (2004, 2013); Mouthereau *et al.* (2014); Tugend *et al.* (2015b) and Leprêtre *et al.* (2018).

3.4.1. *Recognition of a proximal domain: the case study of the Dauphinois Unit in the Western Alps*

Remnants of the proximal domain of the fossil European margin of the Alpine Tethys have been identified in the external parts of the Alps (e.g. Dauphinois/ Helvetic Units, e.g. Mohn *et al.* 2010). The European external part shows low-grade orogeny-related metamorphic overprint (i.e. greenschist facies conditions), associated with weak to moderate deformation during the Cenozoic (Bousquet *et al.* 2008; Dumont *et al.* 2008; Bellanger *et al.* 2014). This limited Alpine overprint allows for the identification of several tectonic units with coherent lithostratigraphy.

The Dauphinois Unit consists of the exposure of the Variscan basement in the so-called external crystalline massif overlain by Mesozoic to Cenozoic successions (for more details, see Dumont *et al.* 2008 and Boutoux *et al.* 2014).

The thin pre-rift Triassic succession is made of a heterogeneous succession of siliciclastic deposits, carbonates and evaporites with a thickness of ~100 m deposited in a fluvial to shallow-marine depositional environment. These features confirm the absence of a major subsidence during the Triassic in this area, which is consistent with a pre-rift setting.

In contrast, the Lower Jurassic calcareous to argillaceous succession ranges from a few dozen meters to 1.5 kilometers in thickness (Chevalier *et al.* 2003). These variations in thickness are related to high normal faults controlling the formation of half-grabens and associated depocenters (low-beta settings, Figure 3.8). Normal fault heaves reach 2 km, affecting the brittle upper crust.

The subsequent Middle Jurassic to Cretaceous sediments were deposited in a post-rift setting showing, by contrast, fewer thickness variations characterized by the deposition of several carbonate platforms (i.e. Tithonian and Urgonian platforms) intercalated with marl deposits. This post-rift stratigraphic record is associated with a moderate subsidence, suggesting a rather limited post-rift thermal subsidence and consequently minor crustal and lithospheric thinning during rifting in this domain. Today, Mesozoic sediments occur generally as variably pinched synclines separated by basement highs that correspond to inverted former half-grabens.

3.4.2. *Recognition of necking zones: the case studies of the Campo-Grosina Units (Alps) and Southern part of Mauléon Basin (Pyrenees)*

The recognition of former necking zones in fossil orogen remains rare (e.g. Mohn *et al.* 2012; Tugend *et al.* 2015a; Ribes *et al.* 2020). Their identification is often hampered by their transitional characteristics located between proximal and distal domains (Figure 3.4). In addition, these zones are typically overprinted by orogeny-related deformation and metamorphism. Here, we document two complementary examples from the Alps and the Pyrenees where fossil necking zones have been recognized. The identification of the Alpine example relies on access to the basement evolution prior to and during rifting; for the Pyrenees, it relies on observations of the stratigraphic record.

The Campo-Grosina Units (Alps): basement record

In the Alps, evidence for the fossil necking zone of the Adriatic rifted margin is reported from the Campo-Grosina Units in Austroalpine nappe system, in southeast Switzerland and northern Italy (Figures 3.5 and 3.9; for further details, see Mohn *et al.* 2012; Petri *et al.* 2019). The Campo-Grosina Units are structurally underlying the Ela-Ortler Units preserving the proximal domain, and are overlying the Err-Bernina-Platta Units preserving remnants of the distal domain (Figure 3.5). The Alpine deformation and metamorphism, remaining weak in these units, makes it possible to unravel specific tectonic, magmatic and metamorphic events preceding the orogeny. More specifically, the Campo-Grosina Units are made of continental basement, which records a poly-phased history spanning the Devono-Carboniferous Variscan orogeny, a Permian post-Variscan extension, Jurassic rifting and eventually the Alpine orogeny.

The overlying Grosina Unit is separated from the Campo Unit by the so-called Eita Shear zone (Figure 3.9). The Campo basement essentially consists of metasediments showing a well-developed pre-Permian (pre-rift) foliation formed at amphibolite-facies conditions (Figure 3.9I). In the Permian, the Sondalo gabbroic complex was emplaced at mid-crustal depth (4–6 kbar) (Petri *et al.* 2016, 2017, 2018; Braga *et al.* 2001) within the Campo basement. The overlying Grosina basement is made of biotite and chlorite orthogneiss associated with metasediments showing a pre-Permian metamorphic fabric formed under amphibole facies conditions (Figure 3.9IV).

The pre-rift Sondalo gabbro is used as a marker to reconstruct the evolution of the Campo middle crust, prior to, during and after rifting. Petrological and structural studies performed on the intrusion and the associated contact aureole show that after its emplacement at mid-crustal depth, an isobaric cooling was recorded (Petri *et al.* 2016). Furthermore, from the Late Permian-Triassic sedimentary deposits, no significant tectonic events and related exhumation of deep basement are recorded. This suggests that the Sondalo gabbroic complex was still at mid-crustal depth at the onset of the Jurassic rifting. By contrast, the depth of the Grosina basement prior to the rifting remains unconstrained.

The evolution of these units was investigated through $^{40}Ar/^{39}Ar$ thermochronology on muscovite and biotite. Two distinct age clusters were identified, with the Campo basement ranging from ~165 to 210 Ma while the Grosina basement ranged from 250 to 270 Ma (Figure 3.9; Petri 2014). This relative age difference confirms that these two units represent distinct portions of the pre-rift continental crust.

The younger $^{40}Ar/^{39}Ar$ age spectra of the Campo basement can be interpreted as cooling ages recording the exhumation of a section of the middle crust during rifting. The time range provided by $^{40}Ar/^{39}Ar$ data are synchronous to the main rifting event related to the opening of the Alpine Tethys. The $^{40}Ar/^{39}Ar$ age spectra of the Grosina basement are consistently older, showing Permian to Middle Triassic ages. Although the exact depth of the Grosina Unit prior to the rifting is unknown, these "older" ages suggest that the Grosina Unit was already in an upper crustal position (<15 km, assuming a thermal gradient of 30°C/km) before rifting.

These two units are limited by the Eita shear zone, which consists of 200 m thick greenschist-facies mylonites (Figure 3.9III). The Eita shear zone was interpreted as a major extensional rift-related structure (more detail can be found in Petri 2014; Mohn *et al.* 2012; Petri *et al.* 2019). Additional rift-related extensional structures were recognized in the Campo-Grosina Units including the Grosina detachment. This later structure is a brittle deformation zone located at top of the Grosina Unit and interpreted as major extensional detachment (Figure 3.10). These results show that that the Campo-Grosina Units recorded the exhumation of mid-crustal levels during the Jurassic rifting accommodated by several extensional structures.

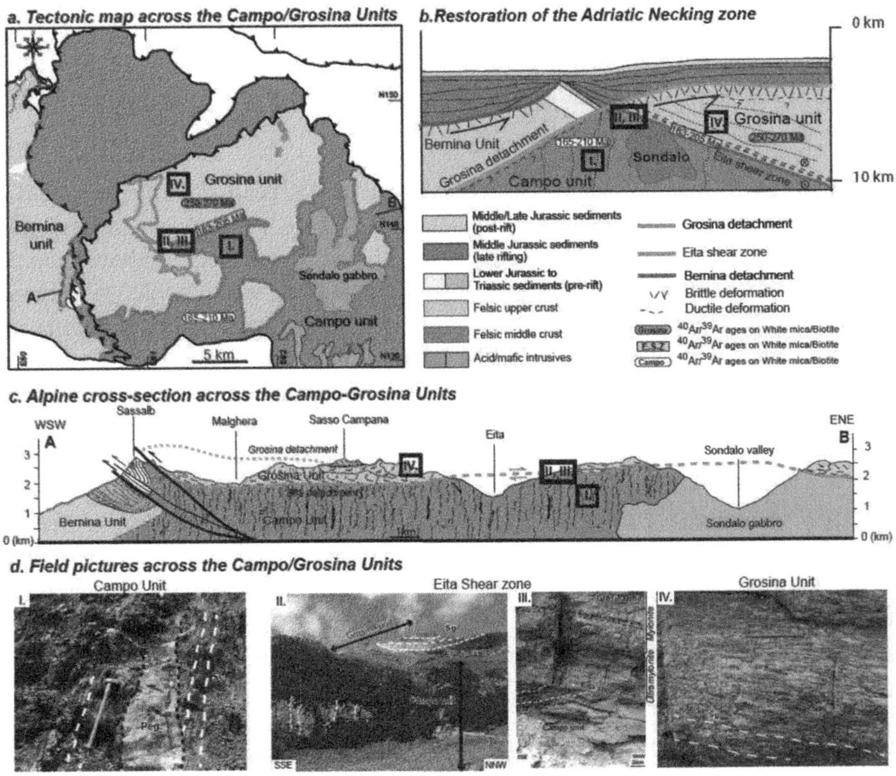

Figure 3.9. *Summary of the observations enabling the identification of the former Adriatic necking zone in the Alps (source: modified from Mohn* et al. *2012; Petri* et al. *2019)*

CONTINUATION OF CAPTION FOR FIGURE 3.9.– *a) The map, c) present-day cross-section and d) field pictures enabled the restoration of b) pre-Alpine geometries. 40Ar/39Ar ages are indicated on the map and restored cross-sections. These ages indicate exhumation of middle crust during Jurassic rifting. Field pictures show the main structures observed in the Campo-Grosina Units. (I) Pegmatite of probable Permian age cross-cutting the steep fabric (Sc) of the Campo Unit mica schist supporting a pre-Permian age for this fabric. (II) Picture showing the relationship between the Grosina and Campo Units separated by the Eita shear zone. An angular discordance is observed between the flat fabric (Sg) of the Grosina Unit and the steep fabric (Sc) of the Campo Unit. (III) Close-up on the Eita shear zone. (IV) Orthogneiss showing an amphibolite facies fabric (Sg) likely of Paleozoic age in the Grosina Unit.*

The Southern part of the Mauléon Basin (Pyrenees): deformation style and stratigraphic record

Remnants of a former necking domain have been identified in the southern part of the Mauléon Basin (Western Pyrenees) based on stratigraphic and structural observations (Figure 3.10; Tugend *et al.* 2014, 2015a). It is now exposed in a large pop-up structure that formed during its incorporation in the Pyrenean orogen in the Late Cretaceous (Velasque *et al.* 1989; Teixell 1998; Jammes *et al.* 2009; Teixell *et al.* 2018; Saspiturry *et al.* 2020). To the north, it is thrust over the Grand Rieu High and southern Arzacq Basin along the Saint Palais and Sainte Suzanne thrust systems (Velasque *et al.* 1989; Daignières *et al.* 1994; Teixell 1998). To the south, it is thrust over the Axial Zone and southern foreland basin along the Lakoura thrust system (Velasque *et al.* 1989; Teixell 1998).

The southern part of the Mauléon Basin exposes thick syn-rift sedimentary sequences deposited onto Palaeozoic metasedimentary rocks, which are part of the pre-rift basement (Mendibelza Massif). The contact between the basement and onlapping sediments is marked by gouges and cataclastic fault zones (Figure 3.10I) evolving laterally into tectono-sedimentary breccias (Figure 3.10II). This sequence of rocks and the low-angle onlap of the sediments are characteristic of *supra-detachment basins* floored by large portions of exhumed basement (Johnson and Hall 1989; Miranda-Avilés *et al.* 2005; Masini *et al.* 2014). Syn-rift sediments consist of their southern proximal part of massive conglomerates (Mendibelza formation, Figure 3.10III) while their distal time equivalents observed to the north consist of turbidites and marls (Figure 3.10IV, "Black Flyschs" formation, Souquet *et al.* 1985). This progressive thickening and deepening of syn-rift sequences along an active detachment fault system is an indicator of significant crustal thinning initiation, consistent with a necking domain (Figure 3.10, Tugend *et al.* 2014, 2015a).

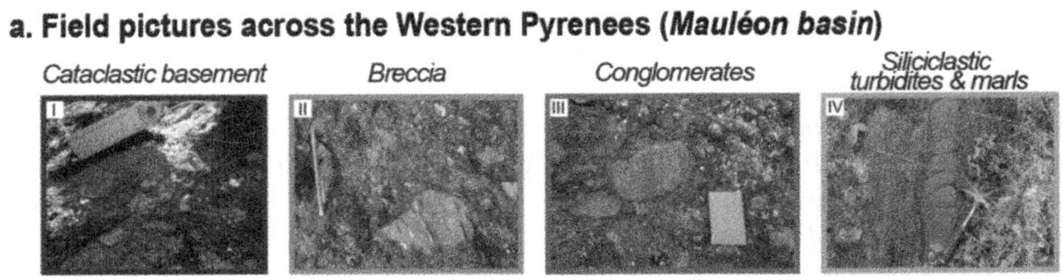

Figure 3.10a. *Summary of the main field observations characterizing the Iberian necking zone cropping out in the Western Pyrenees. (I) Cataclastic basement; (II) tectono sedimentary breccia; (III) syn-rift sediments of the Mendibelza conglomerate formation; (IV) interbedded marls and siliciclastic turbiditic deposits of the syn-rift Flysch noir formation (source: Tugend* et al. *2015a)*

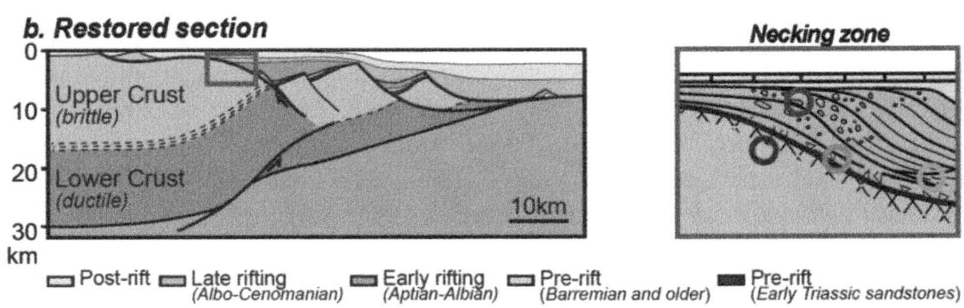

Figure 3.10b. *Restored cross-section across the Iberian margin of the Pyrenean rift system (source: Masini* et al. *2014) and diagram illustrating the tectono–stratigraphic relationships observed in the Iberian necking zone preserved in the Mendibelza Massif (Western Pyrenees, source: Tugend* et al. *2015a)*

3.4.3. *Recognition of hyper-extended domains: the case studies of the Senadhja Nappe (Rif belt)*

Evidence for remnants of hyper-extended domains is now increasingly recognized in orogenic systems. In the following section, we focus on the Rif and more specifically on the External Rif (Suter 1980; Favre *et al.* 1991; Frizon de

Lamotte *et al.* 1991) where recent evidence of a former hyper-extended domain has been reported (Figure 3.2) (Gimeno Vives *et al.* 2019).

The External Rif located in northern Morocco preserves the remnants of the Mesozoic North African rifted margin (Figure 3.11) (Favre *et al.* 1991; Gimeno-Vives *et al.* 2019). This margin was reactivated and deformed during the Cenozoic in the context of the Europe–Iberia–Africa convergence (see Leprêtre *et al.* 2018 and Gimeno Vives *et al.* 2020a and references therein). The External Rif consists of a nappe stack emplaced in a top-to-the-southwest transport direction. Three main units are identified that are, from top to bottom, the Intrarif, the Mesorif and the Prerif, which represent distinct paleogeographic domains. The Mesorif Unit comprises tectonic windows and an allochthonous nappe referred to as the Senadhja nappe. Its internal structure and associated lithostratigraphic associations have been the subject of long-standing controversy and was previously recognized as a chaotic complex in relation to its Miocene emplacement by gravity gliding. However, a recent study attributes this apparent chaotic complex to former rifting processes along the North African distal margin (Gimeno-Vives *et al.* 2019).

The base of this nappe is made of mafic intrusive rocks such as (micro)gabbros representing the Mesorif gabbroic complex (Figure 3.11) (Michard *et al.* 2007; Benzaggagh *et al.* 2014; Gimeno Vives *et al.* 2019). U-Pb geochronology on zircon yielded ages ranging from 200 to 192 Ma (earliest Jurassic). These intusives are likely related to the emplacement of the Central Atlantic Magmatic Province (CAMP; Gimeno Vives *et al.* 2019). Although the country rocks are never exposed, these mafic intrusions are interpreted as being emplaced in continental basement (for further detail, see Gimeno Vives *et al.* 2019, 2020b). The top of these intrusions is a tectonic subhorizontal contact characterized by tectonic breccias and Upper Triassic evaporites (Figure 3.11). Structurally above the latter are dismembered Lower Jurassic carbonate platforms that show steep bedding and a clear cutoff towards the underlying subhorizontal tectonic contact (Figure 3.11II). Lower Jurassic carbonates are limited by a series of extensional structures rooting generally in the Upper Triassic evaporites that represent a major decoupling zone with the Mesorif gabbroic complex. Locally, where the Upper Triassic evaporites were absent, extensional structures rooted in deeper levels, offsetting the top of the Mesorif gabbroic complex (Figure 3.11I). These faults are controlling depocenters of Middle Jurassic sequences showing typical wedge geometries with a thickening towards the faults confirming their syn-rift setting (Figure 3.11II). Above, thick Upper Jurassic deep marine turbiditic sequences seal the extensional structures.

The above-described succession likely reflects an initial Upper Triassic rifting event characterized by the emplacement of mafic intrusions and deposition of Triassic evaporites, followed by a second rifting event during the Middle Jurassic (Gimeno Vives *et al.* 2019). Geometries of this second event were controlled by the previous deposition of Upper Triassic evaporites representing a major decoupling level controlling the formation of "raft" geometries in the Mesozoic successions. The observed complex lithostratigraphic association sampled in the Senhadja Nappe is essentially the result of polyphased rift and magmatic events combined with salt tectonics.

Figure 3.11a. *Different field relationships exposed in the Senhadja Nappe in the External Rif (after Gimeno Vives* et al. *2019, 2020). (I) View of Sof Teirara showing Lower Jurassic carbonates overlying the Mesorif gabbroic complex separated by tectonic breccias. These lithologies are juxtaposed against Middle Jurassic sediments along a major extensional fault (red line). The occurrence of growth strata in Middle Jurassic sediments towards the fault suggests that rifting occurred during this period. (II) Discontinuous blocks of Lower Jurassic carbonates in the Bou Adel area. The carbonate bedding shows a clear cutoff with the underlying Mesorif gabbro. The contact between the two lithologies is characterized by a tectonic breccia corresponding to a major tectonic contact (purple line)*

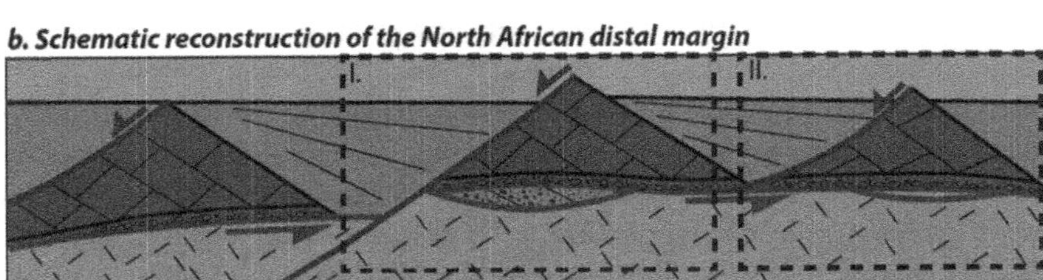

Figure 3.11b. *Diagram illustrating the idealized architecture of the North African distal margin sampled in the Senhadja Nappe and location of field pictures*

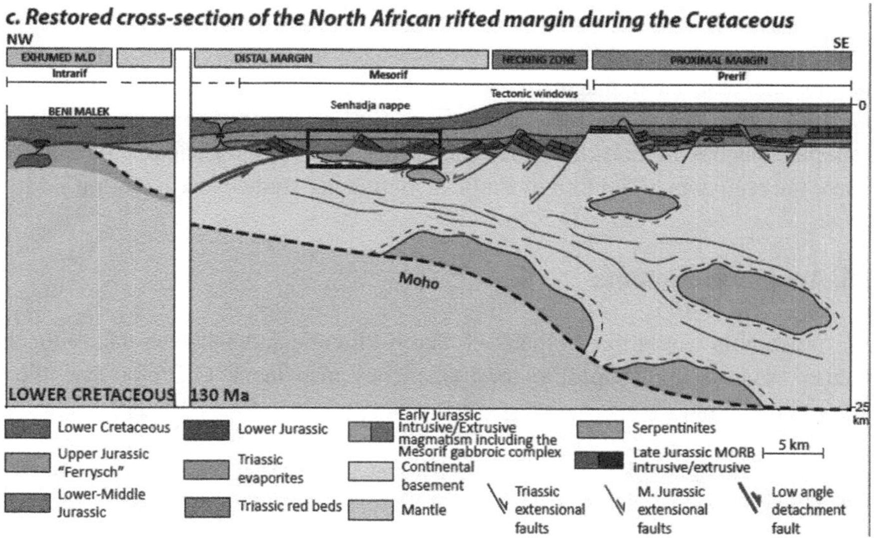

Figure 3.11c. *Diagram illustrating the idealized architecture of the North African distal margin sampled in the Senhadja Nappe and location of field pictures. Restored structure of the North African rifted margin during the Cretaceous (source: Gimeno Vives* et al. *2019 and 2020a)*

3.5. Conclusion

With this chapter, we provide information on the key diagnostic elements which enable the investigation of rifted margin remnants imbricated in orogens. The recognition and characterization of former rift domains in orogenic belts and the comparison with present-day rifted margins remains fundamental for our understanding of continental lithosphere extension mechanisms. However, the study of remnants of rifted margins also challenges our understanding of the formation of orogenic systems. Even in highly deformed internal zones of orogens, pre-orogenic lithostratigraphic associations can be preserved. Taking into account the complexity of pre-orogenic settings (e.g. rift basin architecture and rift-related crustal thickness variations) may have major implications for restoration of balanced cross-sections and estimation of shortening.

Most of these methodologies have been developed from Alpine-type orogens (Alps, Pyrenees) sampling remnants of small and immature rift and oceanic basins. The following chapters will illustrate in greater detail the observations and geometry acquired from this type of orogen. However, applications to other orogenic systems resulting from the closure of wider oceanic domains have not been fully investigated yet. Part of the challenge is related to the potential preservation of former rift-related structures in such orogenic systems characterized by the development of an active margin and associated magmatic arc. However, it also lies in our awareness of potentially complex pre-orogenic systems and in our ability to recognize them. For example, new observations from the Caledonides described in Chapter 8 of Volume 2, where the remnants of the former Baltican margin are studied, are inspiring.

3.6. Acknowledgments

The authors would like to thank G. Peron-Pinvidic and Johannes Jakob for their detailed reviews and helpful suggestions. They also thank G. Peron-Pinvidic for editorial handling.

Further reading. – The above descriptions are abbreviated and often simplified. If interested in reading and learning further, the reader is referred to the following list of publications and references.
– General: (Lemoine *et al.* 1986; Manatschal and Müntener 2009; Mohn *et al.* 2011, 2014; Beltrando *et al.* 2014; Tugend *et al.* 2014, 2015a; Lagabrielle *et al.* 2015; Petri *et al.* 2019; McCarthy *et al.* 2020b).

3.7. References

Agard, P., Omrani, J., Jolivet, L., Whitechurch, H., Vrielynck, B., Spakman, W., Monié, P., Meyer, B., Wortel, R. (2011). Zagros orogeny: A subduction-dominated process. *Geological Magazine*, 148(5–6), 692–725. https://doi.org/10.1017/S001675681100046X.

Alves, T.M., Moita, C., Sandnes, F., Cunha, T., Monteiro, J.H., Pinheiro, L.M. (2006). Mesozoic-Cenozoic evolution of North Atlantic continental-slope basins: The Peniche Basin, western Iberian margin. *AAPG Bulletin*, 90(1), 31–60. https://doi.org/10.1306/08110504138.

Andersen, T.B., Corfu, F., Labrousse, L., Osmundsen, P.T. (2012). Evidence for hyperextension along the pre-Caledonian margin of Baltica. *Journal of the Geological Society*, 169(5), 601–612. https://doi.org/10.1144/0016-76492012-011.

Angiboust, S., Pettke, T., De Hoog, J.C.M., Caron, B., Oncken, O. (2014). Channelized fluid flow and eclogite-facies metasomatism along the subduction shear zone. *Journal of Petrology*, 55(5), 883–916. https://doi.org/10.1093/petrology/egu010.

Anonymous (1972). Penrose field conference on ophiolites. *Geotimes*, 17(12), 24–25.

Bauville, A. and Schmalholz, S.M. (2017). Tectonic inheritance and kinematic strain localization as trigger for the formation of the Helvetic nappes, Switzerland. *Swiss Journal of Geosciences*, 110(2), 523–534.

Bellanger, M., Augier, R., Bellahsen, N., Jolivet, L., Monié, P., Baudin, T., Beyssac, O. (2015). Shortening of the European dauphinois margin (Oisans Massif, Western Alps): New insights from RSCM maximum temperature estimates and 40Ar/39Ar *in situ* dating. *Journal of Geodynamics*, 83, 37–64. https://doi.org/10.1016/j.jog.2014.09.004.

Beltrando, M., Compagnoni, R., Lombardo, B. (2010). (Ultra-) High-pressure metamorphism and orogenesis: An alpine perspective. *Gondwana Research*, 18(1), 147–166. https://doi.org/10.1016/j.gr.2010.01.009.

Beltrando, M., Frasca, G., Compagnoni, R., Vitale-Brovarone, A. (2012). The Valaisan controversy revisited: Multi-stage folding of a Mesozoic hyper-extended margin in the Petit St. Bernard pass area (Western Alps). *Tectonophysics*, 579, 17–36.

Beltrando, M., Manatschal, G., Mohn, G., Dal Piaz, G.V., Vitale Brovarone, A., Masini, E. (2014). Recognizing remnants of magma-poor rifted margins in high-pressure orogenic belts: The Alpine case study. *Earth-Science Reviews*. https://doi.org/10.1016/j.earscirev.2014.01.001.

Beltrando, M., Stockli, D.F., Decarlis, A., Manatschal, G. (2015). A crustal-scale view at rift localization along the fossil Adriatic margin of the Alpine Tethys preserved in NW Italy. *Tectonics*, 34(9), 1927–1951. https://doi.org/10.1002/2015TC003973.

Benzaggagh, M., Mokhtari, A., Rossi, P., Michard, A., El Maz, A., Chalouan, A., Saddigi, O., Rjimati, E.C. (2014). Oceanic units in the core of the External Rif (Morocco): Intramargin hiatus or South-Tethyan remnants? *Journal of Geodynamics*, 77, 4–21 https://doi.org/10.1016/j.jog.2013.10.003.

Berger, A. and Bousquet, R. (2008). Subduction-related metamorphism in the Alps: Review of isotopic ages based on petrology and their geodynamic consequences. *Geological Society Special Publication*, 298(March 2009), 117–144. https://doi.org/10.1144/SP298.7.

Bernoulli, D. (1964). Zur Geologie des Monte Generoso, Beitr. Geol. *Karte Schweiz*, 118, 134.

Bernoulli, D., Manatschal, G., Desmurs, L., Müntener, O. (2003). Where did Gustav Steinmann see the trinity? Back to the roots of an Alpine ophiolite concept. *Special Paper of the Geological Society of America*, 373, 93–110. https://doi.org/10.1130/0-8137-2373-6.93.

Boillot, G., Recq, M., Winterer, E.L., Meyer, A.W., Applegate, J., Baltuck, M., Bergen, J.A., Comas, M.C., Davies, T.A., Dunham, K., Evans, C.A., Girardeau, J., Goldberg, G., Haggerty, J., Jansa, L.F., Johnson, J.A., Kasahara, J., Loreau, J.P., Luna-Sierra, E., Moullade, M., Ogg, J., Sarti, M., Thurow, J., Williamson, M. (1987). Tectonic denudation of the upper mantle along passive margins: A model based on drilling results (ODP leg 103, western Galicia margin, Spain). *Tectonophysics*, 132(4), 335–342. https://doi.org/10.1016/0040-1951(87)90352-0.

Bousquet, R., Oberhänsli, R., Goffé, B., Wiederkehr, M., Koller, F., Schmid, S.M., Schuster, R., Engi, M., Berger, A., Martinotti, G. (2008). Metamorphism of metasediments at the scale of an orogen: A key to the Tertiary geodynamic evolution of the Alps. *Geological Society*, 298(1), 393–411. https://doi.org/10.1144/SP298.18.

Boutoux, A., Bellahsen, N., Lacombe, O., Verlaguet, A., Mouthereau, F. (2014). Inversion of pre-orogenic extensional basins in the external Western Alps: Structure, microstructures and restoration. *Journal of Structural Geology*, 60, 13–29. https://doi.org/10.1016/j.jsg.2013.12.014.

Braga, R., Giacomini, F., Messiga, B., Tribuzio, R. (2001). The Sondalo Gabbroic complex (Central Alps, Northern Italy): Evidence for emplacement of mantle-derived melts into amphibolite-facies metapelites. *Physics and Chemistry of the Earth, Part A: Solid Earth and Geodesy*, 26(4–5), 333–342. https://doi.org/10.1016/S1464-1895(01)00063-1.

Buck, W.R. (1993). Effect of lithospheric thickness on the formation of high- and low-angle normal faults. *Geology*, 21(10). 933–936. https://doi.org/10.1130/0091-7613(1993)021<0933:EOLTOT>2.3.CO;2.

Buiter, S.J.H. and Pfiffner, O.A. (2003). Numerical models of the inversion of half-graben basins. *Tectonics*, 22(5). https://doi.org/10.1029/2002tc001417.

Cannat, M., Sauter, D., Lavier, L., Bickert, M., Momoh, E., Leroy, S. (2019). On spreading modes and magma supply at slow and ultraslow mid-ocean ridges. *Earth and Planetary Science Letters*, 519, 223–233. https://doi.org/10.1016/j.epsl.2019.05.012.

Caron, C., Homewood, P., Wildi, W. (1989). The original Swiss flysch: A reappraisal of the type deposits in the Swiss Prealps. *Earth-Science Reviews*, 26(1–3), 1–45.

de Charpal, O., Guennoc, P., Montadert, L., Roberts, D.G. (1978). Rifting, crustal attenuation and subsidence in the Bay of Biscay. *Nature*, 275(5682), 706–711. https://doi.org/10.1038/275706a0.

Chenin, P., Manatschal, G., Picazo, S., Müntener, O., Karner, G., Johnson, C., Ulrich, M. (2017). Influence of the architecture of magma-poor hyperextended rifted margins on orogens produced by the closure of narrow versus wide oceans. *Geosphere*, 13(2), 559–576. https://doi.org/10.1130/GES01363.1.

Chevalier, F., Guiraud, M., Garcia, J.P., Dommergues, J.L., Quesne, D., Allemand, P., Dumont, T. (2003). Calculating the long-term displacement rates of a normal fault from the high-resolution stratigraphic record (early Tethyan rifting, French Alps). *Terra Nova*, 15(6), 410–416. https://doi.org/10.1046/j.1365-3121.2003.00508.x.

Claringbould, J.S., Bell, R.E., Jackson, C.A.L., Gawthorpe, R.L., Odinsen, T. (2020). Pre-breakup extension in the northern North Sea defined by complex strain partitioning and heterogeneous extension rates. *Tectonics*, 39(8). https://doi.org/10.1029/2019TC005924.

Cluzel, D., Aitchison, J.C., Picard, C. (2001). Tectonic accretion and underplating of mafic terranes in the Late Eocene intraoceanic fore-arc of New Caledonia (Southwest Pacific): Geodynamic implications. *Tectonophysics*, 340(1–2), 23–59.

Coffin, M.F. and Eldholm, O. (1994). Large igneous provinces: Crustal structure, dimensions, and external consequences. *Reviews of Geophysics*, 32(1), 1–36. https://doi.org/10.1029/93RG02508.

Cowie, P.A., Underhill, J.R., Behn, M.D., Lin, J., Gill, C.E. (2005). Spatio-temporal evolution of strain accumulation derived from multi-scale observations of Late Jurassic rifting in the northern North Sea: A critical test of models for lithospheric extension. *Earth and Planetary Science Letters*, 234(3–4), 401–419. https://doi.org/10.1016/j.epsl.2005.01.039.

Daignières, M., Séguret, M., Specht, M., Ecors Team (1994). The Arzacq-western Pyrenees ECORS deep seismic profile. In *Hydrocarbon and Petroleum Geology of France*, Mascle, A. (ed.). Springer, Berlin, Heidelberg.

Dana, J.D. (1873). On some results of the Earth's contraction from cooling, including a discussion of the origin of mountains and the nature of the Earth's interior. *American Journal of Science*. 3, 5, 423–443; 6:6–14; 104–115; 161–171.

Decrausaz, T., Müntener, O., Manzotti, P., Lafay, R., Spandler, C. (2021). Fossil oceanic core complexes in the Alps. New field, geochemical and isotopic constraints from the Tethyan Aiguilles Rouges Ophiolite (Val d'Hérens, Western Alps, Switzerland). *Swiss Journal of Geosciences*, 114(1), 1–27.

Desmurs, L., Manatschal, G., Bernoulli, D. (2001). The Steinmann trinity revisited: Mantle exhumation and magmatism along an ocean-continent transition: The Platta nappe, Eastern Switzerland. *Geological Society Special Publication*, 187, 235–266. https://doi.org/10.1144/GSL.SP.2001.187.01.12.

Dilek, Y. (2003). Ophiolite concept and its evolution. In *Ophiolite Concept and the Evolution of Geological Thought*, Dilek, Y. and Newcomb, S. (eds). Geological Society of America, Boulder.

Dilek, Y. and Furnes, H. (2014). Ophiolites and their origins. *Elements*, 10(2), 93–100. https://doi.org/10.2113/gselements.10.2.93.

Doré, T., and Lundin, E. (2015). Research focus: Hyperextended continental margins – Knowns and unknowns. *Geology*, 43(1), 95–96. https://doi.org/10.1016/0040.

Dumont, T., Champagnac, J.D., Crouzet, C., Rochat, P. (2008). Multistage shortening in the Dauphiné zone (French Alps): The record of Alpine collision and implications for pre-Alpine restoration. *Swiss Journal of Geosciences*, 101(suppl 1), 89–110. https://doi.org/10.1007/s00015-008-1280-2.

Duretz, T., Petri, B., Mohn, G., Schmalholz, S.M., Schenker, F.L., Müntener, O. (2016). The importance of structural softening for the evolution and architecture of passive margins. *Scientific Reports*. https://doi.org/10.1038/srep38704.

Eberli, G. (1988). The evolution of the southern continental margin of the Jurassic Tethys Ocean as recorded in the Allgäu Formation of the Austroalpine Nappes of Graubünden (Switzerland). *Eclogae Geologicae Helvetiae*, 81(1), 175–214. https://doi.org/10.5169/seals-166175.

Elter, P. (1972). La zone ofiolitica del Bracco nel quadro dell'Appennino Settentrionale: Introduzione alla geologia delle Liguridi: Pacini, Pisa, 66° Congresso della Società Geologica Italiana, Guida alle Escursioni, 5–55.

Engi, M., Berger, A., and Roselle, G.T. (2001). Role of the tectonic accretion channel in collisional orogeny. *Geology*, 29(12), 1143–1146.

Etheve, N., Mohn, G., Frizon de Lamotte, D., Roca, E., Tugend, J., Gómez-Romeu, J. (2018). Extreme mesozoic crustal thinning in the Eastern Iberia Margin: The example of the Columbrets Basin (Valencia Trough). *Tectonics*. https://doi.org/10.1002/2017TC004613.

Favre, P., Stampfli, G., Wildi, W. (1991). Jurassic sedimentary record and tectonic evolution of the northwestern corner of Africa. *Palaeogeography, Palaeoclimatology, Palaeoecology*, 87(1–4), 53–73. https://doi.org/10.1016/0031-0182(91)90130-J.

Franke, D. (2013). Rifting, lithosphere breakup and volcanism: Comparison of magma-poor and volcanic rifted margins. *Marine and Petroleum Geology*, 43, 63–87. https://doi.org/10.1016/j.marpetgeo.2012.11.003.

Frizon de Lamotte, D., Andrieux, J., Guezou, J.C. (1991). Cinématique des chevauchements néogènes dans l'Arc bético-rifain : discussion sur les modèles géodynamiques. *Bulletin de la Société géologique de France*, 169, 611–626.

Gillard, M., Autin, J., Manatschal, G., Sauter, D., Munschy, M., Schaming, M. (2015). Tectonomagmatic evolution of the final stages of rifting along the deep conjugate Australian-Antarctic magma-poor rifted margins: Constraints from seismic observations. *Tectonics*, 34(4), 753–783. https://doi.org/10.1002/2015TC003850.

Gillard, M., Sauter, D., Tugend, J., Tomasi, S., Epin, M.E., Manatschal, G. (2017). Birth of an oceanic spreading center at a magma-poor rift system. *Scientific Reports*, 7(1). https://doi.org/10.1038/s41598-017-15522-2.

Gimeno-Vives, O., Mohn, G., Bosse, V., Haissen, F., Zaghloul, M.N., Atouabat, A., Frizon de Lamotte, D. (2019). The mesozoic margin of the Maghrebian Tethys in the Rif Belt (Morocco): Evidence for polyphase rifting and related magmatic activity. *Tectonics*, 38(8), 2894–2918. https://doi.org/10.1029/2019TC005508.

Gimeno-Vives, O., Mohn, G., Bosse, V., Haissen, F., Zaghloul, M.N., Atouabat, A., Frizon de Lamotte, D. (2020a). Reply to comment by Michard *et al.* on "The Mesozoic Margin of the Maghrebian Tethys in the Rif Belt (Morocco): Evidence for polyphase rifting and related magmatic activity". *Tectonics*, 39(4), 1–8. https://doi.org/10.1029/2020TC006165.

Gimeno-Vives, O., Frizon de Lamotte, D., Leprêtre, R., Haissen, F., Atouabat, A., Mohn, G. (2020b). The structure of the Central-Eastern External Rif (Morocco); poly-phased deformation and role of the under-thrusting of the North-West African paleo-margin. *Earth-Science Reviews*, 205(May), 103198. https://doi.org/10.1016/j.earscirev.2020.103198.

Handy, M.R., Schmid, S.M., Bousquet, R., Kissling, E., Bernoulli, D. (2010). Reconciling plate-tectonic reconstructions of Alpine Tethys with the geological-geophysical record of spreading and subduction in the Alps. *Earth-Science Reviews*, 102(3–4), 121–158. https://doi.org/10.1016/j.earscirev.2010.06.002.

Haug, E. (1925) Contribution à une synthèse des Alpes Occidentales. *Bulletin de la Société Géologique de France*, 3(4), 97–244.

Hermann, J., Müntener, O., Trommsdorff, V., Hansmann, W., Piccardo, G.B. (1997). Fossil crust-to-mantle transition, Val Malenco (Italian Alps). *Journal of Geophysical Research: Solid Earth*, 102(B9), 20123–20132.

Jackson, C.A.L. and Lewis, M.M. (2016). Structural style and evolution of a salt-influenced rift basin margin; the impact of variations in salt composition and the role of polyphase extension. *Basin Research*, 28(1), 81–102.

Jakob, J., Alsaif, M., Corfu, F., Andersen, T.B. (2017). Age and origin of thin discontinuous gneiss sheets in the distal domain of the Magma-Poor hyperextended Pre-Caledonian margin of Baltica, Southern Norway. *Journal of the Geological Society*, 174(3), 557–571. https://doi.org/10.1144/jgs2016-049.

Jakob, J., Andersen, T.B., Kjøll, H.J. (2019). A review and reinterpretation of the architecture of the South and South-Central Scandinavian Caledonides – A magma-poor to magma-rich transition and the significance of the reactivation of rift inherited structures. *Earth-Science Reviews*, 192(January), 513–528. https://doi.org/10.1016/j.earscirev.2019.01.004.

Jamieson, R.A. and Beaumont, C. (2013). On the origin of orogens. *Bulletin*, 125(11–12), 1671–1702.

Jammes, S., Manatschal, G., Lavier, L., Masini, E. (2009). Tectonosedimentary evolution related to extreme crustal thinning ahead of a propagating ocean: Example of the western Pyrenees. *Tectonics*, 28(4), 1–24. https://doi.org/10.1029/2008TC002406.

Johnson, J.A. and Hall, C.A. (1989). Tectono-stratigraphic model for the Massif D'Igountze Mendibelza, western Pyrenees. *Journal of the Geological Society*, 146(6), 925–932.

Kjøll, H.J., Galland, O., Labrousse, L., Andersen, T.B. (2019). Emplacement mechanisms of a dyke swarm across the brittle-ductile transition and the geodynamic implications for magma-rich margins. *Earth and Planetary Science Letters*, 518, 223–235. https://doi.org/10.1016/j.epsl.2019.04.016.

Labaume, P. and Teixell, A. (2020). Evolution of salt structures of the Pyrenean rift (Chaînons Béarnais, France): From hyper-extension to tectonic inversion. *Tectonophysics*, 785(July 2019), 228451. https://doi.org/10.1016/j.tecto.2020.228451.

Lagabrielle, Y. (2009). Mantle exhumation and lithospheric spreading: An historical perspective from investigations in the Oceans and in the Alps–Apennines ophiolites. *Ital. J.Geosci. (Boll.Soc.Geol.It.). Sezione Tematica su Alps and Apennines*, 128(2), 279–293. http://dx.doi.org/10.3301/IJG.2009.128.2.279.

Lagabrielle, Y. and Bodinier, J.L. (2008). Submarine reworking of exhumed subcontinental mantle rocks: Field evidence from the Lherz peridotites, French Pyrenees. *Terra Nova*, 20(1), 11–21. https://doi.org/10.1111/j.1365-3121.2007.00781.x.

Lagabrielle, Y. and Cannat, M. (1990). Alpine Jurassic ophiolites resemble the modern central Atlantic Basement. *Geology*, 18, 319–322.

Lagabrielle, Y., Labaume, P., De Saint Blanquat, M. (2010). Mantle exhumation, crustal denudation, and gravity tectonics during Cretaceous rifting in the Pyrenean realm (SW Europe): Insights from the geological setting of the lherzolite bodies. *Tectonics*, 29(4), 1–26. https://doi.org/10.1029/2009TC002588.

Lagabrielle, Y., Vitale Brovarone, A., Ildefonse, B. (2015). Fossil oceanic core complexes recognized in the blueschist metaophiolites of Western Alps and Corsica. *Earth-Science Reviews*, 141, 1–26. https://doi.org/10.1016/j.earscirev.2014.11.004.

Larsen, H.C., Mohn, G., Nirrengarten, M., Sun, Z., Stock, J., Jian, Z., Klaus, A., Alvarez-Zaikian, C.A., Boaga, J., Bowden, S.A., Briais, A., Chen, Y., Cukur, D., Dadd, K., Ding, W., Dorais, M., Ferré, E.C., Ferreira, F., Furusawa, A., Gewecke, A., Hinojosa, J., Höfig, T.W., Hsiung, K.H., Huang, B., Huang, E., Huang, X.L., Jiang, S., Jin, H., Johnson, B.G., Kurzawski, R.M., Lei, C., Li, B., Li, L., Li, Y., Lin, J., Liu, C., Liu, C., Liu, Z., Luna, A.J., Lupi, C., McCarthy, A., Ningthoujam, L., Osono, N., Peate, D.W., Persaud, P., Qiu, N., Robinson, C., Satolli, S., Sauermilch, I., Schindlbeck, J.C., Skinner, S., Straub, S., Su, X., Su, C., Tian, L., van der Zwan, F.M., Wan, S., Wu, H., Xiang, R., Yadav, R., Yi, L., Yu, P.S., Zhang, C., Zhang, J., Zhang, Y., Zhao, N., Zhong, G., Zhong, L. (2018). Rapid transition from continental breakup to igneous oceanic crust in the South China Sea. *Nature Geoscience*, 11(10), 782–789. https://doi.org/10.1038/s41561-018-0198-1.

Lavier, L.L., Buck, W.R., Poliakov, A.N.B. (1999). Self-consistent rolling-hinge model for the evolution of large-offset low-angle normal faults. *Geology*, 27(12), 1127–1130. https://doi.org/10.1130/0091-7613(1999)027.

Lemoine, M., Bas, T., Arnaud-Vanneau, A., Arnaud, H., Dumont, T., Gidon, M., Bourbon, M., de Graciansky, P-C., Rudkiewicz, J-L., Megard-Galli, J., Tricart, P. (1986). The continental margin of the Mesozoic Tethys in the Western Alps. *Marine and Petroleum Geology*, 3(3), 179–199. https://doi.org/10.1016/0264-8172(86)90044-9.

Lemoine, M., Tricart, P., Boillot, G. (1987). Ultramafic and gabbroic ocean floor of the Ligurian Tethys (Alps, Corsica, Apennines). *Search of a Genetic Model. Geology*, 15, 622–625.

Leprêtre, R., Frizon de Lamotte, D., Combier, V., Gimeno-Vives, O., Mohn, G., Eschard, R. (2018). The Tell-Riforogenic system (Morocco, Algeria, Tunisia) and the structural heritage of the southern Tethys margin, BSGF. *BSGF-Earth Sciences Bulletin*, 189(2), 10. https://doi.org/10.1051/bsgf/2018009.

Li, X.H., Faure, M., Lin, W., Manatschal, G. (2013). New isotopic constraints on age and magma genesis of an embryonic oceanic crust: The Chenaillet Ophiolite in the Western Alps. *Lithos*, 160–161(1), 283–291. https://doi.org/10.1016/j.lithos.2012.12.016.

Lombardo, B. and Pognante, U. (1982). Tectonic implications in the evolution of the western Alps ophiolite metagabbro. *Ofioliti*, 7(2/3), 371–394.

Lugeon, M. and Argand, E. (1905). Sur les grandes nappes de recouvrement de la zone du Pirmont. *Comptes rendus de l'Académie des Sciences*, 140, 1364–1367.

Maffione, M., van Hinsbergen, D.J.J., de Gelder, G.I.N.O., van der Goes, F.C., Morris, A. (2017). Kinematics of Late Cretaceous subduction initiation in the Neo-Tethys Ocean reconstructed from ophiolites of Turkey, Cyprus, and Syria. *Journal of Geophysical Research: Solid Earth*, 122(5), 3953–3976. https://doi.org/10.1002/2016JB013821.

Manatschal, G. (2004). New models for evolution of magma-poor rifted margins based on a review of data and concepts from West Iberia and the Alps. *International Journal of Earth Sciences*, 93(3), 432–466. https://doi.org/10.1007/s00531-004-0394-7.

Manatschal, G. and Bernoulli, D. (1998). Rifting and early evolution of ancient ocean basins: The record of the Mesozoic Tethys and of the Galicia-Newfoundland margins. *Marine Geophysical Research*, 20(4), 371–381. https://doi.org/10.1023/A:1004459106686.

Manatschal, G. and Bernoulli D. (1999). Architecture and tectonic evolution of nonvolcanic margins: Present-day Galicia and ancient Adria. *Tectonics*, 18(6), 1099–1119.

Manatschal, G. and Müntener, O. (2009). A type sequence across an ancient magma-poor ocean-continent transition: The example of the western Alpine Tethys ophiolites. *Tectonophysics*, 473(1–2), 4–19. https://doi.org/10.1016/j.tecto.2008.07.021.

Manatschal, G., Froitzheim, N., Rubenach, M., Turrin, B.D. (2001). The role of detachment faulting in the formation of an ocean-continent transition: Insights from the Iberia Abyssal Plain. *Geological Society Special Publication*, 187, 405–428. https://doi.org/10.1144/GSL.SP.2001.187.01.20.

Manatschal, G., Sauter, D., Karpoff, A.M., Masini, E., Mohn, G., Lagabrielle, Y. (2011). The Chenaillet Ophiolite in the French/Italian Alps: An ancient analogue for an Oceanic Core Complex? *Lithos*. https://doi.org/10.1016/j.lithos.2010.10.017.

Manatschal, G., Lavier, L., Chenin, P. (2015). The role of inheritance in structuring hyperextended rift systems: Some considerations based on observations and numerical modeling. *Gondwana Research*, 27(1), 140–164. https://doi.org/10.1016/j.gr.2014.08.006.

Masini, E., Manatschal, G., Mohn, G., Unternehr, P. (2012). Anatomy and tectono-sedimentary evolution of a rift-related detachment system: The example of the Err detachment (central Alps, SE Switzerland). *Bulletin of the Geological Society of America*. https://doi.org/10.1130/B30557.1.

Masini, E., Manatschal, G., Mohn, G., Weissert, H. (2013). The Alpine Tethys rifted margins: Reconciling old and new ideas to understand the stratigraphic architecture of magma-poor rifted margins. *Sedimentology*. https://doi.org/10.1111/sed.12017.

Masini, E., Manatschal, G., Tugend, J., Mohn, G., Flament, J.-M. (2014). The tectono-sedimentary evolution of a hyper-extended rift basin: The example of the Arzacq–Mauléon rift system (Western Pyrenees, SW France). *International Journal of Earth Sciences*. https://doi.org/10.1007/s00531-014-1023-8.

McCarthy, A., Falloon, T.J., Sauermilch, I., Whittaker, J.M., Niida, K., Green, D.H. (2020a). Revisiting the Australian-Antarctic ocean-continent transition zone using petrological and geophysical characterization of exhumed subcontinental mantle. *Geochemistry, Geophysics, Geosystems*, 21, e2020GC009040. https://doi.org/10.1029/2020GC009040.

McCarthy, A., Tugend, J., Mohn, G., Candioti, L., Chelle-Michou, C., Arculus, R., Schmalholz, S., Müntener, O. (2020b). A case of Ampferer-type subduction and consequences for the Alps and the Pyrenees. *American Journal of Science*, 320(4), 313–372. https://doi.org/10.2475/04.2020.01.

McCarthy, A., Tugend, J., Mohn, G. (2021). Formation of the Alpine orogen by amagmatic convergence and assembly of previously rifted lithosphere. *Elements: An International Magazine of Mineralogy, Geochemistry, and Petrology*, 17(1), 29–34.

Michard, A., Feinberg, H., El-Azzab, D., Bouybaouene, M., Saddiqi, O. (1992). A serpentinite ridge in a collisional paleomargin setting: The Beni Malek massif, External Rif, Morocco. *Earth and Planetary Science Letters*, 113(3), 435–442.

Michard, A., Frizon de Lamotte, D., Negro, F., Saddiqi, O. (2007). Serpentinite slivers and metamorphism in the External Maghrebides: Arguments for an intracontinental suture in the African paleomargin (Morocco, Algeria). *Rev. Soc. Geol. España*, 20(3–4), 173–185.

Michard, A., Mokhtari, A., Chalouan, A., Saddiqi, O., Rossi, P., Rjimati, E.C. (2014). New ophiolite slivers in the External Rif belt, and tentative restoration of a dual Tethyan suture in the western Maghrebides. *Bulletin de La Societe Geologique de France*, 185(5), 313–328. https://doi.org/10.2113/gssgfbull.185.5.313; https://doi.org/10.1016/0012-821X(92)90144-K.

Miranda-Avilés, R., Bourrouilh, R., Nava-Sánchez, E.H., Puy-Alquiza, M.J., Bourrouilh-Le Jan, F. (2005). Analyse comparée de bassins sédimentaires transtensionnels : le bassin de Santa Rosalia (Basse Californie du Sud, Mexique) et de Mendibelza (Pyrénées, France). *Estudios geológicos*, 61(3–6), 161–176.

Mohn, G., Manatschal, G., Müntener, O., Beltrando, M., Masini, E. (2010). Unravelling the interaction between tectonic and sedimentary processes during lithospheric thinning in the Alpine Tethys margins. *International Journal of Earth Sciences*. https://doi.org/10.1007/s00531-010-0566-6.

Mohn, G., Manatschal, G., Masini, E., Müntener, O. (2011). Rift-related inheritance in orogens: A case study from the Austroalpine nappes in Central Alps (SE-Switzerland and N-Italy). *International Journal of Earth Sciences*. https://doi.org/10.1007/s00531-010-0630-2.

Mohn, G., Manatschal, G., Beltrando, M., Masini, E., Kusznir, N. (2012). Necking of continental crust in magma-poor rifted margins: Evidence from the fossil Alpine Tethys margins. *Tectonics*. https://doi.org/10.1029/2011TC002961.

Mohn, G., Manatschal, G., Beltrando, M., Haupert, I. (2014). The role of rift-inherited hyper-extension in Alpine-type orogens. *Terra Nova*. https://doi.org/10.1111/ter.12104.

Mohn, G., Karner, G.D., Manatschal, G., Johnson, C.A. (2015). Structural and stratigraphic evolution of the Iberia-Newfoundland hyper-extended rifted margin: A quantitative modelling approach. *Geological Society Special Publication*. https://doi.org/10.1144/SP413.9.

Mouthereau, F., Filleaudeau, P.Y., Vacherat, A., Pik, R., Lacombe, O., Fellin, M.G., Castelltort, S., Christophoul, F., Masini, E. (2014). Placing limits to shortening evolution in the Pyrenees: Role of margin architecture and implications for the Iberia/Europe convergence. *Tectonics*, 33(12), 2283–2314. https://doi.org/10.1002/2014TC003663.

Muñoz, J.A. (2002). The Pyrenees. *The Geology of Spain*, 370–385.

Müntener, O., Hermann, J., Trommsdorff, V. (2000). Cooling history and exhumation of lower-crustal granulite and upper mantle (Malenco, Eastern Central Alps). *Journal of Petrology*, 41(2), 175–200. https://doi.org/10.1093/petrology/41.2.175.

Müntener, O., Manatschal, G., Desmurs, L., Pettke, T. (2010). Plagioclase peridotites in ocean-continent transitions: Refertilized mantle domains generated by melt stagnation in the shallow mantle lithosphere. *Journal of Petrology*, 51(1–2), 255–294. https://doi.org/10.1093/petrology/egp087.

Pantet, A., Epard, J.L., Masson, H. (2020). Mimicking Alpine thrusts by passive deformation of synsedimentary normal faults: A record of the Jurassic extension of the European margin (Mont Fort nappe, Pennine Alps). *Swiss Journal of Geosciences*, 113(1). https://doi.org/10.1186/s00015-020-00366-2.

Peron-Pinvidic, G., Manatschal, G., Osmundsen, P.T. (2013). Structural comparison of archetypal atlantic rifted margins: A review of observations and concepts. *Marine and Petroleum Geology*. https://doi.org/10.1016/j.marpetgeo.2013.02.002.

Petri, B. (2014). Formation et exhumation des granulites permiennes, établir les conditions pré-rift et déterminer l'histoire de l'exhumation syn-rift. PhD Thesis, University of Strasbourg.

Petri, B., Mohn, G., Štípská, P., Schulmann, K., Manatschal, G. (2016). The Sondalo gabbro contact aureole (Campo unit, Eastern Alps): Implications for mid-crustal mafic magma emplacement. *Contributions to Mineralogy and Petrology*. https://doi.org/10.1007/s00410-016-1263-7.

Petri, B., Mohn, G., Skrzypek, E., Mateeva, T., Galster, F., Manatschal, G. (2017). U–Pb geochronology of the Sondalo gabbroic complex (Central Alps) and its position within the Permian post-Variscan extension. *International Journal of Earth Sciences*. https://doi.org/10.1007/s00531-017-1465-x.

Petri, B., Skrzypek, E., Mohn, G., Mateeva, T., Robion, P., Schulmann, K., Manatschal, G., Müntener, O. (2018). Mechanical anisotropies and mechanisms of mafic magma ascent in the middle continental crust: The Sondalo magmatic system (N Italy). *GSA Bulletin*, 130(1–2), 331–352. https://doi.org/10.1130/B31693.1.

Petri, B., Duretz, T., Mohn, G., Schmalholz, S.M., Karner, G.D., Müntener, O. (2019). Thinning mechanisms of heterogeneous continental lithosphere. *Earth and Planetary Science Letters*, 512, 147–162. https://doi.org/10.1016/j.epsl.2019.02.007.

Picazo, S., Müntener, O., Manatschal, G., Bauville, A., Karner, G., Johnson, C. (2016). Mapping the nature of mantle domains in Western and Central Europe based on clinopyroxene and spinel chemistry: Evidence for mantle modification during an extensional cycle. *Lithos*, 266–267(September 2016), 233–263. https://doi.org/10.1016/j.lithos.2016.08.029.

Reston, T.J. (2007). The formation of non-volcanic rifted margins by the progressive extension of the lithosphere: The example of the West Iberian margin. *Geological Society Special Publication*, 282, 77–110. https://doi.org/10.1144/SP282.5.

Reston, T.J. (2009). The structure, evolution and symmetry of the magma-poor rifted margins of the North and Central Atlantic: A synthesis. *Tectonophysics*, 468(1–4), 6–27. https://doi.org/10.1016/j.tecto.2008.09.002.

Reston, T.J., Krawczyk, C.M., Klaeschen, D. (1996). The S reflector west of Galicia (Spain): Evidence from prestack depth migration for detachment faulting during continental breakup. *Journal of Geophysical Research B: Solid Earth*, 101(4), 8075–8091. https://doi.org/10.1029/95jb03466.

Ribes, C., Manatschal, G., Ghienne, J.F., Karner, G.D., Johnson, C.A., Figueredo, P.H., Incerpi, N., and Epin, M.E. (2019). The syn-rift stratigraphic record across a fossil hyper-extended rifted margin: The example of the northwestern Adriatic margin exposed in the Central Alps. *International Journal of Earth Sciences*, 108(6), 2071–2095.

Ribes, C., Petri, B., Ghienne, J.F., Manatschal, G., Galster, F., Karner, G.D., Figueredo, P.H., Johnson C.A., Karpoff, A.M. (2020). Tectono-sedimentary evolution of a fossil ocean-continent transition: Tasna nappe, central Alps (SE Switzerland). *Bulletin*, 132(7–8), 1427–1446.

Ricou, L.E. (1977). Le croissant ophiolitique péri-arabe. Une ceinture de nappes mises en place au Crétacé supérieur. *Rev. Géogr. Phys. Géol. Dynam.*, 327–350.

Rosenbaum, G., Lister, G.S., Duboz, C. (2002). Relative motions of Africa, Iberia and Europe during Alpine orogeny. *Tectonophysics*, 359(1–2), 117–129.

Saspiturry, N., Allanic, C., Razin, P., Issautier, B., Baudin, T., Lasseur, E., Serrano, O., Leleu, S. (2020). Closure of a hyperextended system in an orogenic lithospheric pop-up, Western Pyrenees: The role of mantle buttressing and rift structural inheritance. *Terra Nova*, 32(4), 253–260.

Sauter, D., Cannat, M., Rouméjon, S., Andreani, M., Birot, D., Bronner, A., Brunelli, D., Carlut, J., Delacour, A., Guyader, V., MacLeod, C.J., Manatschal, G., Mendel, V., Ménez, B., Pasini, V., Ruellan, E., Searle, R. (2013). Continuous exhumation of mantle-derived rocks at the Southwest Indian Ridge for 11 million years. *Nature Geoscience*, 6(4), 314–320. https://doi.org/10.1038/ngeo1771.

Schardt, H. (1893). Sur l'origine des Préalpes romandes. *Eclogae Geol. Helv.*, 4, 129–142.

Schmid, S.M., Pfiffner, O.A., Froitzheim, N., Schönborn, G., Kissling, E. (1996). Geophysical-geological transect and tectonic evolution of the Swiss-Italian Alps. *Tectonics*, 15(5), 1036–1064. https://doi.org/10.1029/96TC00433.

Schmid, S.M., Fügenschuh, B., Kissling, E., Schuster, R. (2004). Tectonic map and overall architecture of the Alpine orogen. *Eclogae Geologicae Helvetiae*, 97(1), 93–117. https://doi.org/10.1007/s00015-004-1113-x.

Schmid, S.M., Scharf, A., Handy, M.R., Rosenberg, C.L. (2013). The Tauern Window (Eastern Alps, Austria): A new tectonic map, with cross-sections and a tectonometamorphic synthesis. *Swiss Journal of Geosciences*, 106(1), 1–32.

Schmid, S.M., Kissling, E., Diehl, T., van Hinsbergen, D.J.J., Molli, G. (2017). Ivrea mantle wedge, arc of the Western Alps, and kinematic evolution of the Alps–Apennines orogenic system. *Swiss Journal of Geosciences*, 110(2), 581–612. https://doi.org/10.1007/s00015-016-0237-0.

Şengör, A.M.C. (1990). Plate tectonics and orogenic research after 25 years: A Tethyan perspective. *Earth Science Reviews*, 27(1–2). https://doi.org/10.1016/0012-8252(90)90002-D.

Şengör, A.M.C. (1991). Plate tectonics and orogenic research after 25 years: Synopsis of a Tethyan perspective. *Tectonophysics*, 187(1–3). https://doi.org/10.1016/0040-1951(91)90427-T.

Sibuet, J.-C., Ryan, W.B.F., (1979). Site 398: Evolution of the west Iberian passive continental margin in the framework of the early evolution of the North Atlantic Ocean. *Initial Reports of the Deep Sea Drilling Project*, 47 Pt. 2, (October). https://doi.org/10.2973/dsdp.proc.47-2.138.1979.

Skogseid, J. (2001). Volcanic margins: Geodynamic and exploration aspects. *Marine and Petroleum Geology*, 18(4), 457–461. https://doi.org/10.1016/S0264-8172(00)00070-2.

Souquet, P., Debroas, E.J., Boirie, J.M., Pons, P., Fixari, G., Roux, J.C., Dol, J., Thieuloy, J-P., Bonnemaison, M., Manivit, H., Peybernes, B. (1985). Le groupe du Flysch noir (albo-cénomanien) dans les Pyrénées. *Bull Cent. Rech Exlpo-Prod Elf-Aquitaine Pau*, 9, 183–252.

Steinmann, G. (1905), Geologische Beobachtungen in den Alpen, II. Die Schardtsche Ueberfaltungstheorie und die geologische Bedeutung der Tiefseeabsätze und der ophiolithischen Massengesteine. *Berichte der Naturforschenden Gesellschaft zu Freiburg im Breisgau*, 16, 18–67.

Stern, R.J., Reagan, M., Ishizuka, O., Ohara, Y., Whattam, S. (2012). To understand subduction initiation, study forearc crust: To understand forearc crust, study ophiolites. *Lithosphere*, 4(6), 469–483. https://doi.org/10.1130/L183.1.

Suter, G. (1980). Carte géologique de la chaine rifaine à 1/500 000. *Notes et Mdm. Serv. Geol. Maroc*, 245.

Sutra, E., Manatschal, G., Mohn, G., Unternehr, P. (2013). Quantification and restoration of extensional deformation along the Western Iberia and Newfoundland rifted margins. *Geochemistry, Geophysics, Geosystems*. https://doi.org/10.1002/ggge.20135.

Teixell, A. (1998). Crustal structure and orogenic material budget in the west central Pyrenees. *Tectonics*, 17(3), 395–406.

Teixell, A., Labaume, P., Ayarza, P., Espurt, N., de Saint Blanquat, M., Lagabrielle, Y. (2018). Crustal structure and evolution of the Pyrenean-Cantabrian belt: A review and new interpretations from recent concepts and data. *Tectonophysics*, 724, 146–170.

Thouvenot, F., Paul, A., Frechet, J., Béthoux, N., Jenatton, L., Guiguet, R. (2007). Are there really superposed Mohos in the southwestern Alps? New seismic data from fan-profiling reflections. *Geophysical Journal International*, 170(3), 1180–1194.

Tribuzio, R., Garzetti, F., Corfu, F., Tiepolo, M., Renna, M.R. (2016). U–Pb zircon geochronology of the Ligurian ophiolites (Northern Apennine, Italy): Implications for continental breakup to slow seafloor spreading. *Tectonophysics*, 666, 220–243.

Tricart, P. and Lemoine, M. (1983). Serpentinite oceanic bottom in South Queyras ophiolites (French Western Alps): Record of the incipient oceanic opening of the Mesozoic Ligurian Tethys. *Eclogae Geologicae Helvetiae*, 76(3), 611–629.

Trümpy, R. (1975). Penninic-Austroalpine boundary in the Swiss Alps: A presumed former continental margin and its problems. *American Journal of Science*, 275(A), 209–238.

Tucholke, B.E. and Sibuet, J.-C. (2007). Leg 210 synthesis: Tectonic, magmatic, and sedimentary evolution of the Newfoundland-Iberia Rift. *Proceedings of the Ocean Drilling Program, 210 Scientific Results*, 210(November 2006). https://doi.org/10.2973/odp.proc.sr.210.101.2007.

Tugend, J., Manatschal, G., Kusznir, N.J., Masini, E., Mohn, G., Thinon, I. (2014). Formation and deformation of hyperextended rift systems: Insights from rift domain mapping in the Bay of Biscay-Pyrenees. *Tectonics*. https://doi.org/10.1002/2014TC003529.

Tugend, J., Manatschal, G., Kusznir, N.J., Masini, E. (2015a). Characterizing and identifying structural domains at rifted continental margins: Application to the Bay of Biscay margins and its Western Pyrenean fossil remnants. *Geological Society Special Publication*, 413(1), 171–203. https://doi.org/10.1144/SP413.3.

Tugend, J., Manatschal, G., Kusznir, N.J. (2015b). Spatial and temporal evolution of hyperextended rift systems: Implication for the nature, kinematics, and timing of the Iberian-European plate boundary. *Geology*, 43(1), 15–18. https://doi.org/10.1130/G36072.1.

Tugend, J., Gillard, M., Manatschal, G., Nirrengarten, M., Harkin, C., Epin, M.E., Sauter, D., Autin, J., Kusznir, N., McDermott, K. (2020). Reappraisal of the magma-rich versus magma-poor rifted margin archetypes. *Geological Society, London, Special Publications*, 476(1), 23–47. https://doi.org/10.1144/SP476.9.

Ulrich, M., Picard, C., Guillot, S., Chauvel, C., Cluzel, D., Meffre, S. (2010). Multiple melting stages and refertilization as indicators for ridge to subduction formation: The New Caledonia ophiolite. *Lithos*, 115(1–4), 223–236. https://doi.org/10.1016/j.lithos.2009.12.011.

Van Hinsbergen, D.J., Torsvik, T.H., Schmid, S.M., Maţenco, L.C., Maffione, M., Vissers, R.L., Gürer, D., Spakman, W. (2020). Orogenic architecture of the Mediterranean region and kinematic reconstruction of its tectonic evolution since the Triassic. *Gondwana Research*, 81, 79–229. https://doi.org/10.1016/j.gr.2019.07.009.

Velasque, P.C., Ducasse, L., Muller, J., Scholten, R. (1989). The influence of inherited extensional structures on the tectonic evolution of an intracratonic chain: The example of the Western Pyrenees. *Tectonophysics*, 162(3–4), 243–264.

Whitmarsh, R.B., Manatschal, G., Minshull, T.A. (2001). Evolution of magma-poor continental margins from rifting to seafloor spreading. *Nature*, 413(6852), 150–154. https://doi.org/10.1038/35093085.

Wilson, R.C.L., Manatschal, G., Wise, S. (2001). Rifting along non-volcanic passive margins: Stratigraphic and seismic evidence from the mesozoic successions of the Alps and Western Iberia. *Geological Society Special Publication*, 187, 429–452. https://doi.org/10.1144/GSL.SP.2001.187.01.21.

4

What Can We Learn from Marine Geophysics to Study Rifted Margins?

Julia AUTIN[1] and Louise WATREMEZ[2]

[1] *Université de Strasbourg, CNRS, Institut Terre et Environnement de Strasbourg, UMR 7063, France*

[2] *Université de Lille, CNRS, Université Littoral Côte d'Opale, IRD, UMR 8187 - LOG, Laboratoire d'Océanologie et de Géosciences, Lille, France*

4.1. Geophysical methods

Present-day continental rifted margins are mostly located under water. Access to rocks and structures in the offshore is possible with drillings or dredges; however, the acquisition of such data types is very expensive and time-consuming, and only leads to local constraints. Consequently, such scarce and punctual datasets do not allow us to obtain a global picture of the structure of rifted margins. To acquire such knowledge, researchers may rather use indirect data, such as geophysical methods.

In the following, we list and briefly describe the main geophysical methods used when studying offshore rifts and rifted margins. If interested, and for further details and explanations, the reader is referred to contributions specifically dedicated to the methods. A list of references is proposed at the end of this chapter.

For a color version of all the figures in this chapter, see www.iste.co.uk/peronpinvidic/margins1.zip.

Continental Rifted Margins 1,
coordinated by Gwenn PERON-PINVIDIC. © ISTE Ltd 2022.

4.1.1. *Seismic imaging*

Several methods are based on the properties of seismic waves, that is, acoustic waves, as they propagate through the Earth's layers. However, the observation scale varies with the method. For example, with increasing depth of observation, users can choose between sub-bottom profilers, sparker, boomer, single-channel seismic, multichannel deep seismic, or seismic refraction data. Here, we focus on the deep-penetrating methods, allowing for crustal scale observations: multichannel deep seismic and seismic refraction.

The seismic energy propagates from the source, down to the seafloor and geological layers. It is partly reflected or refracted at each geological boundary (interface), depending on the layers' impedance contrast (Figure 4.1). This impedance contrast corresponds to a change in the physical properties (seismic velocities and densities). However, it is often difficult to determine why such changes occur: is it a lithological change (a different kind of rock media), a discontinuity in rock media (e.g. faults and intrusions), a change in fluid content in the material and so on? Moreover, the imaging can generate artifacts, that is, elements in the underground image that are not real, but are inherent to the behavior of the acoustic waves (lateral echoes, multiple arrivals, slope effects, diffraction hyperbolas, background noise, etc.). Thus, it is of primary importance to separate the observation of the seismic data from their interpretation. Observation should reach a consensus among the researchers, whereas interpretation is more subjective, and is ideally based on discussion and comparison with other datasets.

4.1.1.1. *Multichannel deep seismic*

4.1.1.1.1. Acquisition

The acquisition of offshore deep seismic is done in either 2D (along the path of the ship) or 3D (with multiple receiver arrays along and laterally to the ship path). The geometry of the seismic source and receivers is set to maintain a constant distance between the source and each receiver (offset, Figure 4.1a). This favors the observation of reflected wave arrivals instead of refracted wave arrivals, which usually occur at larger offsets. The receivers are aligned along one (2D) or several streamers (3D), with lengths up to 10–12 km, for the best configurations. The source is chosen depending on the target. Most of the time, scientists make a compromise between penetration (large source volume) and resolution (smaller sources), as they need both energy to penetrate down to the crustal basement and sufficient resolution in the sediments (~10 m).

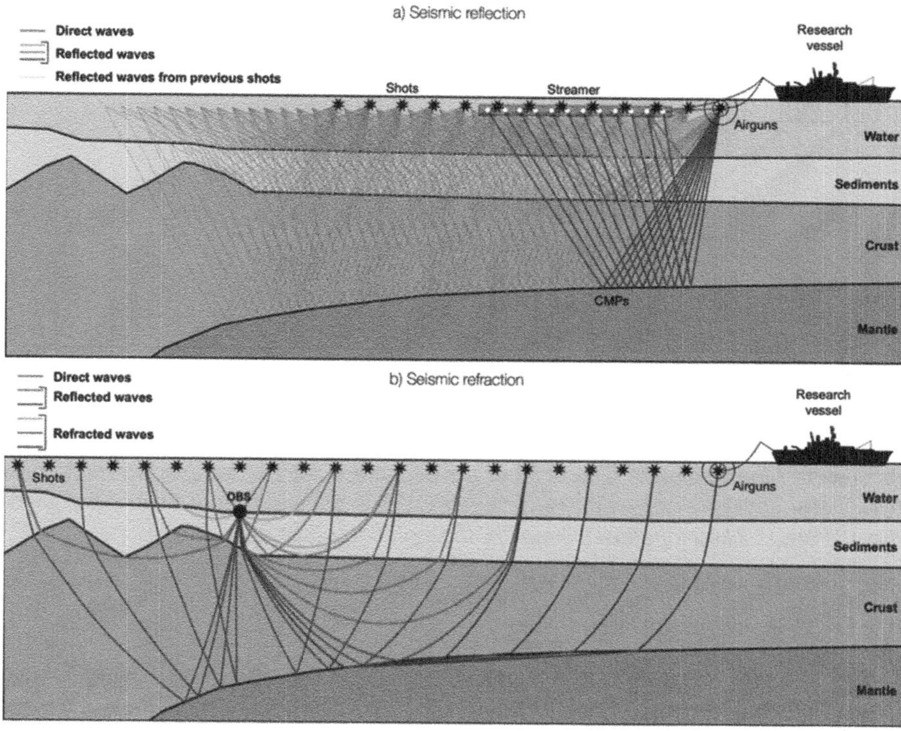

Figure 4.1. *Acquisition principles of seismic reflection and refraction. The research vessel's outline belongs to a French ship, the N/O L'Atalante. a) For seismic reflection, the constant source–receiver distance allows the recording of near-vertical reflected waves and the multiple receivers and shots sample common mid-points (CMPs). See the text for further explanations. b) Seismic refraction is based on increasing source–receiver distances, creating a wider incidence angle and allowing deeper refracted waves to be recorded as the offsets increase*

The quality of the final seismic profiles highly depends on the fold, that is, the theoretical number of times a given point of the seafloor (CMP: common mid-point) is sampled by a reflection and recorded by a receiver (Figure 4.2). The greater the fold, the better the signal to noise ratio. The fold depends on the number of receivers, their spacing and the shot interval distance (e.g. Yilmaz 2001).

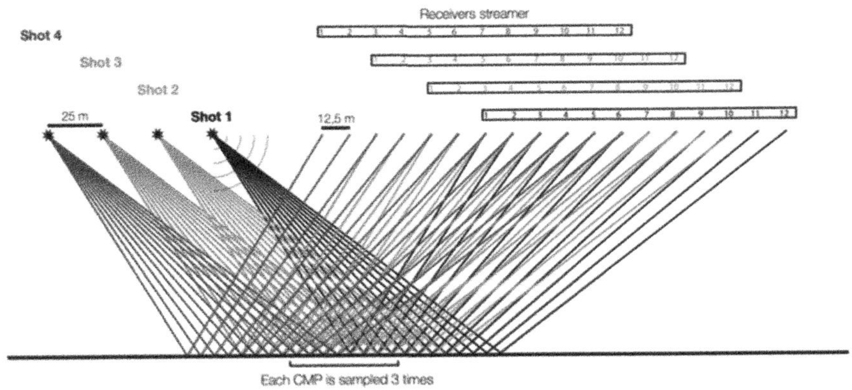

Figure 4.2. *Common mid-points (CMP) acquisition. Each shot is recorded by the 12 receivers of the streamer. One point of the seafloor (CMP) is sampled several times by several shot/receiver pairs. Here, each CMP is sampled three times. The three records are stacked together to improve the signal-to-noise ratio*

4.1.1.1.2. Data processing

Seismic processing aims to reduce the signal-to-noise ratio. Noise can be related to known behavior of the seismic waves (multiples, diffraction hyperbola, etc., see Figure 4.3) or to disturbances during the acquisition (wind, currents, human activity, internal Earth activity, source or receiver issues, etc.).

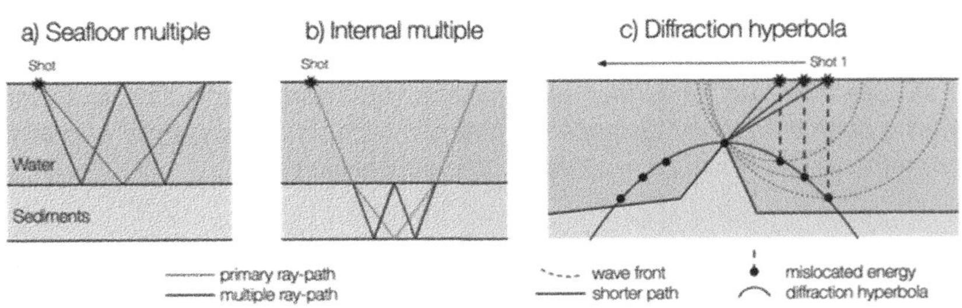

Figure 4.3. *Noises related to seismic waves behavior. a) The seafloor multiple results from waves reflected first on the seafloor, then under the sea surface, and finally on the seafloor again. It produces a ghost of the seafloor with arrival-times doubled compared to the seafloor arrivals. b) Similar behavior within a geological layer can also produce internal multiples. c) Diffraction hyperbola result from diffraction points, such as faults: the seismic energy is mislocated because the diffraction point is reached faster laterally than the actual interface, vertically (the mislocated energy forms a hyperbola)*

The quality control of the traces makes it possible to select or improve the seismic records (filtering and offset- or time-dependent gain applied to the data, either as shot gathers or receiver gathers). The source deconvolution aims to reduce the length of the source wavelet to its minimum, in order to obtain a sharper signal and improve the resolution. Traces are sorted as common midpoint gathers (CMP), that is, all of the traces recording a theoretical point of the seafloor (Figure 4.2). CMP gathers display the arrival times at a common spatial point as a function of the offset (source–receiver distance) and shows a downward hyperbolic shape (Figure 4.4a). Normal move out (NMO) consists of the correction of the arrival times with a theoretical velocity (stacking velocity), in order to align each arrival at the same level (Figure 4.4b). Such a process assumes that the reflector corresponding to those arrivals is horizontal. Stacking consists of summing the energy recorded by each NMO-corrected trace to reinforce the signal in a given point of the underground, thus improving the signal-to-noise ratio (Figure 4.4c). It is particularly efficient to reduce multiples as their arrival-times are corrected with a higher stacking velocity than needed, resulting in downward alignment after NMO, and thus avoiding the stacking of the multiple energy.

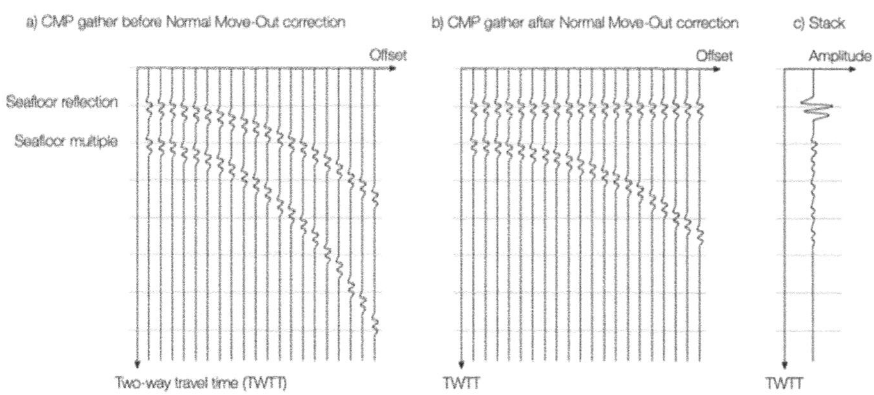

Figure 4.4. *Principle of normal move-out (NMO) correction before stacking. a) CMP gather before NMO, b) CMP gather after NMO and c) stacked CMPs*

After stacking, the seismic section is obtained by juxtaposing the CMP stacked traces along the profile. Further processing is operated to improve the final seismic section. For instance, the migration of the data makes it possible to remove diffraction hyperbolas (Figure 4.3c). The aim of the migration is to restitute the energy at the apex of the hyperbola, in order to place the signal at the right geometrical point (Figure 4.5). Slope migration corrects the effect of non-horizontal reflections: in this case, the theoretical CMP is not focused on one point on the reflector but is scattered around it.

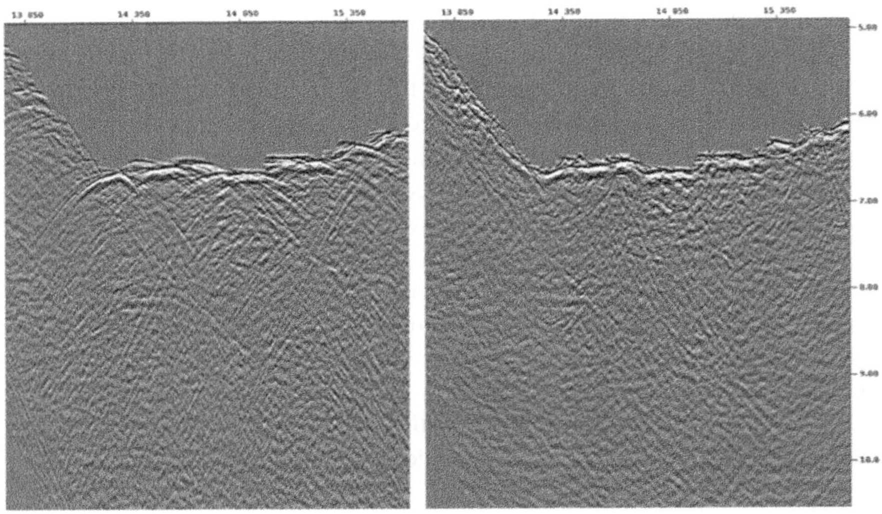

Figure 4.5. *Effect of migration at constant velocity (water velocity, 1500 m/sec) on a stacked section. Left: stacked section. Right: migrated section (source: cruise report, Leroy and Cannat (2014))*

Finally, time-depth conversion is useful to assess the real geometry of the cross-section. It needs a thorough estimation of the propagation velocities and as such, should be considered as a model rather than a result.

We have presented the basic steps of processing. Of course, advanced processing includes more thorough analyses depending on the targets and specificities of the acquisition. For example, more advanced antimultiples will be applied if the target is deeper than the multiple time arrival, Deep Move Out (DMO) is applied where steep slopes are present, and Pre-Stack Depth Migration (PSDM) is a more elaborate way to convert data from time to depth. We encourage the reader to study dedicated literature for further details (e.g. Yilmaz 2001; Dondurur 2018).

4.1.1.1.3. Observation and interpretation

To allow an objective interpretation of seismic data, we follow three main steps during their study: (1) the knowledge of the data acquisition and processing informs us about the quality of the signal-to-noise ratio improvement, the elimination of artifacts and the accuracy of the eventual depth conversion, (2) seismic observation makes it possible to describe the image without preconception on the geological background, (3) seismic interpretation makes it possible to propose a possible architecture of the rifted margin through a scientific argumentation, which will remain theoretical in the absence of *in situ* data.

It is worth saying that beginners should acquire a good understanding of the processing steps required for the production of seismic profiles. Indeed, the observer needs to know whether the seismic section is in time or depth before considering the geometrical aspects (angles, thicknesses, volumes, etc.), if multiples still appear, if a migration was applied and so on. This makes it possible to estimate the degree of confidence in the observations.

Then, seismic observation consists of defining seismic facies (i.e. packages of reflections that display a common seismic pattern) relative to seismic units (Figure 4.6). It can be defined using criteria such as amplitude, frequency, continuity, and also with geometrical observations describing the termination of seismic reflections (onlap, dowlap, toplap, truncation, etc.) or their internal geometry (e.g. parallel, oceanward or continentward dipping, wedge-shaped, divergent, concave, folded reflections, affected by major or minor faults). Nevertheless, the internal geometry of a given seismic unit can vary along a seismic line. The goal is to achieve a coherent description of the seismic facies throughout the entire profile and then the whole seismic dataset.

Finally, the actual interpretation begins. One should be able to recognize a logical pattern in deposition of the sedimentary seismic units from base to top, in order to constrain a depositional and tectonic history. If any well or dredge data are available, they bring further information on the nature and age of the rocks. This makes it possible to point out possible gaps in the sedimentary column. For example, an observed seismic unit in the proximal domain can be missing in the distal domain, due to erosion (erosional surfaces could be observed), sediment bypass, non-deposition (e.g. due to delayed subsidence), or changes in the deposited material at a given time. Thick sand deposits in the proximal domain can evolve into thin condensed intervals in the distal domain. Hence, the evolution of the sedimentation (thicknesses, bypass, erosion, etc.) gives an indication of the available accommodation space for sedimentation and the distance to sediment sources during rifting, and thus of the subsidence history.

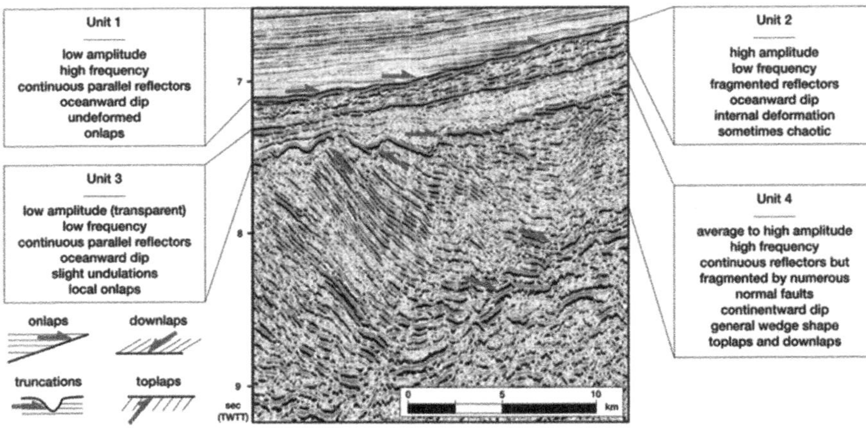

Figure 4.6. *Example of seismic facies description. The red arrows highlight the termination of reflectors in each unit*

We must carefully study the sedimentary units and also the basement seismic units. Indeed, it is sometimes possible to recognize distinct basement seismic facies. For example, an oceanic basement can be distinguished from a continental basement by the strong top reflectivity, clear Moho reflection (crust typically about 2 s TWTT thick) and transparent facies of the magmatic oceanic crust (see the next section). Nevertheless, it will be difficult to distinguish between (1) a thin continental crust that is for instance deformed, highly intruded and covered by magmatic additions, (2) a zone of exhumed mantle that may also be highly intruded and covered by magmatic additions or (3) a slightly tectonised magmatic oceanic crust (also see Chapter 2). In this case, the lack of the basal sedimentary units could help to determine whether the basement is inherited (previous continental surface) or more recently exposed to the surface (exhumation of deeper crust or mantle, oceanic spreading). The presence or absence of Moho reflections is an indicator of the nature of the basement. Indeed, exhumed mantle does not often display Moho reflections, due to progressive increase of the densities and velocities (no impedance contrast) with depth. Lower crust or oceanic crust should display a basal reflection (sharp impedance contrast).

4.1.1.2. *Seismic refraction*

Seismic refraction makes it possible to use the arrival times from different seismic phases (refracted and reflected, Figure 4.1b) in the geological layers, in order to build a velocity model of the subsurface, usually based on P-wave velocities. S-wave arrivals can also be used when P to S conversions occur and such arrivals are recorded (S-waves do not propagate in the water). Different modeling methods exist, showing different styles of results (smooth or layered models) and using the datasets in different ways (first arrivals or more). It is important to understand how a seismic ray travels in the subsurface to be able to fathom the quality of a model, assess which areas are well-resolved (or not), and finally, have an idea of how to interpret these (usually) colorful images.

4.1.1.2.1. Principle

Seismic refraction, also called "refraction and wide-angle reflection" or just "wide-angle seismics" (R-WAR or WAS), uses the ray-tracing theory to assess seismic ray-paths in the subsurface, compute source-to-receiver travel-times and deduce the seismic velocities, as well as the depth and geometry of the geological interfaces. The simplistic theory, as usually taught, assumes simple layers of constant velocities, with head-waves propagating along the interfaces when the seismic velocities increase with depth. However, the head-wave corresponds to a very specific case, where the wave front hits the geological interface at an incidence angle, such as when the refracted angle is equal to 90° (and refracts along the interface). Most of the seismic energy is either refracted in the layer or reflected at the interface. Thus, head-waves show very low energy and the clear arrivals on the seismic refraction sections mostly belong to refracted waves (within layers) and wide-angle reflections. In reality, the seismic velocities in a geological layer tend to gradually increase with depth (vertical velocity gradient) and may also vary laterally (because of different compaction in the sediments, lateral variations in the nature of the geological structures, further lithostatic pressure increase in the crust or mineralogical changes with depth, etc.). We can imagine a geological layer as an infinity of very thin constant-velocity-layers where the seismic ray progressively refracts. Thus, the refracted waves usually turn within a geological layer until it comes back up to the receiver. Consequently, the understanding of the recorded travel-times involves the modeling of these vertical and lateral variations through ray-tracing, using computer software (see the data processing section).

4.1.1.2.2. Acquisition

The acquisition of seismic refraction is achieved by recording seismic shots on multiple instruments, either along the profile (2D) or on a grid (3D). Seismic rays travel deeper in the Earth structures as the source–receiver distance increases (Figure 4.1b). Instruments are deployed from a research vessel at the beginning of the survey and are recovered once the seismic shots are done and recorded by the instruments. Seismic records are organized as receiver-gathers, with one trace per shot, and sorted by offsets (2D) or shot numbers (3D).

The data acquired during the scientific cruise are pre-processed in order to be ready for picking. The main pre-processing step consists of instrument relocation. Indeed, ocean-bottom seismometers (OBS) are deployed from the sea-surface, sink and drift, following currents down to the seafloor. As GPS signal does not transmit underwater, the exact position of the instrument has to be deduced by triangulation, with the help of the arrival-times of the direct waves, which propagated from the source to the instrument, in the water. Note that water depth and acoustic velocities must be known. An instrument that is properly relocated shows a symmetrical seismic record at short offsets (water-wave arrivals). The relocation process is essential to obtain accurate travel-times, and therefore, velocity modeling.

Some data filtering (e.g. low frequencies on hydrophone data) may be applied to the data, but it is best to pick unfiltered data to avoid travel-time shifts due to the processing. Light bandpass filters, offset-dependent gain and deconvolution may be applied to help see far-offset arrivals (be careful, the filter may lead to a shift of the arrival-times) or data display.

4.1.1.2.3. Data processing

Seismic refraction processing aims to build a velocity model of the crustal structures, down to the uppermost mantle, when the source–receiver distances are large enough. The velocity model should account for the observed travel-times on the seismic records. These arrival times are compared to computed arrival times, through ray-tracing in the model (Figure 4.7, middle and lower panels).

Picking on the seismic records provides the seismic travel times used for the modeling. Depending on the modeling method, the data is picked following different strategies. Please note that the method and software examples mentioned below consist of some of the available software used by the authors of this chapter, but do not consist of an exhaustive list of the existing software. The main seismic refraction modeling methods are (see Figure 4.8 for data picking and model examples along the same profile):

1) First arrival-time tomography (e.g. FAST, Zelt and Barton 1998): this inversion method only takes first arrival times into account and uses a velocity grid as input. The output is an updated smooth velocity grid, accounting for the

travel-times. This method is interesting for onboard processing along 2D lines, for example, or 3D modeling.

2) Joint first-arrival time and wide-angle reflection tomography (e.g. Tomo2D, Korenaga *et al.* 2000): this other inversion method also uses first arrival-times (refracted arrivals), but a wide-angle picked dataset can also be added to the modeling, in order to obtain the depth and morphology of an interface (typically, the Moho discontinuity). This scheme may also be used for layered modeling, adding velocity jumps at interfaces and modeling multiple interfaces, layer by layer (e.g. Sallarès *et al.* 2011). The result is a velocity grid (more or less smooth) with one or several interfaces (if modeled).

3) Forward layered modeling (e.g. Rayinvr, Zelt and Smith 1992): This method is used for 2D modeling, layer by layer, usually from top to bottom. It is more time consuming than the tomography methods, but allows for slight geological interpretation during the modeling process. It allows for distinguishing which layer each refracted ray turned, and which interface each reflected ray hit. The aim of this method is to explain the different seismic phases at best, first and secondary arrivals, using the least model parameters (number of layers and interface and velocity nodes) possible (Zelt 1999). It is possible to model the refracted arrival-times that occur at secondary arrivals, as well as S-wave arrivals, multiples and so on. The result consists of a layered velocity model, with sharp velocity discontinuities at model interfaces, generally related to lithological changes.

Figure 4.8 shows an example of data, picking and the output models for Tomo2D and Rayinvr along the same profile, for comparison. The tomography model (Figure 4.8c) was published five years prior to the forward layered model (Figure 4.8e). It consists of preliminary results for the seismic refraction data. However, the authors did not have access to any coincident multi-channel deep seismic data and assumed a very simple starting model. Furthermore, this first processing did not make it possible to observe some very weak P_n arrivals (yellow picks in Figure 4.8d). The layered forward model presents the final results of this study. The authors had access to high-quality coincident multichannel deep seismic data, showing some insights on the sediment layers and basement morphology, as well as clear mid-crustal and Moho reflectivity. To a first order, the two models show similar velocities and Moho depths. However, there are some differences that should be explained. The Moho depths are shallower in the tomography model than in the forward layered model in the south-easternmost part of the model. The picks used as P_mP reflections for the tomography were probably arrivals from the strong mid-crustal reflection, rather than the Moho. The velocity discrepancies between the two models are due to the lack of knowledge of the shallow structures, as well as some smearing. For example, the velocities of the tomography model at a model distance of 80 km are too high compared to the forward layered model. Tomography will compensate by using velocities that are slightly too low underneath, allowing them to fit the picked travel-times. Thus, this comparison shows the importance of data complementarity to help the modeling.

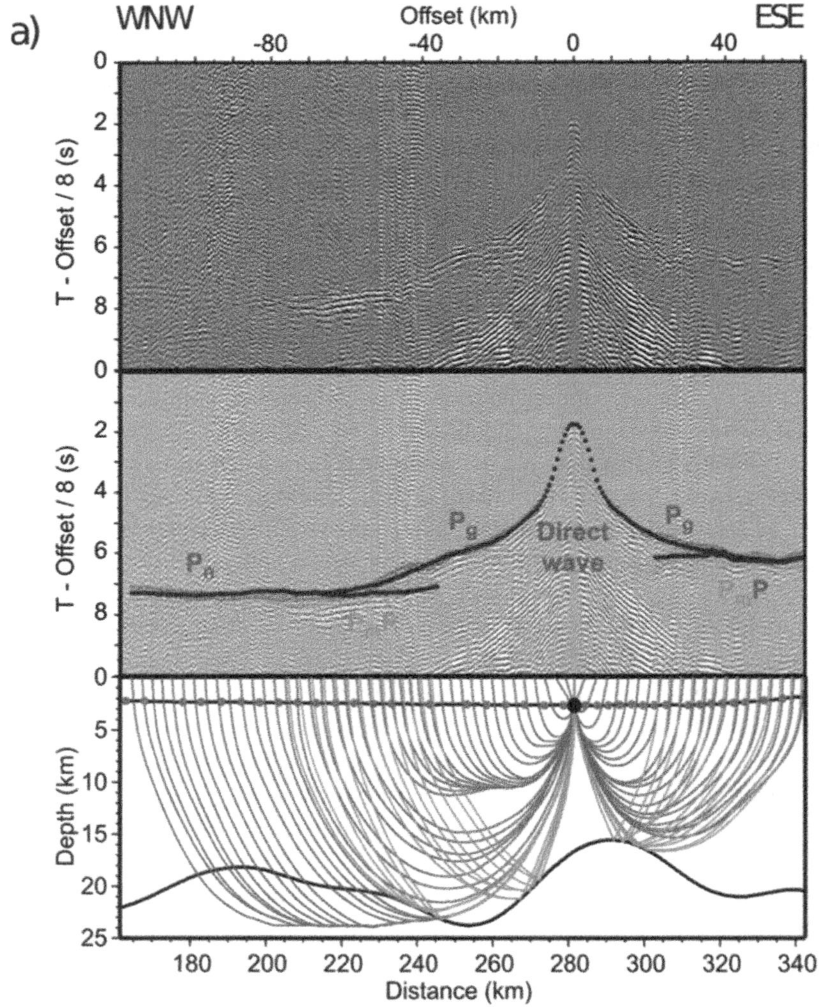

Figure 4.7a. *Seismic record (top panel), picks, computed arrival times (middle panel) and rays (lower panel) for a tomography model (Orphan Basin). All first arrival-times are picked as a unique phase and wide-angle reflections at the Moho are picked separately to model the depth of a reflector (black line) (source: Watremez* et al. *(2015))*

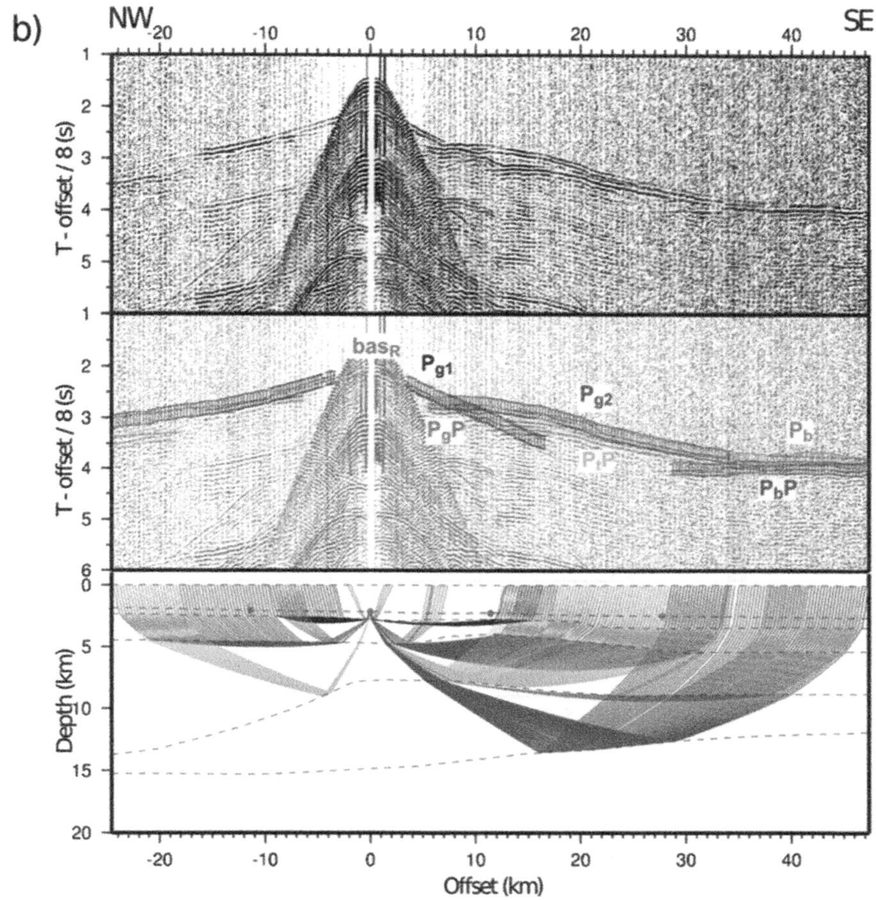

Figure 4.7b. *Same display for a layered forward model (North-Eastern Gulf of Aden). Dashed lines show the model interfaces. Each seismic phase is picked separately, the code allows the layer that the ray travels in to be chosen, and whether it is to be traced as a refraction or a refraction. The color bars in the middle panels correspond to the picked arrival-times, their height being the uncertainty for each pick. The strong amplitude hyperbola at the center of the sections (centered around offset = 0 km) correspond to the direct waves that traveled in the water column (Figure 4.1b). Acronyms:* bas_R *= top of acoustic basement reflection,* P_g, P_{g1}, P_{g2} *= refracted arrivals in the crustal layers,* P_gP *= mid-crustal reflection,* P_mP *= Moho reflection,* P_tP *and* P_bP *= top and bottom of a lower-crustal body, respectively,* P_b *= refraction in a lower-crustal body and* P_n *= upper-most mantle refractions. The purple (a) and black (b) dots correspond to the computed arrival times. The color codes of the picks and rays in the lower panels are the same (source: Watermez* et al. *2011)*

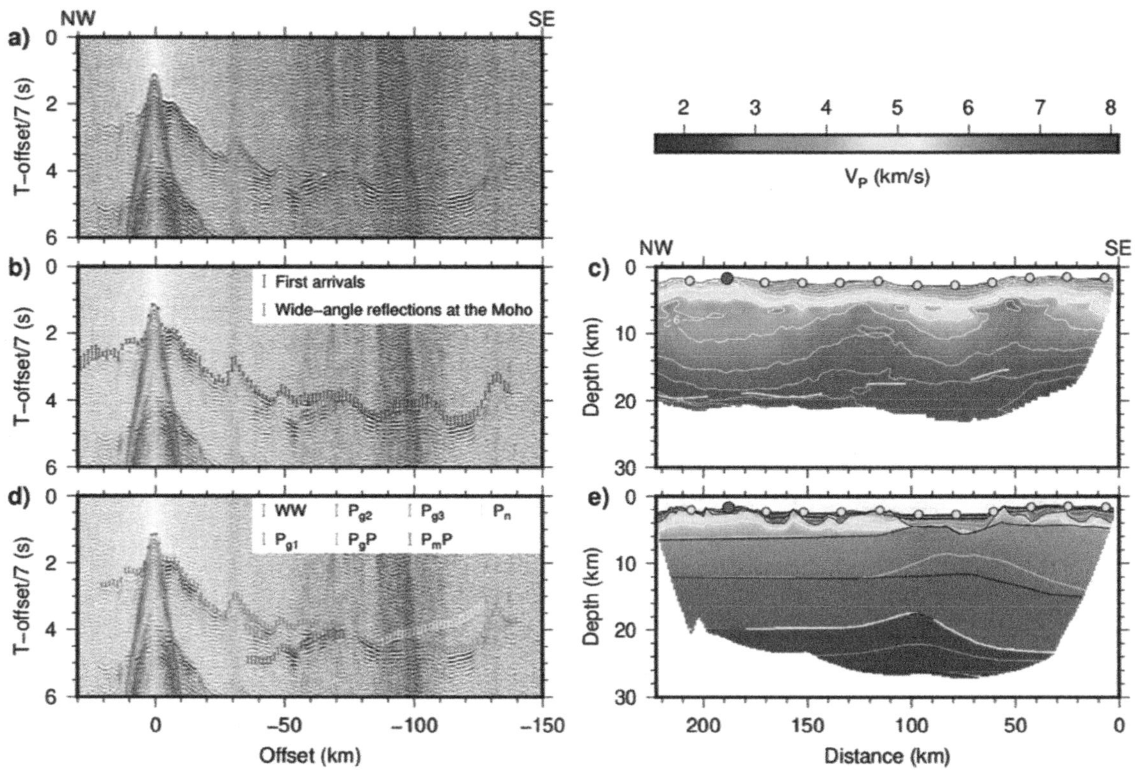

Figure 4.8. *Example of OBS data, picks and results along a same South China Sea profile using Tomo2D and Rayinvr (source: Pichet* et al. *2014 and Liang* et al. *2019, respectively)*

CONTINUATION OF CAPTION FOR FIGURE 4.8.– *a) Example of seismic record along the profile. b) Same data with picks for the travel-time tomography. The color bars correspond to the picks, following the color code in the top-right of the panel. Every 10th pick is shown. Their heights correspond to the picking uncertainty. c) Velocity model resulting from the tomography inversion, using Tomo2D. The velocity color scale is shown in the top-right panel of the figure. The yellow lines correspond to the Moho depth obtained during the joint inversion, showing the interface where some reflected arrivals hit the interface. The white lines show iso-velocity contours, every 0.5 km/s. The yellow and red circles present the OBS locations along the bathymetry line, the red one corresponding to the data presented on the left. d) Same data as in a), with the picks used for forward layered modeling. Picks are presented in the same way as in b). WW = direct wave arrivals ("water wave"),* P_{g1}, P_{g2} *and* P_{g3} *= refracted arrivals in the crustal layers,* P_gP *= intra-crustal reflection,* P_mP *= Moho reflection and* P_n *= upper-most mantle refracted arrivals. e) Velocity model obtained after forward layered modeling, using Rayinvr. Color codes are the same as in c). The black lines present the model interfaces. See the text for further explanation.*

4.1.1.2.4. Resolution assessment

Before moving on to the interpretation step, it is important to assess the resolution of the velocity model. The first two mandatory pieces of information that should be found along each velocity model are the modeling statistics and the fits (or travel-time residuals). The modeling statistics consist of the presentation of number of rays traced (N), normalized misfit (χ^2, average of the travel-time misfits to picking uncertainty ratios) and root-mean squared residual travel-time (t_{RMS}) for each seismic phase, or, at least the whole model. The fit figures show an overlay of the picks and computed arrival-times in the model, which should be as close as possible.

Another important thing to observe is where there is ray coveragence in the model. This is usually shown in the brighter areas of the model, or as a separate plot showing the seismic rays or the derivative weight sum (DWS, a sort of measure of the ray density). Furthermore, areas where rays are traveling in only one direction (i.e. ends of the model) may be affected by non-uniqueness issues, and lower resolution.

Further resolution assessment methods make it possible to identify the resolution of an interface depth, or velocity values, but they are not always shown as the computing time may be very large. Indeed, they are based on realizing a large number of tests, or computing a "new" dataset:

1) Monte-Carlo simulations: the model and picks are randomized before running an inversion. A large number of simulations are necessary to estimate the statistical

resolution of the model. The output shows standard deviations of the velocity field, and model interfaces (if applicable).

2) Checkerboard tests (tomography): travel-times are computed in a perturbed model, following a checkerboard pattern before running a new inversion with these modified travel-times. The result of the inversion is compared to the velocity model, and the areas where the pattern is found are the areas where a velocity anomaly of a similar size is well resolved. Many checkerboard patterns can be tested.

3) Sensitivity tests (forward modeling): this makes it possible to look at the variations of model misfit with systematic changes in chosen model parameters (Moho depth, crustal velocity gradient, upper-most mantle velocities, etc.). The results are usually shown as a plot of model statistics (i.e. χ^2 or tRMS), together with the number of rays traced.

4) Synthetic seismograms (forward modeling): a synthetic wavefield can be computed from the velocity field and impedance contrasts at the interfaces. Synthetic seismograms may be used to justify the choice of a specific velocity gradient, or a feature that is not shown by the travel-times.

4.1.1.2.5. Interpretation

Once the velocity model is finalized, the geological interpretation can be achieved in light of the resolution assessment. The different velocities and bodies highlighted by the model may be interpreted in terms of geological structures, using the seismic velocities and structure morphology. Typical P-wave velocity values are listed below. However, note that any deformation, alteration, hydrothermal and/or geochemical process can lead to profound modifications of these values (see Chapter 2):

– the water (~1.5 km/s);

– sediments (1.6–4.5 km/s);

– continental crust (4.5–7 km/s, with a mid-crust at ~6.5 km/s);

– oceanic crust (oceanic layering with velocities ranging from 4.5 to 7+ km/s, and lower velocity gradient in the gabbros);

– unaltered mantle (7.8–8.3 km/s);

– partially serpentinized mantle (5–8 km/s depending on the degree of serpentinization).

Anomalous velocity bodies are often observed along rifted margins, usually at the crust-mantle boundary. The velocity gradient and eventual reflection observed at the base of the body can help determine the nature of these features (see section 4.2).

4.1.2. *Potential field methods*

Oceanographic vessels are usually equipped with an onboard gravimeter and towed magnetometer. The data acquired using these instruments gives information on the sub-seafloor geology using the density (gravity data) and/or the magnetization (magnetic anomaly data) variations.

4.1.2.1. *Gravity anomalies*

4.1.2.1.1. Acquisition

The gravimeter is installed near the center of motion of the ship to minimize its vertical accelerations, and is mounted on a stabilized platform to reduce the effect of the horizontal accelerations. It measures changes in the vertical component of the gravity field through relative gravity measurements, that is, the gravimeter is not absolute. Relative gravity data needs to be tied to reference data, that is, absolute measures of the gravity field (in harbors), in order to also correct for the system drift. The accuracy of modern marine gravity data is generally less than 1 mGal, thanks to the enhancement of gravimeter and navigational systems. Moreover, those datasets are completed by satellite-derived gravity data (using satellite altimetry) and airborne data. The latter are less accurate, but have a uniform coverage.

4.1.2.1.2. Data processing

In marine geophysics, the free-air gravity anomaly is computed from raw data. The first step is to remove and filter the strong vertical anomalies due to the accelerations of the ship. The acceleration due to the motion of the ship at the Earth's surface is also corrected (Eotvos correction). The free-air gravity anomaly results from the subtraction of the normal gravity field of the Earth (gravity effect of the reference ellipsoid, taking into account the Earth's rotation) from the recorded gravity data. The residual gravity, or Bouguer anomaly, is obtained by subtracting the gravity contrast produced by the water–sediment interface and the sediment–crust interface from the free-air anomaly. Hence, the obtained signal reflects variations in the thickness of the crust or lateral variations of density on the crust.

4.1.2.1.3. Modeling and interpretation

When considering gravity anomalies, it is important to remember that the shape and amplitude of these anomalies are highly dependent on the distance between the measurement location and the considered object. The farther the object, the wider the anomaly and the lower the amplitude (Figure 4.9).

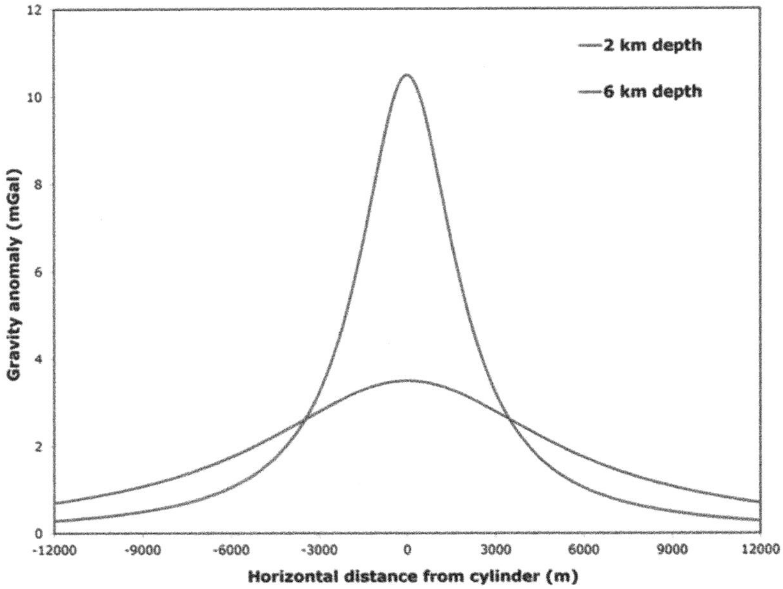

Figure 4.9. *Gravity anomaly for two small cylinders at 2 and 6 km depth. Each cylinder has the same diameter (2 km) and the same density contrast with the surrounding rocks (500 kg.m^{-3})*

Free-air gravity anomalies are often used to perform forward gravity modeling (Figure 4.10). A simple structural model is built from seismic information (depth-converted multi-channel seismic profiles and/or velocity models, converted into densities using empirical relationships). It contains blocks of constant densities that can be interpreted in terms of nature of material. The forward modeling computes the free-air anomaly generated by the structural model. This modeled anomaly is then compared to the measured one. A good fit of modeled and measured data is considered as an indication that the inferred model is possible. However, it is common knowledge that such models are non-unique. Indeed, different structural models can lead to a good fit of the data. It is particularly true if the modeling is achieved along a single seismic profile in 2D. Multiplication of modeled profiles along the same area, or, even better, 3D modeling may be considered as more reliable. Another approach is gravity inversion, that is, the computation of an interface depth and/or of a layer density, to fit the measured data. Here again, the non-uniqueness of the data is a problem. Gravity models can be performed together with isostasy calculation, allowing the models to be better constrained. However, there is a strong initial assumption that the area is in isostatic equilibrium.

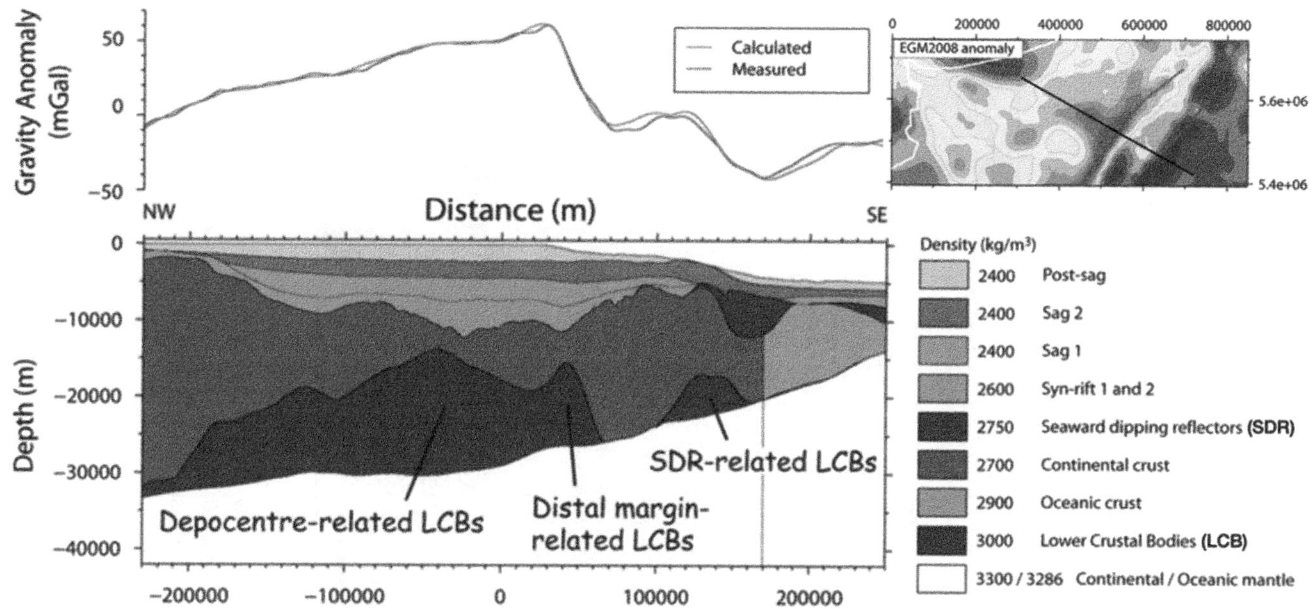

Figure 4.10. *Example of gravity modeling in the Colorado Basin, Atlantic Argentine margin. Each sediment layer or crustal body is assigned a single density value (bottom right). The geometry is known from seismic profiles, except for the lower crustal bodies (bottom left), in which the geometry is inferred from the gravity modeling (top left). Location of the profile on the gravity map (top right) (source: Autin* et al. *2016)*

The gravity signal varies strongly along rifted margins, from the proximal continental crust towards the denser oceanic crust (Figure 4.11). It can be used to localize the ocean–continent transition (OCT), but it is generally more reliable if it is corrected from the influence of the sedimentary thickness. Local high amplitudes in the OCT may be attributed to high-density bodies, related to either (1) underplated or intruded magmatic bodies at volcanic rifted margins, or (2) mantle exhumation, as the dense mantle (partly serpentinized) is located at a shallower depth (<6 km) than below oceanic crust (6–8 km).

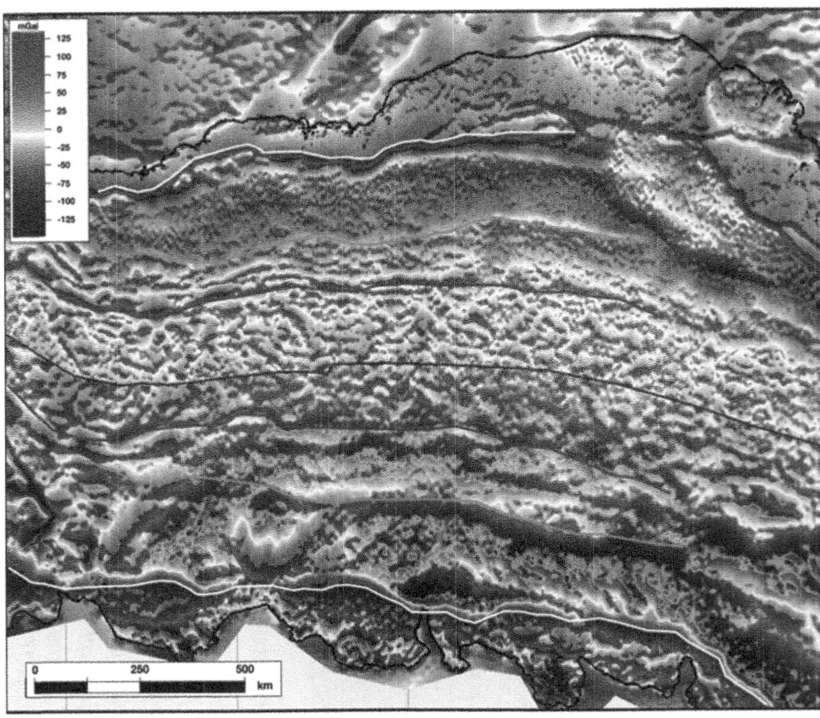

Figure 4.11. *Free-air gravity anomaly map of the conjugated Australia-Antarctica margins (restored to the C18 anomaly, blue line, 38 Ma). Projection: UTM50S. The shadow effect is the north-south horizontal gradient of free-air gravity anomalies derived from Sandwell and Smith (2009). The boundaries are from Gillard* et al. *(2015). White: continental crust necking. Green: limit between hyper-extended continental crust and exhumed mantle domain. Purple: limit between proto-oceanic crust and oceanic crust (source: Sandwell and Smith 2009)*

4.1.2.2. *Magnetic anomalies*

4.1.2.2.1. Acquisition and processing

The intensity of the magnetic field is recorded using an absolute magnetometer towed behind the ship at a distance of at least three times the length of the ship. Magnetic anomalies result from the subtraction of the intensity of the International Geomagnetic Reference Field (IGRF) at the location of measurement from the recorded data. The temporal variations of the geomagnetic field must be considered, generally using the data from a magnetic observatory or complementary measures of a gradient magnetometer. The measurement accuracy is generally of 1 nT (nanoTesla).

4.1.2.2.2. Modeling and interpretation

The shape and amplitude of magnetic anomalies are highly dependent on the distance between the considered object and the measurement location. Similar to gravity anomalies, the farther the object, the wider the anomaly and the lesser the amplitude. Moreover, the anomaly shape is not only linked to the object geometry, but also depends on the magnetization and regional magnetic field directions. As a result, the signal of magnetic anomaly is always formed by a positive and a negative part. The magnetic anomalies mainly result from lateral contrasts of magnetization rather than the magnetization intensity. This magnetic intensity is often considered as a remanent magnetization, for example, for a given rock, the magnetization acquired in the past that is proportional to the reference magnetic field that occurred when the rock cooled.

Also, similarly to gravity anomaly interpretations, forward modeling can be performed to explain magnetic anomalies. However, more parameters have to be taken into account. Indeed, the modeled anomaly depends on the structural model, the magnetic susceptibility assigned to each object and the magnetic field direction. Forward modeling computes the magnetic anomaly generated by the structural model and assigned magnetic parameters. A good fit between the modeled anomaly and the measured one is considered as an indication that the inferred model is possible. Again, such models are non-unique, that is, different structural models/magnetic parameters can lead to a similar good fit of the data. Multiple 2D profiles or 3D modeling may help reduce this non-uniqueness. Another approach is magnetic inversion, that is, the calculation of one interface depth or one magnetic parameter of a layer to fit the measured data. Here again, the non-uniqueness of the model is a problem.

Along rifted margins, magnetic maps are powerful tools to understand the breakup history of the margins (Figure 4.12). Oceanic-crust-related magnetic anomalies record the Earth magnetic field reversal (EMFR), showing isochrones. However, at the OCT, it is not clear whether linear magnetic anomalies are EMFR isochrones or not. Indeed, along volcanic rifted margins, the magma excess that characterizes the OCT could generate a linear magnetic anomaly in contrast with the poorly magnetized continental crust (e.g. the East Coast Magnetic Anomaly (North Atlantic Ocean), Biari *et al.* (2017)). At magma-poor rifted margins, the same contrast-related anomaly could be generated by mantle exhumation (magnetized minerals formed by mantle serpentinization) juxtaposed to the poorly magnetized continental crust. In both cases, the anomaly is not necessarily an isochrone.

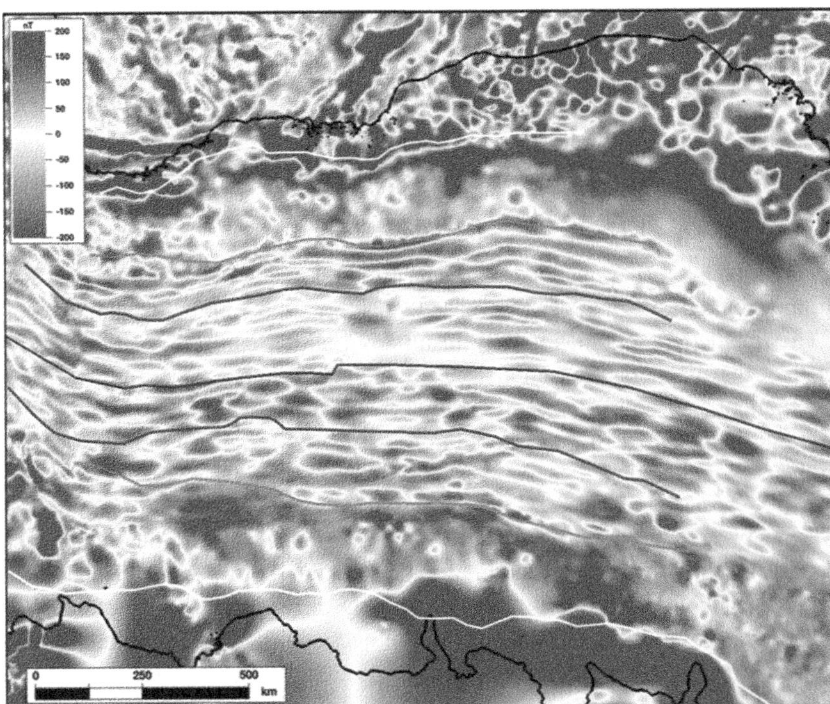

Figure 4.12. *Magnetic anomaly map from the EMAG2 grid model for the conjugated Australia-Antarctica margins (restored to the C18 anomaly, blue line, 38 Ma). Projection: UTM50S. The boundaries are from Gillard* et al. *(2015). The color code of the lines is the same as in Figure 4.10 (source: Maus* et al. *2009)*

4.2. Understanding continental rifted margins using geophysics

4.2.1. *Geological objects interpretation*

There is a large spectrum of interpretations for rifted margins. Sometimes it is linked to the large diversity of rifted margins, sometimes it is linked to different geological scenarios. It is thus important to understand how each geological or geophysical object is recognized and their possible interpretation. Figure 4.13 presents sketches of the two end-member rifted margins: magma-rich or magma-poor rifted margins. Each of the geological objects is described in the following sections and figures.

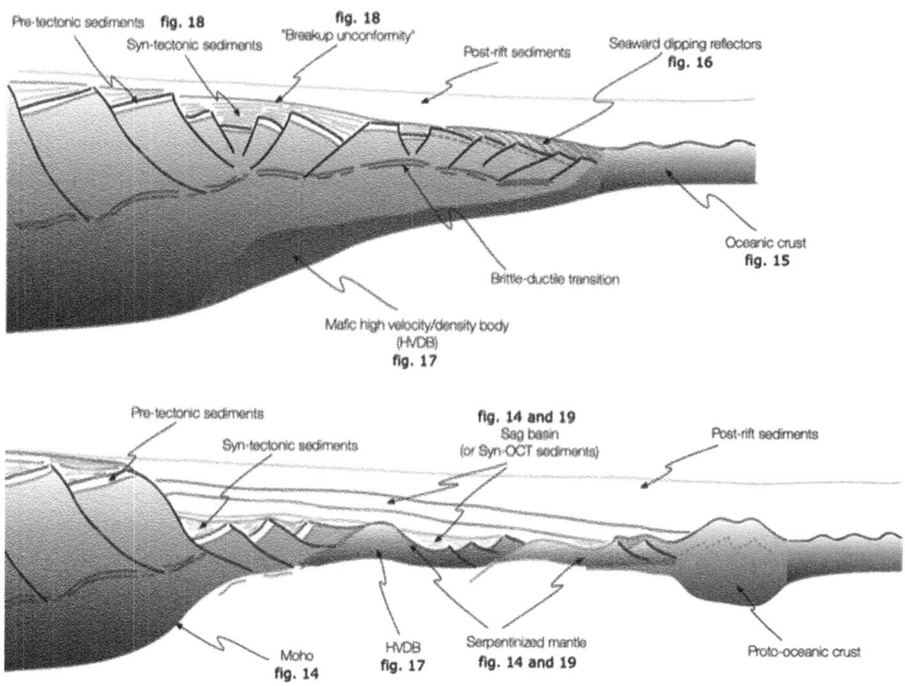

Figure 4.13. *Sketches of the two end-member rifted margins: magma-rich rifted margins (top) and magma-poor rifted margins (base). Each geological object is illustrated by different methods in the indicated figures. "Proto-oceanic crust": basement affected by large magmatic additions before the emplacement of steady-state oceanic crust*

4.2.1.1. *Moho interface*

The Moho is a petrological interface between mantle rocks and crustal rocks. When observed through geophysical methods, the Moho is mainly observed through the supposed density contrast between mantle rocks (3.3) and crustal rocks (2.7–2.9 kg/cm^3). Indeed, gravity models are built considering such density contrasts (modeled gravity Moho). P-wave velocities from seismic refraction modeling are dependent on the densities and lithologies (seismic velocity Moho), as well as reflectivity through the impedance contrast in seismic reflection and seismic refraction (reflective Moho or seismic Moho). It is thus imaged clearly and with high confidence below the continental crust (Figure 4.14) and magmatic oceanic crust (Figure 4.15). However, in the transitional domain, the seismic velocity Moho (8 km/s) is always present, although the petrological Moho may be absent (e.g. below exhumed mantle domains at magma-poor margins). It can also correspond to other interfaces (fresh mantle/serpentinized mantle, mantle/"hot" material in the case of recent underplated material (Watremez *et al.* 2011)). The Moho reflection is present when the impedance contrast is strong enough. Thus, it is absent below exhumed mantle domains as the density gradient is very progressive, following the serpentinization degree (Figure 4.14).

4.2.1.2. *Magmatic additions*

Magmatic objects are frequently observed at rifted margins, in particular, at magma-rich margins. Indeed, the first criteria used to recognize magma-rich margins is the presence of SDR (Seaward Dipping Reflectors, Figure 4.16), that are specifically observed on seismic reflection profiles. They are defined as wedges of stacked, arcuate reflectors that dip seawards with increasing flow thickness down dip (e.g. Mutter *et al.* 1982; Harkin *et al.* 2020). Geologically, they are interpreted as lava flows, interbedded with sediments, that characterize the high magmatic budget of the magma-rich margins during the breakup and formation of a steady-state mid-oceanic ridge (see Chapter 2). Another typical feature observed with deep seismic reflection and/or seismic refraction is an intruded lower crust under the form of a high velocity body (see section 4.2.1.3 for further details).

Other magmatic objects can be recognized as sills, dikes, volcanic seamounts or lava flows and deltas (see Part 1). In seismic reflection profiles, magmatic additions generate strong amplitudes. Sills often appear clearly in the sedimentary succession, with a very high amplitude that can stop abruptly laterally. Volcanic seamounts are frequent and of various sizes. They are dome shaped and generally show no sharp edges (Figure 4.16). When associated with normal faults, they change the faulted block geometry: instead of being asymmetric, they tend to be more symmetric and dome shaped.

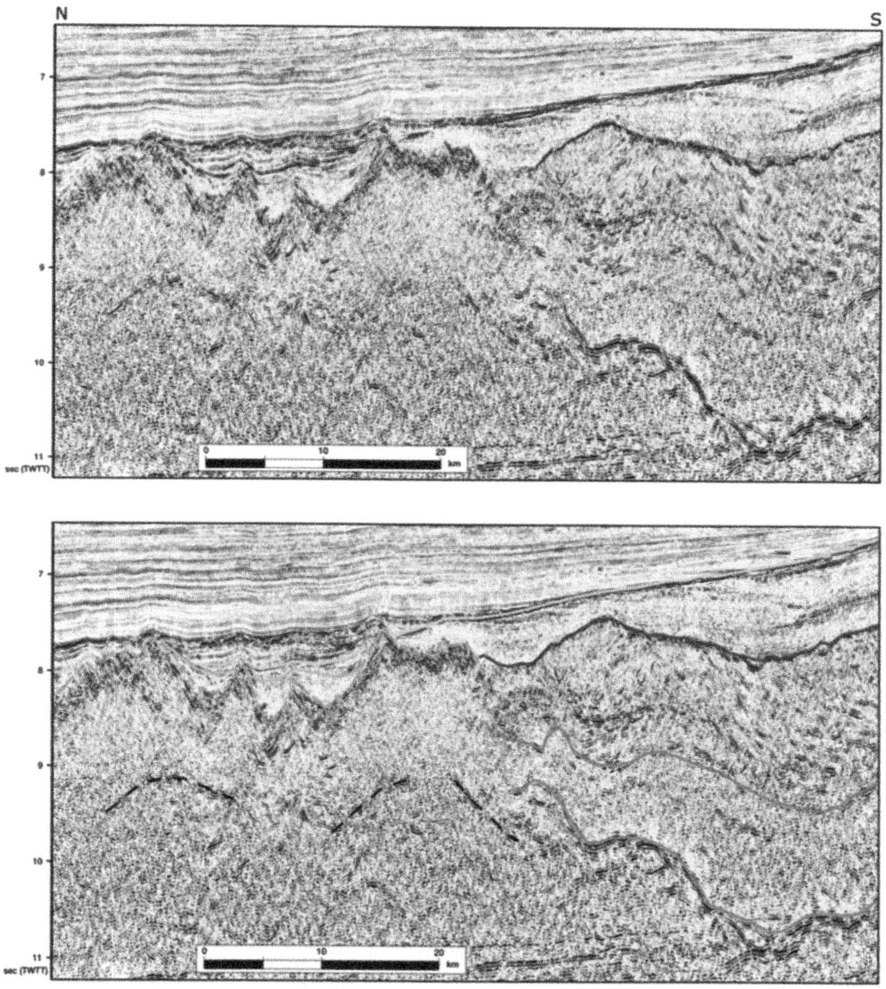

Figure 4.14. *Seismic reflection profile GA228-24 through the transition between continental crust (right) and exhumed mantle (left) without (top) and with interpretation (bottom, Gillard* et al. *2015). Green line: continental Moho. Pink line: top of the continental crust. Brown line: top of the pre-exhumation sediments. Yellow line: top of the sag basin, that is, oceanic breakup unconformity. Blue dashed line: top of the exhumed mantle. Black dashed line: it is not the Moho, but a decollement level that normal faults root into, possibly controlled by the serpentinization degree of the exhumed mantle (source: Gillard* et al. *2019)*

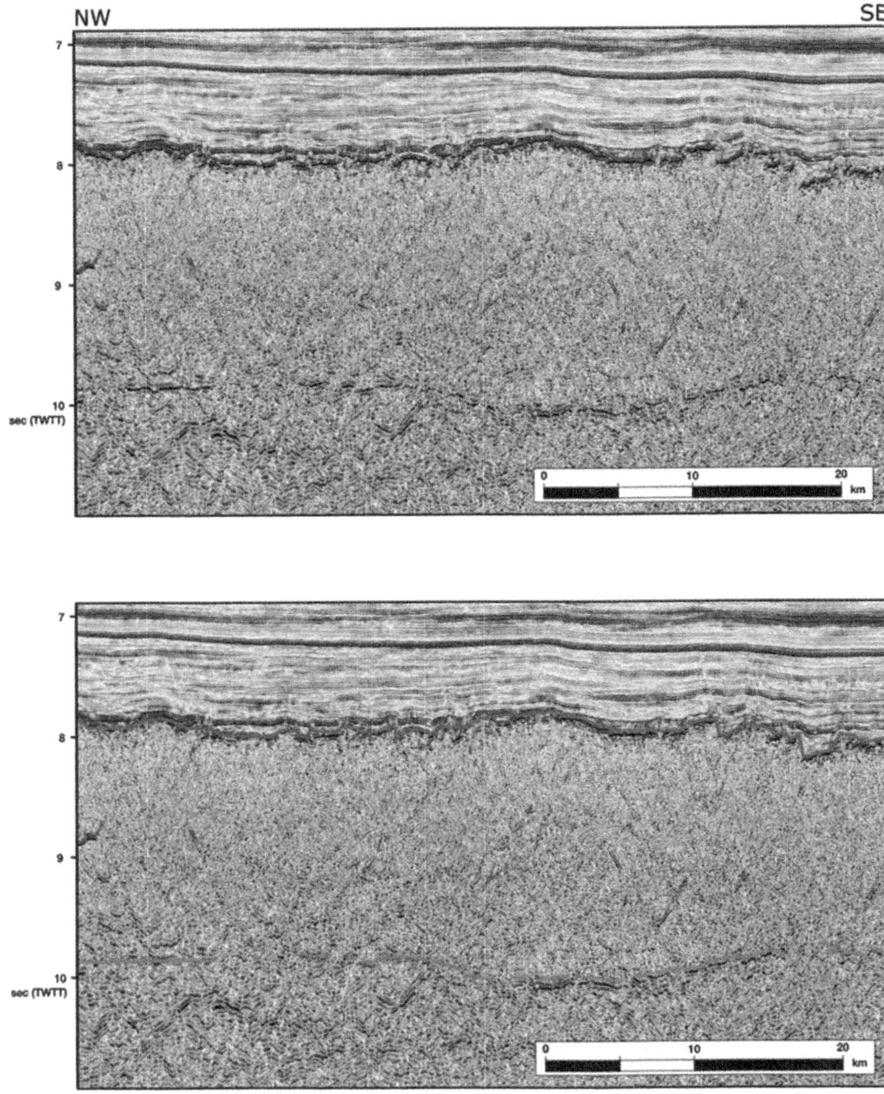

Figure 4.15. *Seismic reflection profile GA229-22 in oceanic crust without (top) and with interpretation (bottom). Green line: oceanic Moho. Blue line: top of the oceanic crust*

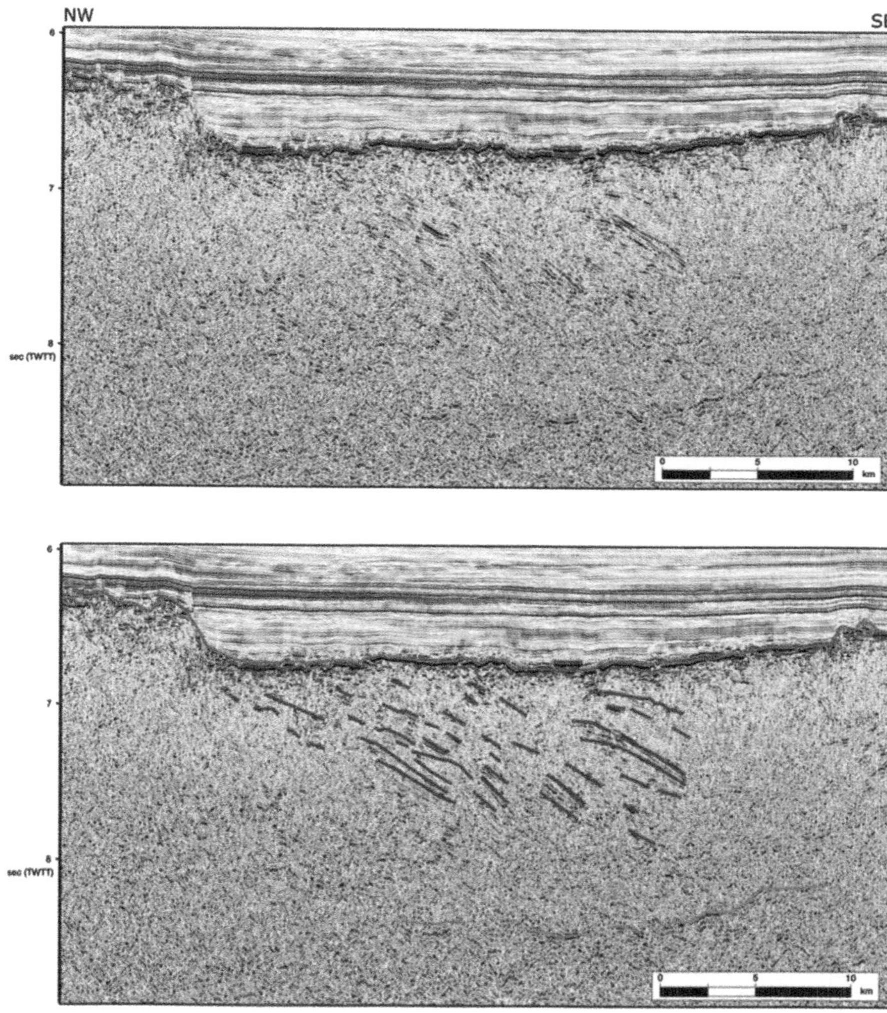

Figure 4.16. *Seismic reflection profile GA228-10 through seaward dipping reflectors (SDR) without (top) and with interpretation (bottom). Green line: transitional Moho. Blue line: top of crust. Red lines: SDR. Please note the volcanic seamount as an acoustic basement, high on the left of the profile*

4.2.1.3. *High velocity/density bodies (HVDB)*

Rifted margins often show the presence of high velocity/density bodies (HVDB) in or at the base of the "crust". They are recognized through seismic refraction (with $V_P > 7.1$ km/s and $V_P < 8$ km/s), or in the gravity signal or modeling. They are also called "lower crustal bodies" (LCB) because of their location at the base of the continental crust or in the OCT deep basement, or "intermediate bodies" as their velocity and density characteristics are higher than typical crust and lower than unaltered mantle. Several hypotheses can explain their lithology and emplacement (see Figure 4.17): mafic intrusions emplaced in the lower crust during the rifting and often related to Seaward Dipping Reflectors (SDRs) at the surface (e.g. Eldholm *et al.* 2000; White and McKenzie 1989); serpentinized mantle exhumed during the final extreme thinning of the crust (e.g. Boillot *et al.* 1987); or high degree metamorphic rocks inherited from a previous orogenic phase (e.g. Ebbing *et al.* 2006).

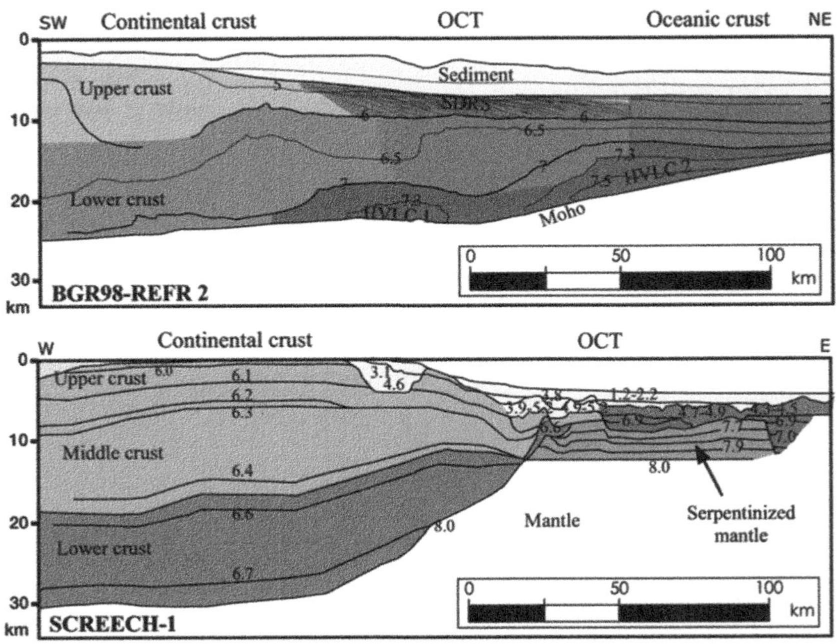

Figure 4.17. *Deep structure of the Argentinian (top) and Newfounland (bottom) continental margins. The magma-rich Argentinian margin displays SDR and deep HVDBs (here HVLC for High Velocity Lower Crust) that are interpreted as mafic intrusions. The magma-poor Newfoundland margin displays a HVDB that is interpreted as serpentinized mantle (source: Biari* et al. *2021, adapted from Schnabel* at al. *2008 and Funck* et al. *2003)*

It is a difficult task to determine the geometry and lithology of a HVDB, but it is most important to understand their role in the development of rifted margins. To discriminate between mafic intrusions and exhumed mantle, the vertical velocity gradient may be relevant: gradients higher than 0.08 s^{-1} are interpreted as typical of serpentinite, while gradients lower than 0.05 s^{-1} may characterize mafic bodies (Watremez *et al.* 2011). Moreover, mafic bodies correlate with greater thicknesses (8–18 km vs. 3–5 km for serpentinite) (Watremez *et al.* 2011). Another method consists of considering the thermal properties of such material (thermal conductivity and radiogenic heat production). It makes it possible to calculate the conductive thermal field and compare it with available measured temperatures (Autin *et al.* 2016). Typical thermal properties are listed in Table 4.1. If mafic intrusions and inherited metamorphic rocks show only slight differences, serpentinized mantle has much lower thermal conductivity and heat production values. These values are derived from rock property records (e.g. Landolt and Börnstein 1983; Vilà *et al.* 2010). The conductivity of serpentinite is considered as half that of typical mantle minerals (White *et al.* 2003).

Material composing HVDB	Bulk thermal conductivity ($mW.K^{-1}$)	Heat production ($\mu W.m^{-3}$)
Mafic intrusions	3.2	0.33
Serpentinized mantle	2	0.007
Inherited high degree metamorphic rocks	3.1	0.25

Table 4.1. *Thermal properties for the possible material composing HVDB*

Of course, at well-known magma-rich margins, such bodies are generally interpreted as intruded magmatic/mafic bodies in the crust, or as underplated and cooled magma below the crust. The difference in the interpretation is important, as the location of the Moho depends on it: either the crust was thickened by intrusions, but the Moho still remains at the base of the HVDB, or the crust was underplated and the paleo-Moho is located at the top of the HVDB. Here, no discriminant method is available.

4.2.1.4. *Sedimentary units*

Originally, the sedimentary seismic packages were systematically referred to as pre-rift, syn-rift and post-rift. However, when describing seismic units, the observer should not use such interpretative terms at first. The observation makes it possible to describe pre-, syn- or post-tectonic units (Figure 4.18). All syn-tectonic units are not coeval and post-tectonic units can either be syn-rift or post-rift (Figure 4.19). In other words, syn-rift sediments can have various shapes and organizations that also depend on the type of fault affecting them: classical normal faults or detachment faults. Ideally, the observer is able to recognize unique seismic facies for each sedimentary unit. This makes it possible to correlate sediments strata of the same age with different tectonic patterns: for example, distal syn-tectonic strata with more proximal post-tectonic ones (Figure 4.19). Such observations highlight the migration of the deformation through the rift and its contribution to subsidence. However, as the sedimentary setting is prone to change laterally from proximal to distal, the seismic facies should change as well. The proximal/distal correlation is thus often difficult and based on strong assumptions.

The following descriptions highlight the main characteristics of pre-syn and post-tectonic units deposited along a classical normal fault (see Figure 4.18).

Pre-tectonic units are not always easily observed, as they generally present a conform relationship with the underlying top basement, that is, the reflectors are parallel to the basement top. It is thus sometimes difficult to differentiate between pre-tectonic sediments and a layered top basement. They are characterized by tilted and parallel reflectors above the tilted blocks in between normal faults. They present toplap terminations, upward, and downlap terminations, downward, where the unit is cut by normal faults.

Syn-tectonic units present typical wedge-shaped organization. The basal reflectors present a conform relationship with the top of the pre-tectonic units. The top reflectors are often sub-horizontal. In between, the reflectors are progressively decreasing their tilted angle to form a sedimentary wedge. The reflector terminations are either downlaps or onlaps against the bounding normal fault.

Post-tectonic units are characterized by parallel reflectors that can either be sub-horizontal or tilted towards the ocean. The basal reflectors often present an unconformity with the top of the syn-tectonic units. The reflector terminations are either onlaps or downlaps above the underlying units.

4.2.1.5. *Unconformities*

Another point of concern is the so-called “breakup unconformity”. It is defined as the interface separating the syn-rift and post-rift strata and is thought to date the breakup as an isochrone timeline. Yet, when looking at interpreted breakup unconformities, it appears that they are often located at the top of syn-tectonic unit, either in the proximal domain or in the distal domain. In such interpretations, it is most likely that the pointed unconformity corresponds to the end of different tectonic events and is thus not an isochrone timeline. Moreover, in its proximal part, it would not date the breakup, but rather the end of the earlier syn-tectonic event (Figures 4.18 and 4.19). Here, it is sometimes referred as the “necking unconformity”. In its distal part, it could date the continental crust breakup and can be referred to as the "crustal breakup unconformity". If there is a sag basin, the eventual unconformity at its top does date the breakup, that is, the onset of spreading, and can be referred to as the "oceanic breakup unconformity" (see Figures 4.14 and 4.19).

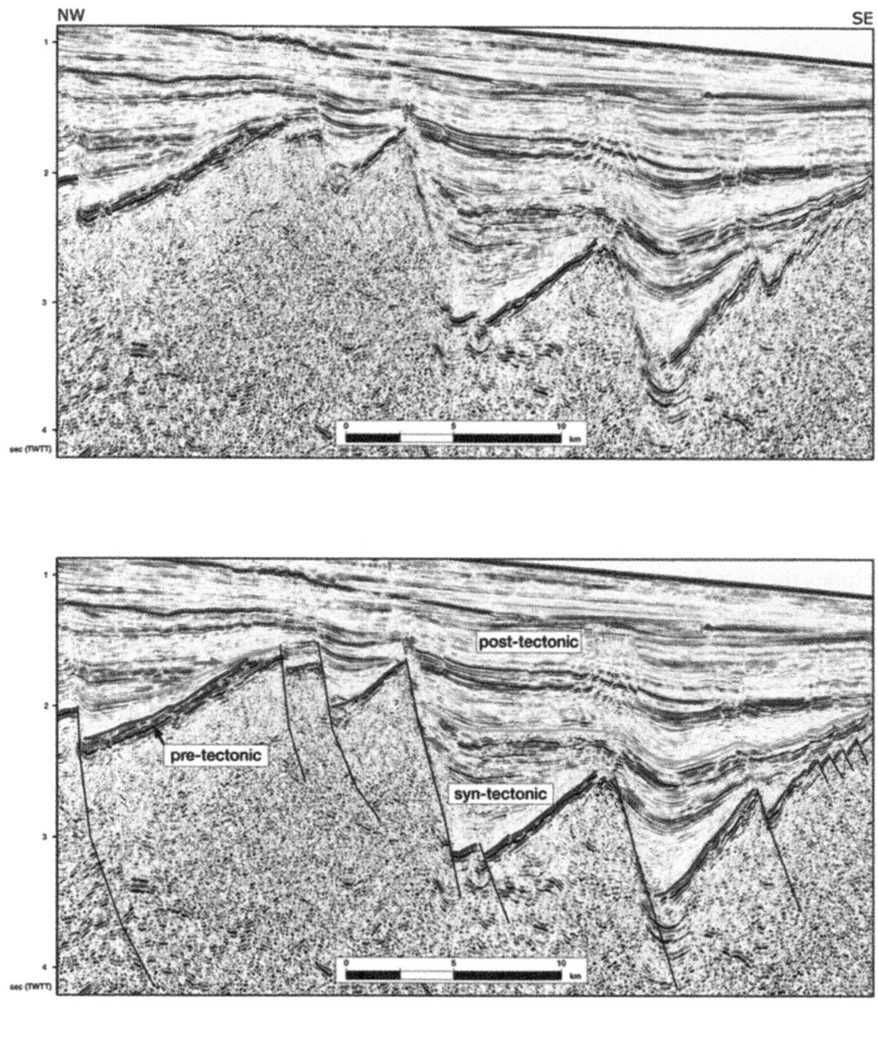

Figure 4.18. *Seismic reflection profile GA199-09 through the proximal margin without (top) and with interpretation (bottom). Blue line: top of pre-tectonic sediments. Orange line: top of syn-tectonic sediments and base of the post-tectonic sediments that onlap above the unconformity. The red arrows highlight the termination of reflectors in each unit (see Figure 4.6)*

4.2.2. Data complementarity

In the absence of rock samples, geophysical methods are the only way to study rifted margins. They provide indirect imaging of the crust and sediments, as well as potential field data. Each method provides information of its own, but the understanding of the geological processes leading to the formation of the rifted margin is best with the combination of these complementary methods. Moreover, onshore/offshore approaches are also an additional bonus, as well as comparisons with field analogs.

To illustrate this powerful complementarity, the difficult location of the OCT boundaries in magma-poor-rifted margins is a good example. The geophysical characteristics of the OCT are neither typical of continental crust nor of oceanic crust. We illustrate this with Figure 4.20, along the North-Eastern margin of the Gulf of Aden, where numerous experiments took place (d'Acremont *et al.* 2005; d'Acremont *et al.* 2006; Lucazeau *et al.* 2008; Autin *et al.* 2010; Leroy *et al.* 2010, 2012; Watremez e*t al.* 2011).

Looking at the magnetic signal, it shows two distinct domains of magnetic pattern: oceanward, the signal shows strong, continuous successive magnetic anomalies whereas, continentward, the signal is weaker, without clear oscillations. In the OCT, there are no clearly identifiable oceanic magnetic anomalies. This excludes the fact that the OCT is composed of a magmatic oceanic crust and makes it possible to locate the oceanward boundary of the OCT along anomaly 5d (d'Acremont *et al.* 2006). As for the wide-angle profile (Leroy *et al.* 2010), it shows a narrow zone of continental crust thinning, with thicknesses decreasing from 30 to less than 10 km thick. In the South, the oceanic crust has typical layers 2 and 3 velocities (5–5.5 and 6.5–7 km/s, respectively) and the thickness of a slow-spreading oceanic crust (ca. 6 km) (Christeson *et al.* 2019). In between, the 5-km-thick OCT shows a high velocity body at the base of the "crust". Such bodies could either result from a mafic body or from serpentinite. To discriminate between the two, vertical velocity gradient seems relevant: gradients higher than 0.08 s^{-1} are typical serpentinite, while gradients lower than 0.05 s^{-1} may characterize mafic bodies (Watremez *et al.* 2011). Moreover, mafic bodies correlate with greater thicknesses (8–18 km vs. 3–5 km for serpentinite) (Watremez *et al.* 2011). Here, the high velocity body is characterized with velocities rapidly decreasing upward from 7.25 km/s to 6.2 km/s, with a gradient of 0.21 s^{-1} and a thickness of 5 km. This typical P-wave velocity increase at the base of the "crust" characterizes the presence of serpentinized mantle: it corresponds to an increasing serpentinization degree upward and thus, less dense material. Further evidence resides in the seismic reflection profile. First, there are no clear Moho reflections in the OCT, which indicate that no sharp velocity contrast allows for such reflection. This could again indicate a gradual downward decrease in the degree of mantle serpentinization. Then, if mantle exhumation occurs, the older sediments deposited on the continental crust, before the beginning of exhumation, are by definition restricted to the continental domain alone. Therefore, it is possible to locate the exhumation point (i.e. the northward boundary of the OCT) where the pre-exhumation sediments stop (Gillard *et al.* 2015). Moreover, the authors observe that a thick horizontal layer of sediments is deposited before the post-rift sediments (which are deposited above the oceanic crust) and after the sedimentary fans that are typical of the syn-rift period. Thus, these sediments were deposited during the continental break-up itself and are referred as "syn-OCT sediments" (d'Acremont *et al.* 2005; Autin *et al.* 2010) (defined as "sag sediments" in other margins). Indeed, the formation of the OCT, here by mantle exhumation along large-offset exhumation faults, does not create syn-kinematic sedimentary wedges, but rather flat or slightly downlapping syn-kinematic units (Péron-Pinvidic *et al.* 2007; e.g. Gillard *et al.* 2015). It is interesting to highlight that the syn-OCT sediments are coeval with both mantle exhumation and on-land major uplift, associated with a phase of erosion, reflecting the general seaward tilting of the whole margin (Leroy *et al.* 2012).

Finally, the presence of major deformation structures (large-offset normal fault) affecting both basement and overlying sediments are more typical of mantle exhumation processes than of oceanic crust formation (e.g. Gillard *et al.* 2015). These observations make it possible to locate the OCT between 30 and 70 km along the wide-angle profile: it must include the high velocity body, as well as the syn-OCT sediments, but no pre-exhumation sediments (Autin *et al.* 2010). Looking at the free-air gravity signal, it shows high positive amplitude in the oceanic domain (40 mGal). There are slightly higher amplitudes in the OCT above the basement highs, then a negative gradient towards strong negative continental amplitudes (-40 mGal). In the OCT, the high velocity body observed on wide-angle data, is associated with higher density than in the surrounding continental and oceanic crust. This lateral contrast can generate a positive gravity anomaly, as observed above the interpreted serpentinized mantle high. The heat flow is also interesting as it shows high values in the conductive oceanic crust and in the OCT (120 $mW.m^{-2}$), and lower values in the less conductive continental crust (40 $mW.m^{-2}$). Such high values in the OCT are not common and may be explained by the fact that the Gulf of Aden is a young basin (less than 20 Ma). Thermal modeling (Lucazeau *et al.* 2008, 2010) indicates that they could correspond to a thermal anomaly in the upper mantle, which persisted after the break-up, possibly through small-scale convection cells below the margin during and after the rifting. If this behavior occurred in older rifted margins, it could explain the shallow water deposits during post rift-time.

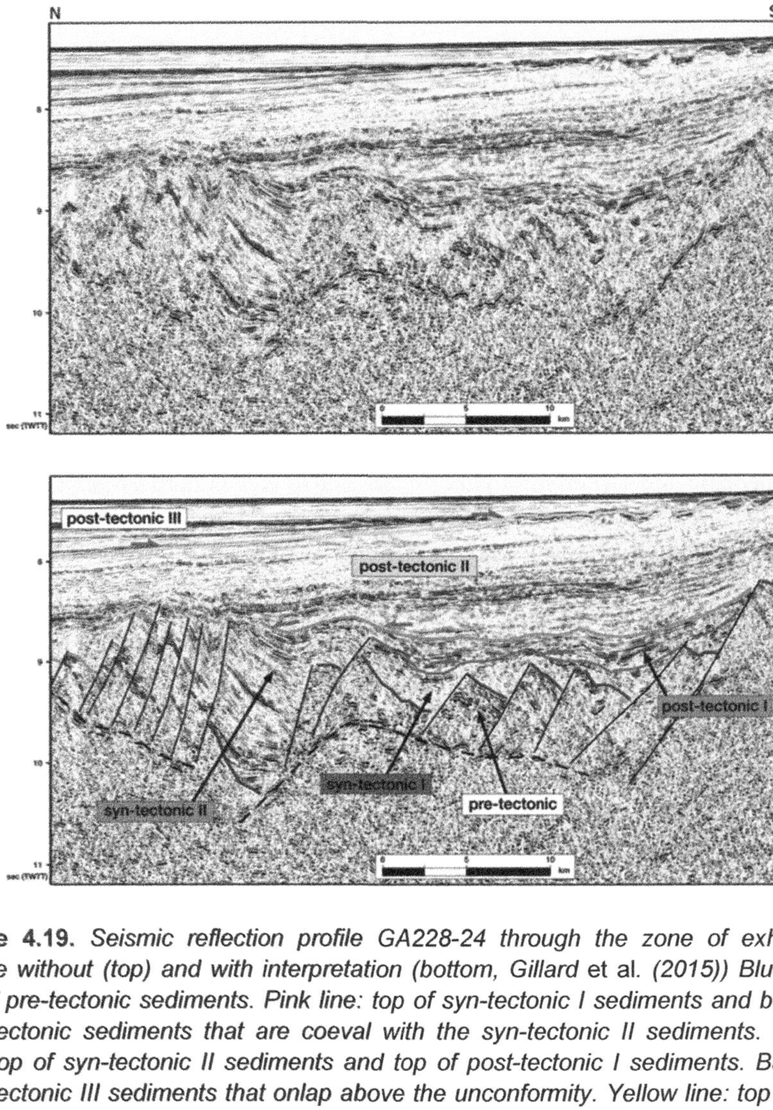

Figure 4.19. *Seismic reflection profile GA228-24 through the zone of exhumed mantle without (top) and with interpretation (bottom, Gillard* et al. *(2015)) Blue line: top of pre-tectonic sediments. Pink line: top of syn-tectonic I sediments and base of post-tectonic sediments that are coeval with the syn-tectonic II sediments. Green line: top of syn-tectonic II sediments and top of post-tectonic I sediments. Base of post-tectonic III sediments that onlap above the unconformity. Yellow line: top of the sag basin and base of post-tectonic III sediments that onlap above the oceanic breakup unconformity. Black dashed line: it is not the Moho, but a decollement level where normal faults root into, possibly controlled by the serpentinization degree of the exhumed mantle. The red arrows highlight the termination of reflectors in each unit (see Figure 4.6) (source: Gillard* et al. *2019)*

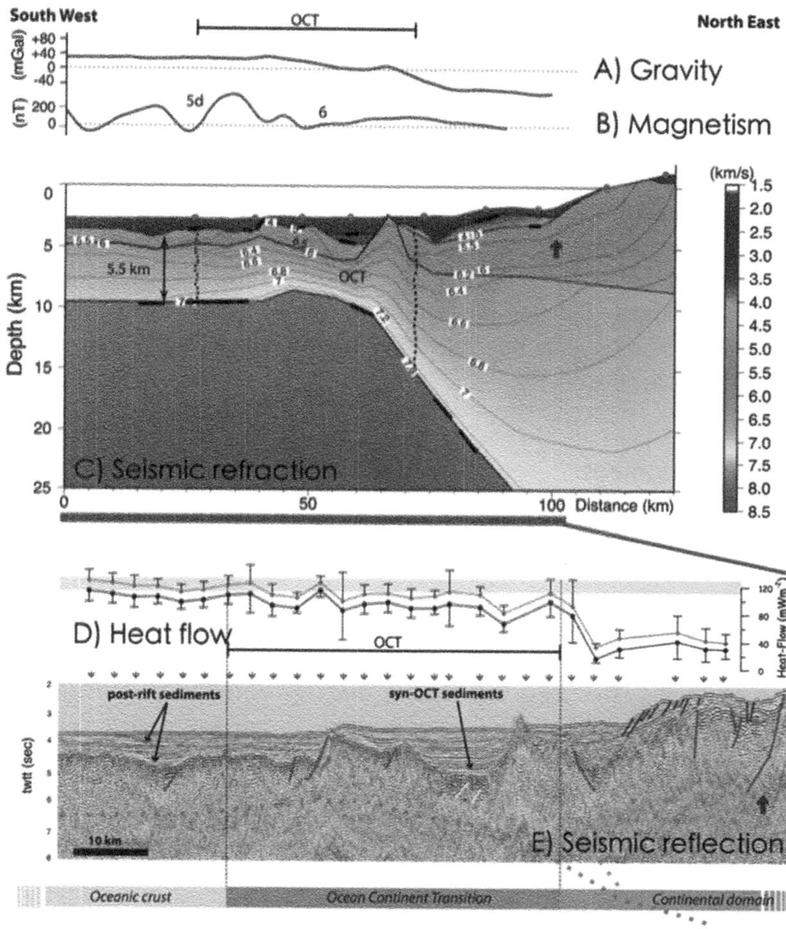

Figure 4.20. *Example of methods complementarity along the Mirbat segment in the North Eastern margin of the Gulf of Aden. a) Gravity signal is high above dense oceanic crust the and low above continental crust. b) Above the oceanic crust, magnetic signal is strong, with continuous successive magnetic anomalies oscillations, whereas continentward, the signal is weaker, without clear oscillations. c) The 5-km-thick OCT shows a high velocity body at the base of the "crust" i with velocities rapidly decreasing upward characterizing the presence of serpentinized mantle. d) Heat flow is high above the conductive oceanic crust and low above continental crust. e) On the seismic reflection profile, faults active at the beginning of the syn-rift period are in red, faults active during the exhumation phase are in yellow, those active during the post-rift phase are in orange and those associated with active submarine slides are in violet. The colored dotted lines are derived from the velocity model, converted into two-way travel time; they correspond to the main interfaces underneath the acoustic basement (source: Leroy* et al. *2010, 2012)*

4.3. Conclusion

Geophysical data are often the only constraints available for interpretation of offshore rifted continental margins. Seismic reflection, refraction and potential fields are all valuable datasets carrying relevant information. However, numerous uncertainties are inherent to the methods, which often hamper a proper validation of the geological model. The complementarity of the dataset, as well as the large amount of data are favored to reach the most accurate interpretation. Nevertheless, even the most complete set of data can lead to various – if not contradictory – interpretations, due to the non-uniqueness of the models. The interpreter must convince their reader by gathering scientific proofs, as well as giving arguments for a consistent geological scenario.

4.4. Acknowledgment

The seismic lines of the AGSO surveys GA199, GA228 and GA229 (Figures 4.6, 4.14, 4.15, 4.16, 4.18 and 4.19) have been provided by Geoscience Australia after personal request and are published with the permission of Geoscience Australia.

Further reading.– The above descriptions are abbreviated and often simplified. If interested in reading and learning further, the reader is referred to the following list of publications and references:
– Reflection: (Yilmaz 2001; Dondurur 2018; Gillard *et al.* 2017).
– Refraction: (Zelt 1999; Korenaga *et al.* 2000; Biari *et al.* 2021).
– MAG: (Szameitat *et al.* 2020).
– HVDB: (Autin *et al.* 2016).
– Thermal approach: (Gholamrezaie *et al.* 2018)

4.5. References

d'Acremont, E., Leroy, S., Beslier, M.-O., Bellahsen, N., Fournier, M., Robin, C., Maia, M., Gente, P. (2005). Structure and evolution of the eastern Gulf of Aden conjugate margins from seismic reflection data. *Geophysical Journal International*, 160(3), 869–890 [Online]. Available at: http://onlinelibrary.wiley.com.biblioplanets.gate.inist.fr/doi/10.1111/j.1365-246X.2005.02524.x/abstract [Accessed 25 February 2013].

d'Acremont, E., Leroy, S., Maia, M., Patriat, P., Beslier, M.-O., Bellahsen, N., Fournier, M., Gente, P. (2006). Structure and evolution of the eastern Gulf of Aden: Insights from magnetic and gravity data (Encens-Sheba MD117 cruise). *Geophysical Journal International*, 165(3), 786–803 [Online]. Available at: http://onlinelibrary.wiley.com.biblioplanets.gate.inist.fr/doi/10.1111/j.1365-246X.2006.02950.x/abstract [Accessed 25 February 2013].

Autin, J.A., Leroy, S., Beslier, M.-O., d'Acremont, E., Razin, P., Ribodetti, A., Bellahsen, N., Robin, C., Al Toubi, K. (2010). Continental break-up history of a deep magma-poor margin based on seismic reflection data (northeastern Gulf of Aden margin, offshore Oman). *Geophysical Journal International*, 180(2), 501–519 [Online]. Available at: https://academic.oup.com/gji/article-lookup/doi/10.1111/j.1365-246X.2009.04424.x [Accessed 12 April 2017].

Autin, J.A., Scheck-Wenderoth, M., Götze, H.-J., Reichert, C., Marchal, D. (2016). Deep structure of the Argentine margin inferred from 3D gravity and temperature modelling, Colorado Basin. *Tectonophysics*, 676, 198–210 [Online]. Available at: http://linkinghub.elsevier.com/retrieve/pii/S0040195115006526 [Accessed 12 April 2017].

Biari, Y., Klingelhoefer, F., Sahabi, M., Funck, T., Benabdellouahed, M., Schnabel, M., Reichert, C., Gutscher, M.-A., Bronner, A., Austin, J.A. (2017). Opening of the central Atlantic Ocean: Implications for geometric rifting and asymmetric initial seafloor spreading after continental breakup: Opening of the Central Atlantic Ocean. *Tectonics*, June [Online]. Available at: http://doi.wiley.com/10.1002/2017TC004596 [Accessed 7 July 2017].

Biari, Y., Klingelhoefer, F., Franke, D., Funck, T., Loncke, L., Sibuet, J.-C., Basile, C., Austin, J.A., Rigoti, C.A., Sahabi, M., Benabdellouahed, M., Roest, W.R. (2021). Structure and evolution of the Atlantic passive margins: A review of existing rifting models from wide-angle seismic data and kinematic reconstruction. *Marine and Petroleum Geology*, 126, 104898 [Online]. Available at: https://www.sciencedirect.com/science/article/pii/S0264817221000027 [Accessed 18 February 2021].

Boillot, G., Recq, M., Winterer, E.L., Meyer, A.W., Applegate, J., Baltuck, M., Bergen, J.A., Comas, M.C., Davies, T.A., Dunham, K., Evans, C.A., Girardeau, J., Goldberg, G., Haggerty, J., Jansa, L.F., Johnson, J.A., Kasahara, J., Loreau, J.P., Luna-Sierra, E., Moullade, M., Ogg, J., Sarti, M., Thurow, J., Williamson, M. (1987). Tectonic denudation of the upper mantle along passive margins: A model based on drilling results (ODP leg 103, western Galicia margin, Spain). *Tectonophysics*, 132(4), 335–342 [Online]. Available at: http://www.sciencedirect.com/science/article/B6V72-488953N-4H/2/857bb8b2715606703820d8036cf14afd.

Christeson, G.L., Goff, J.A., Reece, R.S. (2019). Synthesis of oceanic crustal structure from two-dimensional seismic profiles. *Reviews of Geophysics*, 57(2), 504–529 [Online]. Available at: http://agupubs.onlinelibrary.wiley.com/doi/abs/10.1029/2019RG000641 [Accessed 22 February 2021].

Dondurur, D. (2018). *Acquisition and Processing of Marine Seismic Data*. Elsevier, Amsterdam.

Ebbing, J., Lundin, E., Olesen, O., Hansen, E.K. (2006). The mid-Norwegian margin: A discussion of crustal lineaments, mafic intrusions, and remnants of the Caledonian root by 3D density modelling and structural interpretation. *Journal of the Geological Society*, 163(1), 47–59.

Eldholm, O., Gladczenko, T.P., Skogseid, J., Planke, S. (2000). Atlantic volcanic margins: A comparative study. *Geological Society, London, Special Publications*, 167(1), 411–428 [Online]. Available at: http://sp.lyellcollection.org/content/167/1/411.abstract [Accessed 20 December 2011].

Funck, T., Hopper, J.R., Larsen, H.C., Louden, K.E., Tucholke, B.E., Holbrook, W.S. (2003). Crustal structure of the ocean-continent transition at Flemish Cap: Seismic refraction results. *Journal of Geophysical Research*, 108(B11), 20.

Gernigon, L., Ringenbach, J.-C., Planke, S., Le Gall, B. (2004). Deep structures and breakup along volcanic rifted margins: Insights from integrated studies along the outer Vøring Basin (Norway). *Marine and Petroleum Geology*, 21(3), 363–372.

Gholamrezaie, E., Scheck-Wenderoth, M., Sippel, J., Strecker, M.R. (2018). Variability of the geothermal gradient across two differently aged magma-rich continental rifted margins of the Atlantic Ocean: The Southwest African and the Norwegian margins. *Solid Earth*, 9(1), 139–158 [Online]. Available at: https://www.solid-earth.net/9/139/2018/ [Accessed 22 February 2018].

Gillard, M., Autin, J., Manatschal, G., Sauter, D., Munschy, M., Schaming, M. (2015). Tectonomagmatic evolution of the final stages of rifting along the deep conjugate Australian-Antarctic magma-poor rifted margins: Constraints from seismic observations. *Tectonics*, 34(4), 753–783 [Online]. Available at: http://doi.wiley.com/10.1002/2015TC003850 [Accessed 17 August 2018].

Gillard, M., Sauter, D., Tugend, J., Tomasi, S., Epin, M.-E., Manatschal, G. (2017). Birth of an oceanic spreading center at a magma-poor rift system. *Scientific Reports*, 7(1), 15072 [Online]. Available at: http://www.nature.com/articles/s41598-017-15522-2 [Accessed 3 September 2018].

Gillard, M., Tugend, J., Müntener, O., Manatschal, G., Karner, G.D., Autin, J., Sauter, D., Figueredo, P.H., Ulrich, M. (2019). The role of serpentinization and magmatism in the formation of decoupling interfaces at magma-poor rifted margins. *Earth-Science Reviews*, 196, 102882 [Online]. Available at: https://linkinghub.elsevier.com/retrieve/pii/S0012825218307013 [Accessed 18 February 2021].

Harkin, C., Kusznir, N., Roberts, A., Manatschal, G., Horn, B. (2020). Origin, composition and relative timing of seaward dipping reflectors on the Pelotas rifted margin. *Marine and Petroleum Geology*, 114, 104235 [Online]. Available at: http://www.sciencedirect.com/science/article/pii/S0264817220300180 [Accessed 15 December 2020].

Korenaga, J., Holbrook, W.S., Kent, G.M., Kelemen, P.B., Detrick, R.S., Larsen, H.-C., Hopper, J.R., Dahl-Jensen, T. (2000). Crustal structure of the southeast Greenland margin from joint refraction and reflection seismic tomography. *Journal of Geophysical Research: Solid Earth*, 105(B9), 21591–21614 [Online]. Available at: http://agupubs.onlinelibrary.wiley.com/doi/abs/10.1029/2000JB900188 [Accessed 16 December 2020].

Landolt, H.H. and Börnstein, R. (1983). *Landolt-Börnstein: Zahlenwerte und Funktionen aus Naturwissenschaften und Technik: Neue Serie*. Springer, Berlin.

Leroy, S. and Cannat, M. (2014). MD 199 / SISMO-SMOOTH cruise, Marion Dufresne R/V [Online]. Available at: https://campagnes.flotteoceanographique.fr/campagnes/14003300/ [Accessed 18 February 2021].

Leroy, S., Lucazeau, F., d'Acremont, E., Watremez, L., Autin, J., Rouzo, S., Bellahsen, N., Tiberi, C., Ebinger, C., Beslier, M.-O., Perrot, J., Razin, P., Rolandone, F., Sloan, H., Stuart, G., Al Lazki, A., Al-Toubi, K., Bache, F., Bonneville, A., Goutorbe, B., Huchon, P., Unternehr, P., Khanbari, K. (2010). Contrasted styles of rifting in the eastern Gulf of Aden: A combined wide-angle, multichannel seismic, and heat flow survey. *Geochemistry, Geophysics, Geosystems*, 11(7) [Online]. Available at: http://doi.wiley.com/10.1029/2009GC002963 [Accessed 12 April 2017].

Leroy, S., Razin, P., Autin, J.A., Bache, F., d'Acremont, E., Watremez, L., Robinet, J., Baurion, C., Denèle, Y., Bellahsen, N., Lucazeau, F., Rolandone, F., Rouzo, S., Kiel, J.S., Robin, C., Guillocheau, F., Tiberi, C., Basuyau, C., Beslier, M.-O., Ebinger, C., Stuart, G., Ahmed, A., Khanbari, K., Al Ganad, I., de Clarens, P., Unternehr, P., Al Toubi, K., Al Lazki, A. (2012). From rifting to oceanic spreading in the Gulf of Aden: A synthesis. *Arabian Journal of Geosciences*, 5(5), 859–901 [Online]. Available at: http://link.springer.com/10.1007/s12517-011-0475-4 [Accessed 12 April 2017].

Liang, Y., Delescluse, M., Qiu, Y., Pubellier, M., Chamot-Rooke, N., Wang, J., Nie, X., Watremez, L., Chang, S.-P., Pichot, T., Savva, D., Meresse, F. (2019). Décollements, detachments, and rafts in the extended crust of dangerous ground, South China Sea: The role of inherited contacts. *Tectonics*, 38(6), 1863–1883 [Online]. Available at: http://agupubs.onlinelibrary.wiley.com/doi/abs/10.1029/2018TC005418 [Accessed 16 December 2020].

Lucazeau, F., Leroy, S., Bonneville, A., Goutorbe, B., Rolandone, F., d'Acremont, E., Watremez, L., Dusunur, D., Tuchais, P., Huchon, P., Bellahsen, N., Al-Toubi, K. (2008). Persistent thermal activity at the Eastern Gulf of Aden after continental break-up. *Nature Geoscience*, 1(12), 854–858 [Online]. Available at: http://dx.doi.org/10.1038/ngeo359 http://www.nature.com/ngeo/journal/v1/n12/suppinfo/ngeo359_S1.html.

Lucazeau, F., Leroy, S., Rolandone, F., d'Acremont, E., Watremez, L., Bonneville, A., Goutorbe, B., Düsünur, D. (2010). Heat-flow and hydrothermal circulation at the ocean–continent transition of the eastern gulf of Aden. *Earth and Planetary Science Letters*, 295(3–4), 554–570 [Online]. Available at: http://www.sciencedirect.com/science/article/pii/S0012821X10002852 [Accessed 6 September 2012].

Maus, S., Barckhausen, U., Berkenbosch, H., Bournas, N., Brozena, J., Childers, V., Dostaler, F., Fairhead, J.D., Finn, C., von Frese, R.R.B., Gaina, C., Golynsky, S., Kucks, R., Lühr, H., Milligan, P., Mogren, S., Müller, R.D., Olesen, O., Pilkington, M., Saltus, R., Schreckenberger, B., Thébault, E., Caratori Tontini, F. (2009). EMAG2: A 2-arc min resolution Earth Magnetic Anomaly Grid compiled from satellite, airborne, and marine magnetic measurements: EARTH MAGNETIC ANOMALY GRID. *Geochemistry, Geophysics, Geosystems*, 10(8) [Online]. Available at: http://doi.wiley.com/10.1029/2009GC002471 [Accessed 20 October 2021].

Mutter, J.C., Talwani, M., Stoffa, P.L. (1982). Origin of seaward-dipping reflectors in oceanic crust off the Norwegian margin by "subaerial sea-floor spreading". *Geology*, 10(7), 353–357 [Online]. Available at: https://doi.org/10.1130/0091-7613(1982)10<353:OOSRIO>2.0.CO;2 [Accessed 15 December 2020].

Peron-Pinvidic, G., Manatschal, G., Minshull, T.A., Sawyer, D.S. (2007). Tectonosedimentary evolution of the deep Iberia-Newfoundland margins: Evidence for a complex breakup history. *Tectonics*, 26.

Pichot, T., Delescluse, M., Chamot-Rooke, N., Pubellier, M., Qiu, Y., Meresse, F., Sun, G., Savva, D., Wong, K.P., Watremez, L., Auxiètre, J.-L. (2014). Deep crustal structure of the conjugate margins of the SW South China Sea from wide-angle refraction seismic data. *Marine and Petroleum Geology*, 58, 627–643 [Online]. Available at: http://www.sciencedirect.com/science/article/pii/S0264817213002602 [Accessed 16 December 2020].

Sallarès, V., Gailler, A., Gutscher, M.-A., Graindorge, D., Bartolomé, R., Gràcia, E., Díaz, J., Dañobeitia, J.J., Zitellini, N. (2011). Seismic evidence for the presence of Jurassic oceanic crust in the central Gulf of Cadiz (SW Iberian margin). *Earth and Planetary Science Letters*, 311(1), 112–123 [Online]. Available at: http://www.sciencedirect.com/science/article/pii/S0012821X11005140 [Accessed 16 December 2020].

Sandwell, D.T. and Smith, W.H.F. (2009). Global marine gravity from retracked Geosat and ERS-1 altimetry: Ridge segmentation versus spreading rate. *Journal of Geophysical Research: Solid Earth*, 114(B1) [Online]. Available at: http://agupubs.onlinelibrary.wiley.com/doi/abs/10.1029/2008JB006008 [Accessed 16 February 2021].

Schnabel, M., Franke, D., Engels, M., Hinz, K., Neben, S., Damm, V., Grassmann, S., Pelliza, H., Dos Santos, P.R. (2008). The structure of the lower crust at the Argentine continental margin, South Atlantic at 44°S. *Tectonophysics*, 454(1–4), 14–22 [Online]. Available at: http://www.sciencedirect.com/science/article/pii/S0040195108000759 [Accessed 22 February 2013].

Szameitat, L.S.A., Manatschal, G., Nirrengarten, M., Ferreira, F.J.F., Heilbron, M. (2020). Magnetic characterization of the zigzag shaped J-anomaly: Implications for kinematics and breakup processes at the Iberia–Newfoundland margins. *Terra Nova*, 32(5), 369–380 [Online]. Available at: https://onlinelibrary.wiley.com/doi/10.1111/ter.12466 [Accessed 19 November 2020].

Vilà, M., Fernández, M., Jiménez-Munt, I. (2010). Radiogenic heat production variability of some common lithological groups and its significance to lithospheric thermal modeling. *Tectonophysics*, 490(3–4), 152–164 [Online]. Available at: http://www.sciencedirect.com/science/article/pii/S0040195110002064 [Accessed 22 February 2013].

Watremez, L., Leroy, S., Rouzo, S., d'Acremont, E., Unternehr, P., Ebinger, C., Lucazeau, F., Al-Lazki, A. (2011). The crustal structure of the north-eastern Gulf of Aden continental margin: Insights from wide-angle seismic data. *Geophysical Journal International*, 184(2), 575–594 [Online]. Available at: http://onlinelibrary.wiley.com.biblioplanets.gate.inist.fr/doi/10.1111/j.1365-246X.2010.04881.x/abstract [Accessed 25 February 2013].

Watremez, L., Helen Lau, K.W., Nedimović, M.R., Louden, K.E. (2015). Traveltime tomography of a dense wide-angle profile across Orphan Basin. *GEOPHYSICS*, 80(3), B69–B82 [Online]. Available at: http://library.seg.org/doi/10.1190/geo2014-0377.1 [Accessed 22 February 2021].

White, R. and McKenzie, D. (1989). Magmatism at rift zones: The generation of volcanic continental margins and flood basalts. *Journal of Geophysical Research*, 94(B6), 7685–7729.

White, N., Thompson, M., Barwise, T. (2003). Understanding the thermal evolution of deep-water continental margins. *Nature*, 426(6964), 334–343.

Yilmaz, Ö. (2001). *Seismic Data Analysis*. Society of Exploration Geophysicists [Online]. Available at: https://library.seg.org/doi/book/10.1190/1.9781560801580 [Accessed 22 February 2021].

Zelt, C.A. (1999). Modelling strategies and model assessment for wide-angle seismic traveltime data. *Geophysical Journal International*, 139(1), 183–204 [Online]. Available at: https://doi.org/10.1046/j.1365-246X.1999.00934.x [Accessed 22 February 2021].

Zelt, C.A. and Barton, P.J. (1998). Three-dimensional seismic refraction tomography: A comparison of two methods applied to data from the Faeroe Basin. *Journal of Geophysical Research: Solid Earth*, 103(B4), 7187–7210 [Online]. Available at: http://agupubs.onlinelibrary.wiley.com/doi/abs/10.1029/97JB03536 [Accessed 16 December 2020].

Zelt, C.A. and Smith, R.B. (1992). Seismic traveltime inversion for 2-D crustal velocity structure. *Geophysical Journal International*, 108(1), 16–34 [Online]. Available at: https://academic.oup.com/gji/article/108/1/16/661014/Seismic-traveltime-inversion-for-2-D-crustal [Accessed 7 February 2017].

5

Numerical Modeling of Rifting: An Overview

Marta PÉREZ-GUSSINYÉ and Zhonglan LIU

MARUM – Center for Marine Environmental Sciences, University of Bremen, Germany

5.1. Introduction

Our understanding of rifted margin formation has been typically built on observations of their structure, obtained from multichannel seismic reflection and wide-angle velocity data, drilling and field geology. These observations only provide their present-day structure and/or are local. Therefore, they need to be interpreted in order to unravel the spatio-temporal evolution that the margins underwent. Unfortunately, the link between present-day margin configuration and its evolution is difficult to establish because most geological processes occur on time scales that are much larger than human observation ones and are influenced by nonlinear interactions between the different variables guiding deformation. **Geodynamic modeling has become an increasingly important tool in rifted margin evolution studies, as it fills this gap by quantitatively simulating rifting processes and their interactions.** Computer simulations provide the link between final rifted margin architecture and the processes involved in shaping it. Computational geodynamic modeling has bloomed in all branches of geosciences in the last 40 years. In studies of extensional environments, it is increasingly used to analyze fundamental processes that shape rifted margin's architecture, such as

For a color version of all the figures in this chapter, see www.iste.co.uk/peronpinvidic/margins1.zip.

Continental Rifted Margins 1,
coordinated by Gwenn PERON-PINVIDIC.

lithospheric deformation, melting and sedimentation, and guide interpretations of available observations.

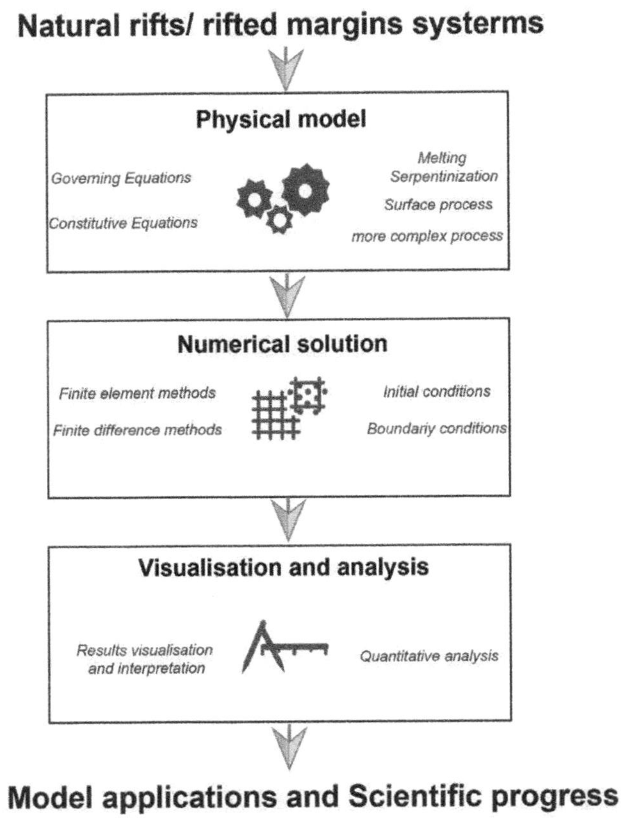

Figure 5.1. *General workflow of numerical modeling of rifting. For the numerical simulation of geological processes, it is first required to conceptualize its physics and describe the fundamental mathematical equations that guide it. These governing equations can then be solved using different numerical methods. Once the equations have been solved numerically, the results need to be visualized, interpreted and quantitatively analyzed. The schematic icon in the figure has been adapted from van Zelst* et al. *(2021)*

The first step towards numerically simulating a geological process is to conceptualize its physics and describe the fundamental mathematical equations that guide it (Figure 5.1). In geodynamics, there are three main equations that describe the conservation of momentum, mass and energy and will be described in the following sections. Most of these equations only have analytical solutions for very simple cases, which do not describe the complexity of geological systems. Although these analytical solutions for simple approximations were used in the past, today, we use modern computational methods to solve them. These methods require transforming the space-time from a continuous framework to a discrete one, where functions and variables are defined at given spatial locations and time intervals, that is, in a spatio-temporal grid or mesh. Meshes can be 1D, 2D or 3D. Transforming a function from the continuous spatio-temporal framework to a discrete one is known as **discretization**. A great variety of discretization and numerical methods to solve the required equations exist. The most widely used include **finite differences**, **finite elements**, **spectral methods**, **discrete elements** and **finite volume**. The reader is referred to Ismail-Zadeh and Tackley (2010) and Gerya (2019) for an exhaustive description of these methods. Once the equations have been solved numerically, the results need to be visualized. This is typically done as a post-processing step. Visualization and interpretation of the results are crucial to connect the model with observations.

Numerical models can be used to test and generate hypotheses, as well as to analyze the feedback between different processes. They have undergone a huge advancement in the last 40 years, and a large number of codes, some of which are available to the scientific community, have been developed (see the list in Table 5.1). Model progress has been driven by the need to incorporate more realistic and complex processes to reproduce increasingly higher-quality observations, and by the concomitant exponential increase in computing power. Numerical models constitute not only technical advances, but also their theoretical conception, and results lead to conceptual advances in our understanding of rifting. They advance hand in hand with the acquisition of newer, higher-resolution data, as we will see in the following section.

5.2. A brief historical introduction to numerical modeling of rifting

At the start of the 1970s, it was recognized that post-rift subsidence at margins resembled that of oceanic crust, that is, subsidence decayed exponentially with time (Sleep 1971). By analogy to oceanic plates, the subsidence was thought to be the consequence of thermal contraction after the heating of the lithosphere during rifting (Sleep 1971). The mechanisms by which the crust would heat and thin during rifting were not well understood. Some authors suggested that a sudden rifting event would lead to heating and broad uplift of the rift region, accompanied by extensive erosion of the crust, resulting in the observed crustal thinning (e.g. Sleep (1971) among others). Others, instead, supported the idea that the thinning of the crust was mainly caused by extensional stresses instead of erosion, and that this would mainly lead to subsidence (e.g. Bott (1971) among others). The breakthrough to this controversy arrived with the acquisition of new seismic and drill data in the Bay of Biscay and Rockall Trough in the eastern North Atlantic (Montadert *et al.* 1977; De Charpal *et al.* 1978). Seismic sections in this area showed rotated basement blocks bounded by listric faults and no evidence for extensive sub-areal erosion. Hence, extensional stresses were thought to be responsible for crustal thinning through upper crustal faulting and ductile extension of the lower crust and mantle. The conceptual idea that emerged is that the crust thins due to far-field extensional stresses: during extension, the crust subsides, the mantle rises and then melts as it decompresses during uplift. Finally, breakup occurs and a new oceanic crust is formed. Montardet *et al.* (1977) used drill data to show that post-rift subsidence was also matched by an exponential decay similar to that of oceanic plates as previously shown by Sleep (1971). Quantitative predictions for this conceptual model were first made by McKenzie (1978) and are commonly known as the "**pure-shear model**". These predictions are based on simple assumptions that lead to analytical solutions for the temperature and subsidence evolution (Figure 5.2). The simplicity of this model and its solutions have made it extraordinarily popular, and they are still regularly used in basin analysis for industrial and academic applications.

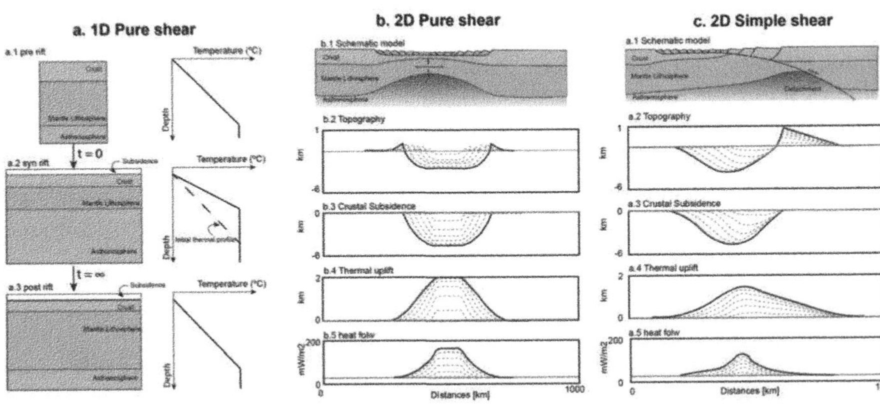

Figure 5.2. *Modes of deformations in rifts (a) The three rifting phases of the 1D pure-shear model proposed by McKenzie (1978). The pre-rift phase depicts the crustal and lithospheric thickness and its thermal state at the start of rifting. The syn-rift is the phase of active extension, during which the lithosphere thins, leading to subsidence and asthenospheric upwelling. Once extension ceases, during the post-rift phase, the lithosphere thermally relaxes and cools and the geothermal gradient returns to its initial value. Cooling during the post-rift phase leads to an increase in the density of the mantle, which results in post-rift thermal subsidence. (b, c) 2D pure-shear/simple-shear model and associated topography, crustal subsidence, thermal uplift and heat flow from Buck* et al. *(1988)*

McKenzie (1978) envisaged three phases that describe the rifting process (Figure 5.2). The pre-rift phase depicts the crustal and lithospheric thickness and its thermal state at the start of rifting. The syn-rift is the phase of active extension, during which the lithosphere thins, leading to subsidence and asthenospheric upwelling. As the asthenosphere's temperature is considered to be constant, its upwelling leads to an increase in the geothermal gradient. Once extension ceases, during the post-rift phase, the lithosphere thermally relaxes and cools and the geothermal gradient returns to its initial value (Figure 5.2a). Cooling during the post-rift phase leads to an increase in the density of the mantle, which results in post-rift thermal subsidence. In the McKenzie (1978) model, some crucial assumptions about rift development are included: 1) thinning is instantaneous in geological time scales, 2) the upper and lower crusts and mantle extend by the same amount, β, (where β, the stretching factor, is calculated by dividing the final thickness of the crust at a given location along the margin by its initial thickness before rifting), 3) lateral heat transport is not significant during rifting and 4) the lithosphere has no strength. Together, these assumptions imply that vertical, as well as lateral heat conduction are negligible and the different rheology of lithospheric layers does not play a role in the amount of extension they experience. Subsidence and uplift are calculated assuming Airy isostasy. As we will see, these assumptions are all violated to a certain extent, but they provide a first quantitative estimation of rifting processes. Given these assumptions, McKenzie (1978), estimated that the syn-rift, the heat flow for a given stretching factor β, would be equal to βQ_0, where Q_0 is the pre-rift heat flow value (Figure 5.2). Hence, in this model, to estimate the heat flow and subsidence, only the stretching factor must be known. This can be obtained from: 1) wide-angle seismic data profiles, 2) measuring the extension on faults, 3) subsidence analysis or 4) heat flow data when the rift is still active. All of these estimates should be equal if extension is assumed to be uniform within the lithosphere, but this is often not the case (see below). The pure-shear model also predicts symmetric conjugate margins, displaying the same tectonic architecture, subsidence, heat flow and sedimentation histories, with no shoulder uplift and an amount of syn-rift subsidence, which is similar to post-rift subsidence (Figure 5.2).

A series of subsequent modifications to the McKenzie (1978) model work in a similar fashion, in the sense that the stretching factor is given as an input. Subsidence is calculated by either assuming no lithospheric strength, for example, Airy isostatic equilibrium, or assuming flexural strength, in which case the flexural equation is used to calculate subsidence and uplift. We call these models kinematic since the amount of extension, that is, the deformation, which results from the rifting process, is given as an input to the model. This contrasts with more modern dynamic models in which the deformation and thus the amount of extension, together with temperature, heat flow and subsidence/uplift, are an output of the models, as we will see later. Although all models are simplifications of geological processes, the degree of simplification in kinematic models is larger than in dynamic ones, as realistic nonlinear lithospheric rheologies are not considered in the former. Nevertheless, kinematic models are still extensively used, especially by the oil industry, since they are much quicker to run than dynamic ones. Additionally, by giving the amount of extension as an input, they can be used to study the heat flow, temperature and subsidence along a given seismic profile of a margin. Instead, in dynamic models, the amount of extension is an output to the model. Dynamic models are typically used to investigate the range of initial, boundary conditions and extension velocities that lead to a general margin architecture, but cannot be used to model a given margin in particular. With this in mind, the next section gives a broad overview of the main modifications to the pure-shear model within the framework of kinematic models. Subsequently, we focus on dynamic models, which is the main topic of this chapter.

Modifications to the pure-shear model with kinematic models

First-order modifications of the pure-shear model have looked at the effects of a **finite rift duration** (Cochran 1983; Jarvis and McKenzie 1980), **lateral heat conduction** (Cochran 1983; Steckler 1985), **non-uniform thinning of crust and mantle** (Royden and Keen 1980), **lithospheric strength** on basin configuration, subsidence and uplift (Beaumont *et al.* 1982; Weissel and Karner 1989; Kooi *et al.* 1992) and the possibility that the lithosphere does not extend by pure shear, but **simple shear** (Wernicke 1985; Buck 1991; Lister *et al.* 1991). Although the general mechanism of lithospheric extension followed by post-rift subsidence has remained accepted, these modifications impart significant differences on the timing, magnitudes and spatial distribution of vertical movements and temperatures during the syn- and post-rift phases, as well as on final margin tectonic architecture.

A **finite duration of rifting** results in heat lost by conduction during the rifting event (Jarvis and McKenzie 1980; Cochran 1983). Heat can be lost in the vertical direction, that is, from depth towards the surface, as well as in the lateral direction, from the center of the rift towards the sides (Steckler 1985). Jarvis and McKenzie (1980) compared the predictions of temperature and subsidence for a finite rift event to those of instantaneous rifting. They used a simple model in which only vertical heat conduction was allowed and concluded that the predictions of the instantaneous rifting model could be used, provided the rift duration was less than 20 Myr. However, Cochran (1983) showed that if **lateral heat conduction** was also considered, heat loss would be much larger, resulting in a consequent increase of the syn-rift subsidence at the expense of the pos-trift subsidence by 25% for a rift duration of 20 Myr.

One important observation guiding the conception of rift models has been the subsidence of basins determined from deep, exploratory wells. If the timings and magnitude of the subsidence are known, implications can be drawn about the timing and amount of extension. Models can be conceived, whose predictions match those observations. The comparison of observations of model predictions is typically done after removing the influence of the sediment load to the subsidence history, a process known as "**backstripping**" (Watts and Ryan 1976). The observation in some margins (e.g. Labrador, Canada and Northwest Australia), of a smaller proportion of syn-rift to post-rift subsidence than that predicted by McKenzie (1978), was attributed to a larger amount of extension in the deeper parts of the lithosphere compared to that in the shallower parts, a modification to the pure-shear model, which is commonly known as a "**non-uniform**" **thinning** model (Royden and Keen 1980). The precise depth level at which the amount of extension becomes different was not specified by Royden and Keen (1980). Different amounts of extension in shallow brittle and deeper ductile layers have also been interpreted from seismic profiles at margins (e.g. de Charpal 1978). The non-uniform extension model often assumes that mantle thinning occurs over a broader area than crustal thinning. It predicts uplift in the rift shoulders, where crustal thinning is much less than mantle thinning, and subsidence over the center of the basin where crustal thinning is typically larger and more similar to the mantle thinning (Royden and Keen 1980). Uplift and erosion of the graben shoulders is often interpreted from the lack of pre-rift sediments in wells located in the coastal plains. These sediments are interpreted to have been uplifted and then eroded at the start of rifting (Royden and Keen 1980). A problem with the formulation of Royden and Keen's (1980) "non-uniform" model is that it needs a surface of decoupling within the lithosphere to separate regions with different stretching factors. This requirement is removed if stretching is allowed to vary continuously with depth (Rowley and Sahagian 1986).

The **strength** of the lithosphere during rifting also plays a crucial role in shaping basin and margin architecture. Upon rifting, the shoulders of a strong lithosphere will tend to flexural rebound, producing shoulder uplift (Weissel and Karner 1989).

Opposite to thermal mechanisms for rift shoulder uplift, e.g. lateral heat conduction (Cochran 1983), vertical heat transfer through small-scale convection (Buck 1985; Keen 1985) and non-uniform extension (Royden and Keen 1980), flexural uplift allows permanently elevated topography in the rift shoulders, even after rifting has ceased, as observed in many margins of the Gondwana continents (Weissel and Karner 1989). Lithospheric flexural rigidity also plays a crucial role in shaping margin architecture with initially strong lithospheres, for example, those derived from rifting a craton, predicted to form deep basins with large shoulder uplift, and weak lithospheres, those derived from rifting post-orogenic lithosphere, exhibiting shallower basins with smaller shoulder uplifts (Kooi *et al.* 1992).

The prediction of **tectonic symmetry** of conjugate margins in the McKenzie (1978) model has also been widely disputed. For example, Wernicke and Burchfield (1982) proposed a model by which the lithosphere extends in a **simple-shear** manner, through a lithospheric-scale detachment fault that cross-cuts the lithosphere and results in asymmetric margins, the upper plate and the lower plate to the detachment. Wernicke (1985) proposed that the upper plate to the detachment was characterized by little faulting compared to lithospheric thinning, a higher flank topography and larger heat flow and post-rift subsidence; while the lower plate was characterized by more abundant normal faulting, less lithospheric thinning and post-rift subsidence (Figure 5.2). Low-angle detachment-like faults have been interpreted in the oceanwards edge of the thinned continental crust at the West Iberia margin (De Charpal *et al.* 1978; Boillot and Winterer 1988; Krawczyk *et al.* 1996), the Bay of Biscay (De Charpal *et al.* 1978; Le Pichon and Sibuet 1981), the Møre and Vøring basins (Osmundsen and Ebbing 2008) and also at the Camamu basin in Brazil (Blaich *et al.* 2011). Other observations in support of large-scale simple-shear extension include the observation of asymmetric topography on opposite rift shoulders (e.g. the Arabian side of the Red Sea is much higher than the Egyptian side, Buck *et al.* (1988)) and the contrasting widths of some conjugate margins, such as those of the central South Atlantic (e.g. Santos-Namibe and Davison 1997). Buck *et al.* (1988) made quantitative predictions of the simple-shear model (Figure 5.2), comparing them to topography and heat flow over the Red Sea, and concluded that while topography better fitted a simple-shear model of extension, heat flow was better predicted by a pure-shear one. As a result of this, some models were built, which were a combination of pure shear in the ductile part and simple shear in the brittle part (Weissel and Karner 1989; Kusznir and Ziegler 1992). The extension mode of the lithosphere and the contribution of detachment-like faults to the overall asymmetry of conjugate margins is still debated today, as we will see in the later sections.

In summary, towards the end of the 1980s and throughout the 1990s, the basic mechanisms by which the lithosphere rifted were clear. Rifting consisted of crustal and lithospheric extension followed by thermal relaxation. It had been shown that various mechanisms that were linked to the lateral variation in temperature field, finite rift duration and lithospheric strength during rifting could strongly affect the temperature and subsidence evolution of margins, as well as their final tectonic architecture. At the same time, the increasing number of seismic data along margins showed that margin architectural style varies and that a given model may not explain the wide variety of tectonic architectures observed. Modeling efforts shifted towards the use and development of more complex dynamic models that could consider the spatio-temporal changes in lithospheric rheology during rifting, and explored the range of conditions that led to the variety of observed tectonic styles. In the following sections, we introduce the basic principles of dynamic modeling and then give a broad perspective on some aspects of rifting that have been illuminated by these models.

5.3. Dynamic modeling: description and formulation

5.3.1. *Eulerian and Lagrangian frameworks*

By dynamic models of rifting, we encapsulate a wide variety of numerical models that seek to understand rift dynamics by quantitatively modeling the interactions between the temperature and deformation fields during extension, given an initial lithospheric rheology and geotherm. To do this, **the models solve the momentum, energy and mass conservation equations** (see below). These equations can be written in a Eulerian or Lagrangian framework (Figure 5.3). In the Lagrangian framework, the observer that describes the flow or deformation field sits on a particle that moves with the flow. Plotting the position of a given particle through time shows the trajectory of the particle. In the Eulerian framework, the observer that describes the flow is located at a fixed point in space and describes how the flow changes at that point with time. This can be visualized by thinking of a river flow. While the Lagrangian description of the river flow would be that of an observer sitting in a boat on the river, the Eulerian description would be that of an observer sitting on the riverbank. In modeling, when the deformation is described in a Lagrangian framework, the mesh deforms with time, while in the Eulerian framework, the mesh nodes are fixed in space (Figure 5.3). The reader is referred here to Gerya (2019) for a detailed description of the Eulerian and Lagrangian frameworks.

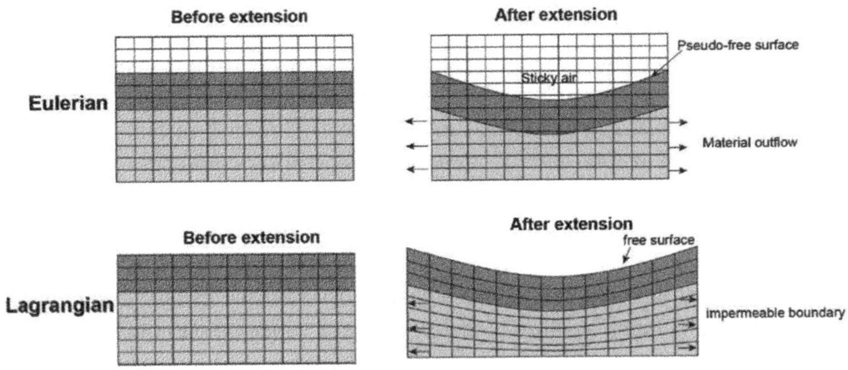

Figure 5.3. *Eulerian and Lagrangian frameworks. In the Lagrangian framework, the observer that describes the flow or deformation field sits on a particle that moves with the flow. Plotting the position of a given particle through time shows the trajectory of the particle. In the Eulerian framework, the observer that describes the flow is located at a fixed point in space and describes how the flow changes at that point with time*

5.3.2. *Governing equations*

The momentum equation is the analogue of Newton's second law of motion, $f = ma$, where m is the mass, a is the acceleration and f is the force, in a continuum mechanics framework. In a Lagrangian framework, the momentum equation is expressed by considering each node/point of the grid as a very small, infinitesimal volume, dV, then adding the forces acting over this volume and equating them to the mass times the acceleration experienced by the infinitesimal volume. This is done in each of the spatial directions, considering gravity in the vertical direction and normalizing by the infinitesimal volume $dV = \Delta x \Delta y \Delta z$. The resulting form of the equation for the x-direction is:

$$\frac{\partial \sigma_{xx}}{\partial x} + \frac{\partial \sigma_{xy}}{\partial y} + \frac{\partial \sigma_{xz}}{\partial z} + \rho g_x = \rho a_x \qquad [5.1]$$

And generalized for all directions is:

$$\frac{\partial \sigma_{ij}}{\partial x_j} + \rho g_i = \rho a_i = \rho \frac{Dv_i}{Dt} \qquad [5.2]$$

where i and j are coordinate indexes that span the spatial directions. As usual, the summation is over the indexes that are repeated, that is, the j index in equation [5.2], as shown in the equation for the x-direction, equation [5.1]. σ_{ij} are the stresses in the direction i acting over the face perpendicular to the direction j, v_i are the velocity and gravity acceleration in the direction i, t is the time and ρ is the density.

We often express the total stress, σ_{ij}, in terms of the deviatoric stress, σ_{ij}' (also often referred to as τ_{ij}), which is the deviation of the stress from the hydrostatic state:

$$\sigma_{ij} = \sigma_{ij}' - P\delta_{ij} \quad [5.3]$$

where P is the hydrostatic pressure and δ_{ij} is the Kronecker delta, which is equal to 1 for $i = j$, and 0 for $i \neq j$. By using the deviatoric stress in equation [5.2], we get the Navier–Stokes equation:

$$\frac{\partial \sigma'_{ij}}{\partial x_j} - \frac{\partial P}{\partial x_i} + \rho g_i = \rho \frac{Dv_i}{Dt} \quad [5.4]$$

The term $\rho \frac{Dv_i}{Dt}$ represents the inertial force, which is very small in comparison to ρg_i when we deal with highly viscous flow, as in geodynamics, so we neglect the term $\rho \frac{Dv_i}{Dt}$ from equation [5.4] (the reader is again referred to Gerya (2019) to see the order of magnitudes of these two terms). This approximation is called the infinite Prandtl number approximation, and by using it, we get the Stokes equation that is typically used in geodynamic codes for modeling lithospheric deformation:

$$\frac{\partial \sigma'_{ij}}{\partial x_j} - \frac{\partial P}{\partial x_i} + \rho g_i = 0 \quad [5.5]$$

The deformation is calculated by solving the momentum equation, equation [5.5], and the mass conservation equation [5.6], also called the continuity equation (Figure 5.4). The mass conservation is typically solved assuming the incompressibility approximation. Using this assumption, we get the mass conservation equation:

$$\frac{\partial v_i}{\partial x_i} = 0 \quad [5.6]$$

where i are again coordinate indexes that span the spatial directions and v is the velocity. Pressure and velocity are determined from equations [5.5] and [5.6], respectively, often using an iterative Uzawa penalty method (Pelletier *et al.* 1989).

The momentum equation is strongly dependent on the viscosity, and this is, in turn, dependent on the temperature. The temperature is calculated by solving the energy conservation equation. This can be derived by relating the change of internal energy in an infinitesimal volume to the heat flow in and out of the volume, and yields:

$$\rho C_p \frac{DT}{Dt} = \nabla \cdot (\kappa \nabla T) + H \qquad [5.7]$$

where ρ is the density, C_p is the heat capacity, κ is the thermal conductivity and H are additional heat sources, which can include radioactive heating, shear heating, etc.

5.3.3. *Rheological models*

In modern rifting codes, the constitutive relation between deviatoric stress and deviatoric strain rate, $\dot{\varepsilon}_{ij}'$, is introduced into the Stokes equation [5.5], in order to consider materials with visco-elasto-plastic rheology (Figure 5.4):

$$\sigma'_{ij} = \eta_{eff} \left(2\dot{\varepsilon}_{ij}' + \frac{\sigma'^{oldJ}_{ij}}{\mu \Delta t} \right) \qquad [5.8]$$

where η_{eff} is the effective viscosity, σ'^{oldJ}_{ij}, is the Jaumann corrotational stress, μ is the shear modulus and Δt is the time step. The effective viscosity formulation depends on the deformation regime, namely whether the material undergoes plastic yielding or deforms in the visco-elastic regime. To recognize where plasticity occurs, different criteria are employed by different authors. As an example, Andrés-Martínez *et al.* (2019) employ a Drucker–Prager yield criterion for plasticity (Drucker and Prager 1952), that is, material is considered to deform plastically when the square root of the second invariant of the deviatoric stress (σ_{II}) is larger than or equal to the yield stress (σ_y) (Ranalli 1995), where σ_y is:

$$\sigma_y = P sin\phi + C cos\phi \qquad [5.9]$$

where ϕ is the friction angle, C is the cohesion and P is pressure. To include plasticity into the viscous formulation, a Prandtl–Reuss flow law may be used (Andrés-Martínez *et al.* 2019), which brings τ_{II} to the yield envelope ($\sigma_{II} = \sigma_y$) for yielding materials, so that when $\tau_{II} > \sigma_y$, the effective viscosity is defined as:

$$\eta_{eff} = \frac{\sigma_y}{\left(2\dot{\varepsilon} + \frac{\sigma'^{oldJ}_{ij}}{\mu \Delta t} \right)_{II}} \qquad [5.10]$$

In the case of $\tau_{II} < \sigma_y$, the material flows visco-elastically and the effective viscosity is:

$$\eta_{eff} = \left(\frac{1}{\eta_{dis}} + \frac{1}{\eta_{dif}} + \frac{1}{\Delta t\mu}\right)^{-1} \quad [5.11]$$

where η_{dis} and η_{dif} are the dislocation creep viscosity and the diffusion creep viscosity, respectively, given by the general equation:

$$\eta_{dis/dif} = FB^{\frac{1}{n}} \dot{\varepsilon}_{II}^{\frac{1-n}{n}} exp\left(\frac{E+PV}{nRT}\right) \quad [5.12]$$

where F is a factor for scaling parameters obtained from uniaxial/triaxial experiments to the second-invariant-based formulation of the viscosity (Andrés-Martínez *et al.* 2019), B is the dislocation or diffusion creep constant, $\dot{\varepsilon}_{II}$ is the square root of the second invariant of the deviatoric strain rate, n is the power law exponent, E is the activation energy, V is the activation volume, R is the universal gas constant and T is the absolute temperature.

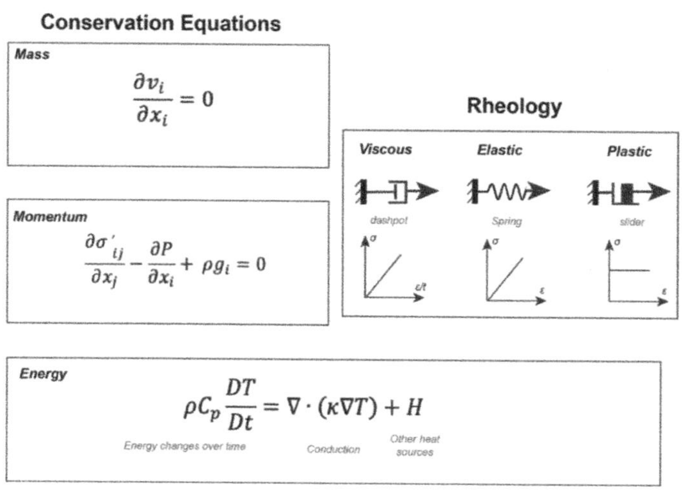

Figure 5.4. *Conservation equations and rheology model. The momentum equation (equation [5.5]) is the analogue of Newton's second law of motion. Velocities and pressures are solved using the momentum and mass conservation equation (equation [5.6]), given a rheological model. The temperature is obtained by solving the equation for the conservation of energy (equation [5.7]). In the viscous model, the deviatoric stress (σ) is proportional to the strain rate (ε̇) (equation [5.12]); in the elastic model, the stress is proportional to the strain (ε) (equation [5.8]), while plastic is used in geodynamic models to simulate the formation of brittle ruptures, where the stress does not change with strain once it reaches yield stress (equation [5.9])*

5.3.4. *Strain softening*

To simulate strain localization processes on faults and shear zones in nature, many models include strain weakening (also known as strain softening), which leads to the weakening of previously deformed elements within the mesh. Hence, an area where deformation has occurred will acquire a lower effective viscosity and will tend to focus deformation in the next time step, leading to localization of strain and formation of fault-like and shear zone-like structures. In nature, weakening mechanisms along faults include the loss in rock cohesion with accumulated strain along faults and fractures (Lavier *et al.* 2000), the increase in fluid pressure along fault planes and mineral transformations that replace strong minerals by weak minerals, both of which reduce the effective friction angle (Bos and Spiers 2002). Hence, some authors implement strain softening by reducing cohesion (e.g. Lavier *et al.* 2000), while others only reduce the friction angle. This is because, at depth, cohesion only contributes a small amount to the yield strength reduction in comparison to the friction angle, which is multiplied by pressure (equation [5.9]). Friction angle or cohesion reduction occurs from an initial to a final value, as a function of the second invariant of the finite plastic strain (Huismans and Beaumont 2007). Some models also include weakening of the ductile deformation field. This mimics weakening due to grain size reduction and crystallographic preferred orientation (Hansen *et al.* 2012). The viscous strain weakening is implemented by linearly increasing the pre-exponential factor, B, in equation [5.12] until a given factor Wmax (see Ros *et al.* 2017; Andrés-Martínez *et al.* 2019 for details).

5.3.5. *Initial and boundary conditions*

To start the model run, the thermal and rheological state of the modeled lithospheric/asthenospheric section needs to be defined. As we will see later, this profoundly influences the rift evolution and final margin architecture. The continental lithosphere is characterized by a layered rheology typically containing two strong layers in the upper crust and mantle (although the mantle may be weak if the temperature is high enough, if it is wet, or if the crust is thick). As shown in Figure 5.5, initial lithospheric strength is modified by changing the lower crustal rheology, crustal thickness and initial geotherms. In the models shown, the upper crust is assumed to have a wet quartzite rheology. Typical options for lower crustal strength include mafic granulite, wet anorthite and wet quartzite flow laws, which exhibit decreasing lower crustal strengths. These flow laws are determined in a laboratory and extrapolated to geological strain rates (e.g. Wilks and Carter 1990;

Gleason and Tullis 1995; Hirth and Kohlstedt 2003). For any given rheology, increasing crustal thickness decreases total lithospheric strength, because strong mantle is substituted by a weaker lower crust. Likewise, because rheology is inversely proportional to temperature (equation [5.12]), the effect of increasing geothermal gradient is to decrease the integrated lithospheric strength (Figure 5.5; see also Kusznir and Park 1987).

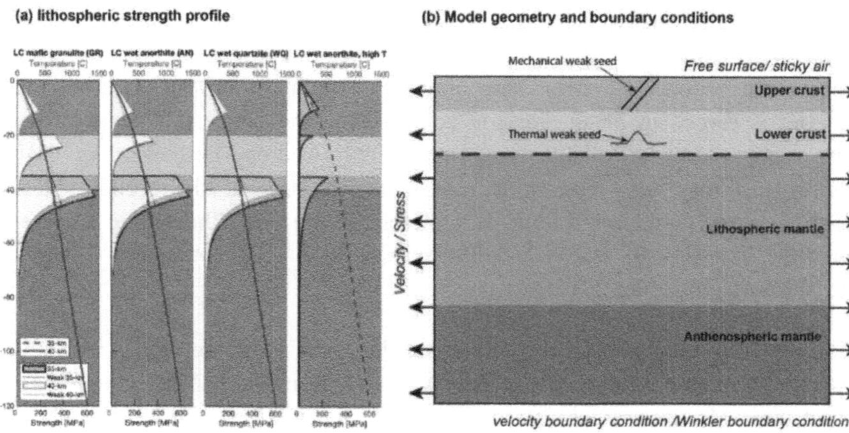

Figure 5.5. *a) Typical initial strength and temperature profiles used in geodynamic simulations of rifting. The lower crustal rheologies shown are from strong to weak, mafic granulite, GR, wet anorthite, AN, and wet quartzite, WQ. The blue envelopes are for 35 km thick crust, and the orange ones are for 40 km thick crust. The red solid and dashed lines are the geotherm in each case. The first three columns correspond to a model initialized with a thermal weak seed, TWS, at the center of the model and a geotherm with 600 °C at 35 km (Types 1a–c in Figure 5.8). The fourth column corresponds to a model initialized with a random weak seed, RWS, anorthite in the lower crust, a 35 km thick crust and a temperature of 700 °C at 35 km (Type 3 in Figure 5.8). Figure 5.8 shows the final architectures of some of these models. The strength profiles are computed with a strain rate of 10^{-15} s^{-1}. The effect of the initial weak seed on the final margin architecture is shown in Figure 5.6. Note that the total lithospheric strength decreases by 1) changing the rheology, 2) increasing the crustal thickness or 3) increasing the geothermal gradient. b) Lithospheric cross section showing typical boundary conditions choices (see sections 5.3.5 and 5.3.6) (source: adapted from Pérez-Gussinyé* et al. *2020)*

Boundary conditions are imposed at the top, bottom and sides of the model (Figure 5.5). These conditions control the temperature and velocities fields on the sides of the modeled region. They are chosen in such a way that the modeled region, that is, a cross section (in 2D models) or rectangular volume (in 3D models) of the lithosphere and underlying asthenosphere, behaves as it would in a real natural scenario of extension. Different authors use different boundary conditions depending on their understanding of the extensional system. For example, some authors use pure-shear far-field boundary conditions (BCs) that consist of prescribing the horizontal velocities on the model sides as half the extension velocities and the bottom vertical velocities as full extension velocities. The horizontal velocities on the bottom and the vertical velocities on the sides are unconstrained, that is, they are allowed to free slip (e.g. Ros *et al.* 2017). Other authors use a Winkler BC to determine the bottom velocities. This simulates a situation in which the bottom of the model is forced to be in isostatic equilibrium, and hence no lateral pressure gradients are allowed (Poliakov and Buck 1998; Andrés-Martínez *et al.* 2019). Velocity BCs may also consist of constant velocities at the model sides at lithospheric levels and an inflow of material in the sub-lithospheric mantle to balance the exit flux of material at lithospheric levels (Huismans and Beaumont 2011; Naliboff and Buiter 2015).

The top of the model intends to simulate the Earth's surface, which is a stress-free surface, where normal and shear stress vanish (Kaus *et al.* 2010; Crameri *et al.* 2012; Andrés-Martínez *et al.* 2015). This type of situation generates instabilities when the time step for computation of the velocities is larger than the viscous relaxation time (Zhong 1996; Zhong and Gurnis 1996). The reason for this is that in most geodynamic codes, the velocity is calculated at the nodes at the start of the time step and assumed to be constant throughout the time step. However, if the time step is larger than the viscous relaxation time, significant relaxation of the velocities should occur throughout the time step, which is not considered. This leads to an overestimation of the velocity throughout the time step and oscillations of the topography (see Andrés-Martínez *et al.* 2015). Several approaches are used to remediate this problem. One is the use of the "sticky air" approximation, which consists of adding a low-viscosity, low-density layer on top of the model free-surface, which is a proxy for air; although its viscosity is much larger than air (see the review in Crameri *et al.* 2012). In a benchmark to test the effect of the sticky air approximation on the geometry of the slab, Schmeling *et al.* (2008) found that the properties of the sticky-air layer control the geometry of the slab. Kaus *et al.* (2010) and Andrés-Martínez *et al.* (2015) propose a scheme in which the predicted displacement at the start of the time step is damped in order to avoid oscillations.

To solve the heat conservation equation, BCs are also needed on the temperature field (Figure 5.5). These are typically taken to be a constant temperature at the surface and a constant temperature at the initial base of the lithosphere and below it (typical values are 0°C and 1300°C/1350°C for the lithosphere's bottom). On both sides, the temperature boundary condition is generally set to have the heat flux equal to zero to ensure energy conservation throughout the system.

5.3.6. *Initial weaknesses*

Dynamic models are often designed so that the rheological structure at the start of rifting is laterally homogeneous at the start of the model run. If the model is left to run without an additional perturbation, deformation may localize on the sides of the model. In order to localize deformation in the center of the model and be free from side effects, a thermomechanical heterogeneity is introduced in the center of the model. Weak seeds can be of thermal origin, in which case they represent regions with anomalously higher temperatures than the surroundings (e.g. Ros *et al.* 2017; Andrés-Martínez *et al.* 2019), or mechanical origin, in which case a seed of reduced effective viscosity is defined (Huismans and Beaumont 2003; Naliboff and Buiter 2015). Weak seeds can be placed at different depth levels within the lithosphere. Thermal weak seeds located in weaker layers, for example, the deeper part of the mantle (e.g. Brune *et al.* 2014), tend to localize deformation much more slowly than those located in stronger layers (e.g. Huismans and Beaumont 2003). Thus, models with weak seeds in weak layers exhibit a more prolonged phase of distributed deformation at the start of rifting. Initial weaknesses can also be localized at a point or be spread over an area (Naliboff *et al.* 2017). Although weak seeds are only meant to localize the deformation at the rift start, Dyksterhuis *et al.* (2007) have shown that they have a significant influence on margin architecture. This is also observed in more modern codes, for example, Pérez-Gussinyé *et al.* (2020) show how for a given initial rheological and thermal setup, a thermal weak seed (TWS) of 100°C at the crust base leads to asymmetric margins; however, when a mechanical, spatially distributed, randomized weak seed (RWS) is used, wide, symmetric margins are produced (Figure 5.6). In the case shown in Figure 5.6, while the TWS quickly promotes deformation in the area where it is located, the RWS tends to inhibit strong spatial localization until shortly before breakup.

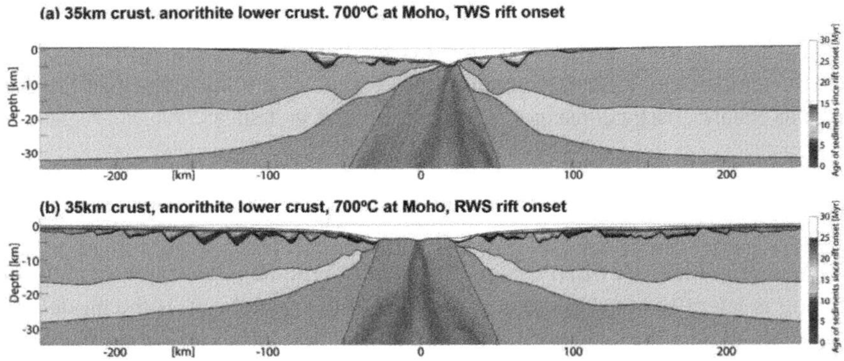

Figure 5.6. *Effect of changing the initial weak seed used to focus the deformation at the start of rifting. Both models have the same initial rheological and thermal configurations at the start of the rift. Their initial thermal and rheological profiles are shown in Figure 5.5. Model a) is initialized with a thermal weak seed (TWS). Model b) is initialized with a mechanical random weak seed (RWS). The mechanical random weak seed distributes deformation at rift start and results in symmetric wide margins (see the Type 3 margins in Figure 5.8). The thermal weak seed localizes deformation at the start of the rift and results in asymmetric margins (see section 5.3.6) (source: adapted from Pérez-Gussinyé* et al. *2020)*

5.4. Dynamic modeling: applications

This section gives a broad overview on how numerical models have helped to unravel the main controls on margin and sediment architecture. We first broadly review some early dynamic models that led to significant advances, and then we focus on results with more modern codes.

5.4.1 *Margin width and extension mode*

Rifts and rifted margins display a large variability in their width. While some rifts are very narrow and intensively thinned over ~100 km lateral distances, others exhibit much gentler thinning profiles and extend over very large lateral distances. Rifts such as the Gulf of Suez, the Rhine Graben, the east African rift system, the Baikal rift and the Rio Grande Rift are examples of narrow rifting. The Basin and Range, with more than 800 km of width, is one of the classical examples of wide rift (see Buck 1991, and the references therein). Likewise, some conjugate margins, like the South China Sea, are extremely wide and display crustal thinning from ~26 km to break up over ~200 km lateral distance on each margin (Pichot *et al.* 2014; Sun *et al.* 2018; Cameselle *et al.* 2020).

One of the main results of dynamic modeling over the last 40 years has been to show the influence of the initial lithosphere's rheology on margin tectonic architecture (Figure 5.8). Rifts tend to form along pre-existing weak trends around cratonic regions, that is, young orogenic belts, suture zones and other mobile belts, which are extensively weakened by various processes (e.g. Dunbar and Sawyer 1989). These include intensive faulting within the upper crust, increased volatile content within the mantle, locally elevated geotherms, increased crustal thickness and increased lateral heterogeneity within the crust (see Dunbar and Sawyer (1989), and the references therein). Each of these factors weakens one of the strong layers of the lithosphere, for example, an increase in crustal thickness weakens the mantle lithosphere (Figure 5.5), while an increase in quartz content in the upper crust, as in granitic arcs, reduces the strength of the upper crust. Dunbar and Sawyer (1989) used finite element method (FEM) dynamic models to understand the effect on rift architecture of an initial mantle weakness (by increasing the crustal thickness over a broad area) and of an initial crustal weaknesses (by introducing a weak granite layer in the upper crust). They found that a crustal weakness leads to localized extension in the crust and to spatially distributed extension in the mantle. This leads to initial graben formation, cooling in the lower crust and upper mantle and broad extended margins. Instead, a mantle weakness localizes extension in the mantle and spatially distributes it in the crust, leading to early melting, margin uplift during the last stages of extension and relatively narrow symmetric margins. These results are reproduced and extended in modern dynamic models, where it is shown that rifting, which initiates in a crust, which is thicker than the surrounding regions, leads to the emergence above sea level of some portions of the margin in advanced stages of thinning (Chenin *et al.* 2019).

In a closely related work, Bassi (1991) and Bassi *et al.* (1993) investigated the factors that control margin width, given the same initial weakness. They ran and compared models with relatively cold and warm initial geotherms and "wet" and "dry" rheologies. Cold and/or dry conditions lead to larger initial lithospheric strength, while warm and/or wet conditions lead to smaller initial lithospheric strength. They found that plasticity plays a central role in the evolution of lithospheric necking. The more the lithosphere behaves plastically, the earlier the necking takes place and the sharper the strain localization will be, that is, leading to the formation of narrow margins. Conversely, they observed progressive thinning across the margin when ductile deformation is dominant. They found that strengthening due to cooling during extension at slow rates could lead to the lateral migration of deformation and the formation of wide rifts. This idea had been already proposed by Kusznir and Park (1987), who suggested that for very slow extension, the strengthening effect due to the replacement of weak crust by strong mantle may be larger than the weakening arising from the increased geothermal gradient during extension (see also England 1983). The main point suggested in Bassi *et al.* (1993) is that lateral rift migration may lead to rupture on one side of the wide rift, leading to an asymmetric wide–narrow margin pair. This has been corroborated by later, more sophisticated models, which incorporate strain softening (Brune *et al.* 2014; Svartman-Dias *et al.* 2015; Tetreault and Buiter 2017).

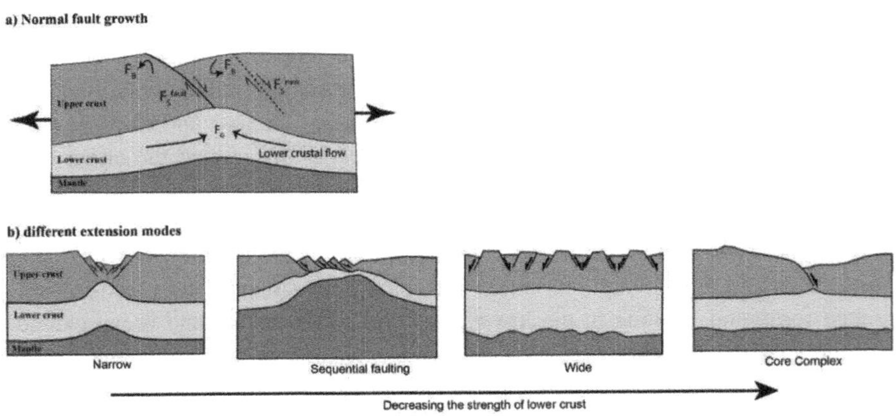

Figure 5.7. *a) Force balance during fault slip. The force required for a fault to continue to slip is made up of two components, F_B: bending stress; F_S^{fault}: the force to continue to slip on the weakened fault surface. F_S^{new} is the force needed to start a new fault. F_G is the force that induces lower crustal flow towards fault tips and results from lateral pressure gradients (see section 5.4.2 for a detailed description). b) The rifting modes are a function of the initial lower crustal strength (see section 5.4.1). Note that, in addition, rifting modes may change during extension. For example, a thick and warm continental crust, which has a very weak lower crust at rift start, may undergo core-complex mode and then transition to a wide, sequential and narrow mode when localization finally leads to breakup (see Ros* et al. *2017; Buck 1991; Andrés-Martínez 2019 for details)*

In an influential paper, Buck (1991) focused not only on the final margin architecture, but also on the modes of extension that the lithosphere experiences during rifting (Figure 5.7). Both Kusznir and Park's (1987) and Buck's (1991) calculations were not fully dynamic but are cited here because of their conceptual significance. Those works were based on the calculation of the lithospheric strength using brittle rheologies (equation [5.9]), non-Newtonian flow laws (equation [5.12]), an imposed strain rate and a 1D estimation of the temperature during rifting (equation [5.7] in 1D). Additionally, lower crustal flow due to lateral pressure gradients towards the extending area was modeled by Buck (1991). He demonstrated that, for initial rheological profiles, including very weak lower crustal viscosities, the lower crust would flow towards the extending area, erasing lateral crustal thickness variations and leading to the formation of core complexes. This flow is driven by lateral pressure gradients from the less extended areas towards the more extended areas, where active faulting occurs (Figure 5.7). Buck (1991) predicted that the initial lower crustal rheology controlled the extension mode. He proposed that with increasing extension, deformation modes would switch to those favored by stronger lower crustal rheologies, as the lower crust progressively cools during slow extension. This switch can explain the wide variety of final margin architectures we see today and their changing tectonic style from the proximal to the distal domains (Figure 5.8; see also Ros *et al.* 2017 and Pérez-Gussinyé *et al.* 2020).

In the dynamic models cited above (e.g. Bassi 1991; Dunbar and Sawyer 1989; Bassi 1991; Bassi *et al.* 1993), strain softening was not incorporated. This did not allow the simulation of faults and shear zones, nor the analysis of the evolution of the grabens and half-grabens in detail. Also, because the code solutions did not converge for large mesh deformations, the modeled lithospheric sections were only extended by a small amount and breakup was not achieved in the simulations. Still, these pioneering works already suggested that the extension of initially strong lithospheres results in narrow margins, while the extension of weak lithospheres results in wide margins, and that, for a very weak lower crust, core complexes would be formed (Figure 5.7). The later generation of numerical models incorporated higher mesh resolutions, strain softening, melting, serpentinization, surface processes and could breakup. Also, more realistic boundary conditions were implemented so that subsidence and uplift could be tracked with more confidence. In the next sections, we discuss the results when using these more modern models.

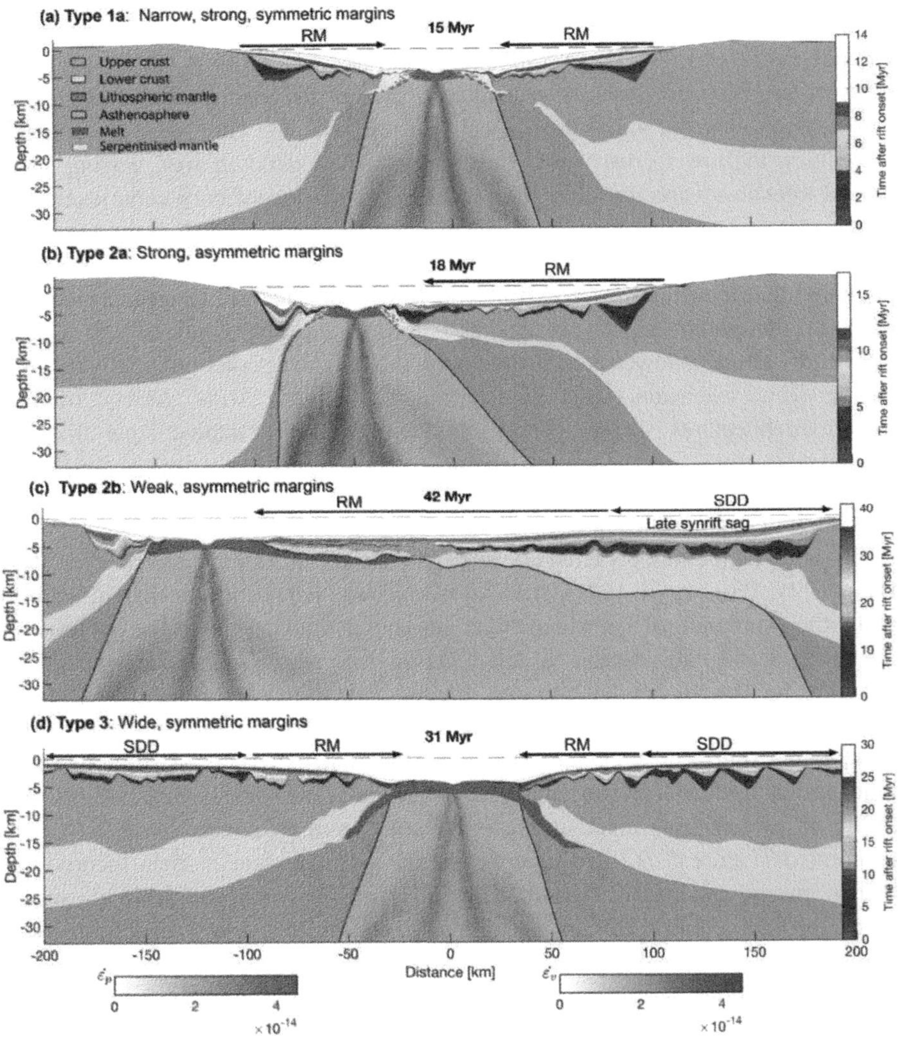

Figure 5.8. *Effect of lithospheric strength on final margin and sedimentary architecture (source: adapted from Pérez-Gussinyé* et al. *2020)*

CONTINUATION OF CAPTION FOR FIGURE 5.8.– *The integrated lithospheric strength decreases from a) to d). Models show margin configuration at 5 million years after breakup for four different rheological configurations. a) Type 1a: Narrow, symmetric margins: mafic granulite lower crust, 35 km crust, 600°C at 35 km at rift start. b) Type 2a: Strong, asymmetric margins: mafic granulite lower crust, 40 km crust, 600°C at 35 km at rift start. c) Type 2b: Weak asymmetric margins: wet quartzite and 40 km crust, 600°C at 35 km at rift start. d) Type 3: Wide symmetric margins: wet anorthite and 35 km crust, 700°C at 35 km at rift start. Models a) to c) are initialized with a thermal weak seed (TWS), and model d) is initialized with a random weak seed (RWS) (see section 5.3.6 and Figure 5.6). Sediment age is shown in a rainbow color scale for the syn-rift period, and in white for the post-rift period. The dashed blue line is the isostatically calculated sea level, assuming that a 30-km-thick crust is at sea level. The blue and red color scales and shading indicate plastic and ductile strain rate. Unidirectional arrows show the direction of rift migration (RM). The bidirectional arrows show areas of spatially distributed deformation (SDD). In areas of RM, the syn-tectonic sediments become younger oceanward. The nature of the continent-ocean transition, as predicted by results of Ros et al. (2017), is shown schematically. Yellow: serpentinized mantle; red: magmatic material. This prediction is conceptual and is based on the model results from Ros et al. (2017) (see Figure 5.10).*

5.4.2. *Normal fault geometry*

Faulting style varies widely among different rift systems and, also, from the proximal to the distal sections of margins. Normal faults may occur at a high angle, (65–30°) or a low angle (<30°). In the East African rift, where extension is far away from breakup, faults tend to occur at a high angle. Some areas of wide margins/basins, such as the South China Sea and the Basin and Range, are characterized by alternating horsts and grabens bounded by high-angle normal faults (Pichot *et al.* 2014; Cameselle *et al.* 2020), while other areas of the same wide margins exhibit faults that are tens of kilometers long, also known as core complexes, that exhume deep lower crustal rocks on their footwall and dip at very low angles at the surface (Wernicke 1985; Deng *et al.* 2020).

The relationship between extension, fault pattern and lithospheric strength may be understood from the force balance acting on a fault using a simple model (see Figure 5.7). This theoretical model was initially developed for a single elastic layer (Forsyth 1992), then for a single elasto-plastic layer (Buck 1993) and finally with fully dynamic models that simulated a single elasto-visco-plastic layer (Behn and Ito 2008; Lavier *et al.* 2000). The conceptual model assumes that the tensional force required for a fault to continue to slip is made up of two components, one to keep the fault slipping on the plane with already weakened strength (F_S^{fault}) and the other to allow the plate around the fault to bend in response to fault slip (F_B). If the fault is to continue to slip, these two components must be less than the stress required to create a new fault (F_S^{new}), that is, $F_S^{fault} + F_B < F_S^{new}$. If we modify the form of the equation, we find that $F_B < F_S^{new} - F_S^{fault}$. On the left-hand side of the equation, the bending stress increases with fault offset and the square of the lithosphere thickness according to elastic-plastic model (Buck 1993), while the right-hand side of the equation depends on the initial cohesion and friction angle, F_S^{new}, and the amount of strain weakening along the already existing fault plane, F_S^{fault} (see section 5.3.4). Therefore, both increasing the magnitude of strain weakening (Figure 5.9b) and decreasing the strength of the lithosphere (Figure 5.9a) can facilitate the formation of faults with larger offset. This is consistent with the observation that core complexes associated with large-offset faults are observed in areas where the heat flow is higher than average, implying a relatively thin brittle layer and low lithospheric strength (Lavier and Buck 2002).

In the analysis above, the effect of lower crustal strength is not directly addressed. Block and Royden (1990) noted that in order to maintain the topography observed in core complexes in the Basin and Range, large-scale lower crustal flow towards the area of extension was required. Lateral lower crustal flow towards the area of active faulting is controlled by a competition between extensional stresses that drive the lower crust out of the basin, and gravitational stresses that pulls the lower crust towards the area of active extension (Figure 5.7a). Thus, this flow may be facilitated in areas of thick continental crust, hot, very ductile and/or partially molten lower crust lower crust, which has a low viscosity and can easily flow (Rey *et al.* 2009). Dynamic models that include temperature and nonlinear relationship of strain rate with viscosity have been used to understand the relationship between lower crustal viscosity and core complex formation. As predicted by several authors, when the lower crust is very weak, either because the geothermal gradient is very high or the crust is very thick (45 km or more), lower crust tends to strongly flow towards the region of upper crustal faulting, generating core complex structures (Block and Royden 1990 and Buck 1991, among many others).

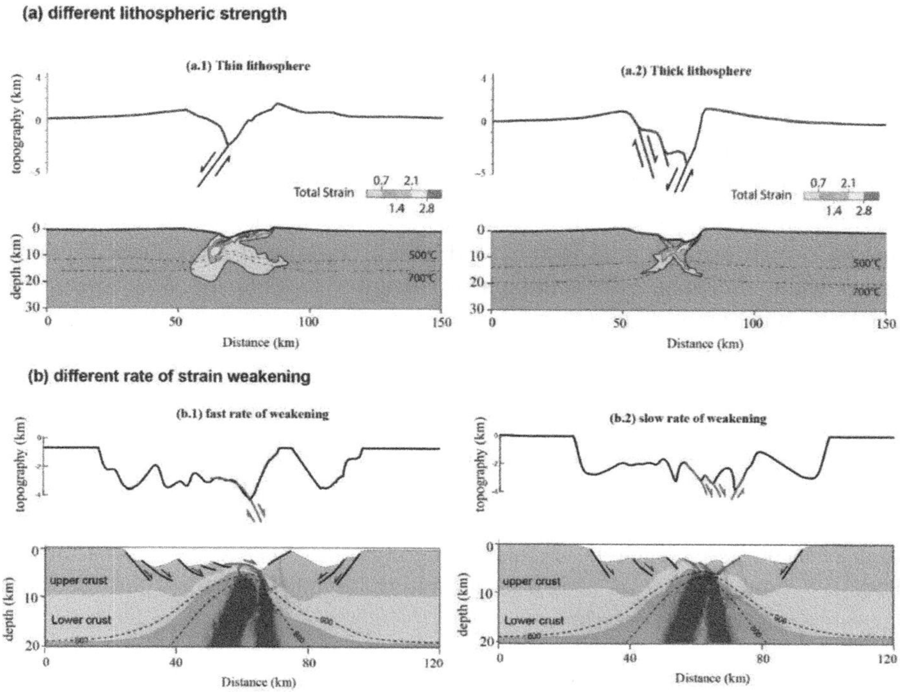

Figure 5.9. *Controls on normal fault growth during rifting. a) Effect of lithospheric thicknesses on the fault model. The results indicate that for the same choice of strain-weakening parameters, thinner lithospheres (a.1) tend to develop single, relatively large offset faults, while thicker lithospheres tend to multiple smaller faults. b) Effect of different strain-weakening parameters on the geometry of the normal faults. The model shown in b.1 shows that the material undergoes full strain weakening (internal friction angle decreases from 30° to 15°, cohesion decreases from 40 MPa to 4 MPa) when the cumulative strain reaches 0.2, whereas the model shown in b.2 requires the cumulative strain to reach 1.0 for full strain weakening to occur. The results of those models show that with thicker lithospheric thickness (e.g. early rifting), both models create multiple faults, while for thin lithospheric thickness (e.g. late rifting), a single low-angle fault is created in the faster strain weakening model and multiple high-angle faults are created in the slower strain weakening model (source: adapted from Lavier and Buck 2002)*

5.4.3. *Symmetric and asymmetric margins*

Margins display different degrees of asymmetry between their conjugates. While some display a rather symmetric conjugate tectonic structure, for example, the South China Sea, others display an asymmetric tectonic structure, with one side of the margin being much wider than the conjugate one, for example, Angola-southern Espirito Santo conjugates in the South Atlantic (Unternehr *et al.* 2010; Pérez-Gussinyé *et al.* 2020).

Early conceptual models proposed that asymmetric extension occurs when deformation is focused on a lithospheric-scale detachment fault/shear zone, which is active from the start of the rift (Figure 5.2). This results in an upper plate, narrow margin, which exhibits little faulting and abrupt crustal and lithospheric thinning, and a wide, conjugate margin, which is the lower plate to the detachment, and exhibits much more faulting and less abrupt crustal and lithospheric thinning (e.g. Lister *et al.* 1986, 1989). Multichannel seismic sections (MCS) show evidence for wide–narrow conjugate margins pairs, whose patterns of upper crustal faulting and crustal thinning resemble those predicted by the simple-shear model (Figure 5.2). However, the evidence for lithospheric-scale detachments has remained elusive in MCS. Numerical models have investigated the parameter space under which such lithospheric scale faults could be formed, and alternative mechanisms have been proposed, such as lateral rift migration (see section 5.4.1, Kusznir and Park 1987 and Bassi *et al.* 1993), in generating asymmetry without the need of a large-lithospheric scale detachment.

In a series of papers, Huismans *et al.* (2005) raised the importance of strain softening for generating margin asymmetry. They included strain softening of the brittle and ductile layers. Their coupled models, where the lower crust is strong enough to effectively connect upper crustal and mantle deformation since rift start, result in the formation of large asymmetry. The models are initiated by a plastic weak seed in the center of the model, which leads to the formation of two big faults/shear zones dipping towards it. One of them randomly accumulates a bit more strain, and due to strain softening, its strength decreases and becomes dominant. This leads to asymmetric upwelling of the asthenosphere towards its hanging wall, which becomes the weakest area, and where the next fault/shear zone will occur. This process continues, and soon after that, breakup occurs, leaving two relatively narrow, but asymmetric margins. Because the spacing between faults is a function of the strength of the lithosphere, these strong, coupled models exhibit a relatively large block size in comparison to natural observations. Decoupled models instead lead to symmetric margins. With a similar setup, but using only strain softening in the brittle domain, Huismans and Beaumont (2011) simulated narrow, symmetric margins, exhibiting large shoulder uplift and syn-rift subsidence at the center of the basin. Instead, very wide, symmetric conjugate margins arise, with little accompanying subsidence, and no shoulder uplift, when the crust is extremely weak. These results confirm earlier predictions regarding the relationship between initial rheology and margin architecture (Dunbar and Sawyer 1987, Buck *et al.* 1991,

Kooi *et al.* 1992, Bassi *et al.* 1993 among others). Additionally, Huismans and Beaumont (2011) linked margin geometry to the nature of the continent–ocean transition (section 5.4.4).

In nature, asymmetric conjugate margins, such as the magma-poor West Iberia-Newfoundland margins, often exhibit a distal section where faults consistently dip oceanward and culminate in a low-angle, detachment-like fault located at breakup location in the wide margin (e.g. Reston 1996). Pérez-Gussinyé and Reston (2001) noticed that for the very low extension velocities typical of magma-poor margins (i.e. less than 20 mm/year full extension), the lower crust would become progressively cooler. Whole crustal embrittlement would occur after a β of 3–4, which coincides with the crustal thickness observed before the detachment breakaway. They hypothesized that after embrittlement, water could reach the mantle through brittle faults, serpentinize it, and weaken it enough to allow detachment formation, crustal breakup and mantle exhumation. Although their calculations were only 1D, and deformations were not computed, these results have been widely used to explain magma-poor margin evolution (e.g. Péron-Pinvidic *et al.* 2013). Using 2D fully dynamic models of rifting and through detailed comparison and use of observations made in the ancient Tethys margin exposed in the Alps, Lavier and Manatschal (2006) proposed that deformation at magma-poor margins was guided first by the strain softening of a mid-crustal layer, which would decouple the upper crust from the lower crust resulting in symmetric extension, their "stretching mode", followed by extension affecting the distal margin, their "thinning mode" and finally their "exhumation mode". Their exhumation mode was facilitated by concave-downward detachment faults that led to crustal breakup, but, in their models, the final geometry of the margins shows a very small degree of asymmetry.

Kinematic reconstructions of the West Iberia-Newfoundland margins suggest that the asymmetry in faulting pattern and crustal thinning can be well reproduced by the emergence of an oceanward dipping and younging fault array, which is sequential in time and is only active in the future wide margin (Ranero and Pérez-Gussinyé 2010). According to this interpretation, the narrow conjugate margin is the hanging wall to the sequential fault array, whose activity laterally migrates from the wide towards the future narrow margin. Using geodynamic simulations, Brune *et al.* (2014) uncovered the dynamics by which lateral migration of deformation may occur. This migration is maintained by a lower crust, which is strong enough to allow deformation to be localized in a single fault zone at a given time, but weak and ductile enough to prevent crustal breakup by faulting (Brune *et al.* 2014, Figure 5.7). In this case, localization of deformation in a single fault leads to asymmetric uplift of the mantle, heating and weakening the hanging wall of the large active fault, where the next fault will form (Brune *et al.* 2014). The mechanism for asymmetry formation is similar to that of Huismans and Beaumont (2003). The crucial difference is that Brune *et al.* (2014) included an initial thermal weak seed in the mantle, which allowed for distributed deformation at the start of rifting. Additionally, they used a wet anorthite rheology for the lower crust, which led to much more progressive coupling of the upper crust and mantle deformation. These two factors result in a much larger degree of asymmetry of conjugate margins, a longer phase of lateral rift migration and a wider wide margin (Brune *et al.* 2014).

In summary, numerical modeling results indicate that for a very strong lower crust, upper crustal faulting is highly coupled to mantle deformation by narrow shear zones within the lower crust. These produce very effective crustal thinning, resulting in narrow symmetric margins characterized by large oceanward dipping faults that produce large syn-rift subsidence (narrow rift mode, Figures 5.7 and 5.8). However, for a very weak lower crust, the deformation is distributed over a large area, leading to the formation of symmetric margins, with small offset faults dipping both landward and oceanward and small syn-rift subsidence (wide rift mode: see Figures 5.7 and 5.8). For intermediate rheologies, margins become asymmetric and are characterized by a wide margin with a gentle crustal thinning profile and large oceanward dipping faults in its distal section, and a narrow margin with little faulting and abrupt crustal thinning (sequential fault mode: see Figures 5.7 and 5.8). The start and duration of the rift migration phase depend on the underlying crustal rheology, with strong rheologies going into asymmetric rift migration earlier, resulting in shorter wide margins than weaker rheologies (Brune *et al.* 2017).

5.4.4. *Melting and serpentinization*

Modern modeling efforts have been focused on understanding the factors that control the nature of the continent–ocean transition (COT) before normal seafloor spreading occurs. While some margins present an abrupt transition to normal magmatic, oceanic crust after the continental crust breaks up (Larsen *et al.* 2018; Sun *et al.* 2018), others, the so-called "magma-poor", exhibit a very wide, up to 200 km long, transition between the thinned continental and the oceanic crust, where the mantle is exhumed and serpentinized (e.g. West Iberia-Newfoundland margins, see Whitmarsh *et al.* 2001). Magma-poor margins where the mantle is exposed at the seafloor before steady-state seafloor spreading starts have intrigued the geological community since they were first discovered (Boillot *et al.* 1987). The reason for this is that simple 1D modeling of the thermal and magmatic evolution during rifting shows that, at normal extension rates, ~6 km of melt should be produced at crustal breakup (Bown and White 1995). Fully dynamic models that also consider the influence of lithospheric rheology on the magmatic and temperature evolution of rifting have been used to understand the nature of the continent–ocean transition (COT) at margins. The main lessons learned from those works is that extension velocity (Pérez-Gussinyé *et al.* 2006), initial mantle composition (Pérez-Gussinyé *et al.* 2006) and lower crustal rheology (Ros *et al.* 2017; Armitage *et al.* 2018) control when melting starts during rifting and how much melting is produced. For sufficiently slow extension velocities (i.e. < ~ 7 mm/year), the mantle cools during uplift so that less melt is produced. Simultaneously, cooling at shallow levels leads to crustal embrittlement and mantle serpentinization, so that once melting starts, melt is underplated beneath the already exposed and serpentinized mantle, generating a poorly magmatic COT with an exposed and serpentinized mantle, such as the West Iberia-Newfoundland margins (see Figure 5.10 and the Type 1a and 2a margins in Figure 5.8, Pérez-Gussinyé *et al.* 2006 and Ros *et al.* 2017). On the other hand, in rifts that initiate in regions with very weak lower crustal rheologies, the crust does not become brittle before breakup, even at very low extension velocities (Ros *et al.* 2017). Here, melting starts earlier than serpentinization, and the wide margins are underplated with a thin magmatic layer (see Figure 5.10, and the Type 2a and 3 margins in Figure 5.8). After breakup, the transition to normal, magmatic oceanic crust is abrupt in time and space (see Figure 5.8). This is the case for the South China Sea or the asymmetric margins of Angola-Espirito Santo (Ros *et al.* 2017). Even though melting was not included in the calculations of Huismans and Beaumont (2011), similar conclusions were reached. They associated wide margins, which developed in very weak lower crust conditions, with an abrupt transition to magmatic oceanic spreading (their Type II margins). Instead, narrow margins were implied to exhibit mantle exhumation (their Type I margins). However, the last premise only occurs when the velocity is sufficiently slow or the initial mantle is depleted (see also Manatschal *et al.* 2007).

The very same reason that allows us to explain the formation of a magma-poor COT prevents the formation of a steady state, robustly magmatic mid-ocean ridge thereafter. The solution to this may be to visualize the rift as a system bounded by constant force boundary conditions (BC), as previously shown by Brune *et al.* (2016). When a lithospheric section is subjected to constant-force BCs, the system self-accelerates as it approaches breakup, as the viscosities in the center of the model are small due to asthenospheric uplift. This explains recent compilations that show rift velocities accelerating towards breakup (Brune *et al.* 2016). However, it does not explain how the transition from a self-accelerating system to one in which the oceans spread at a constant rate occurs. Morgan and Reston (2005) instead proposed a hypothesis in which magma-poor margins only go to steady-state spreading when plume fed hot and fertile material invades them from out of the plane of extension. The effect of feeding plume material into an extending rift has been tested in 3D by various authors (see below).

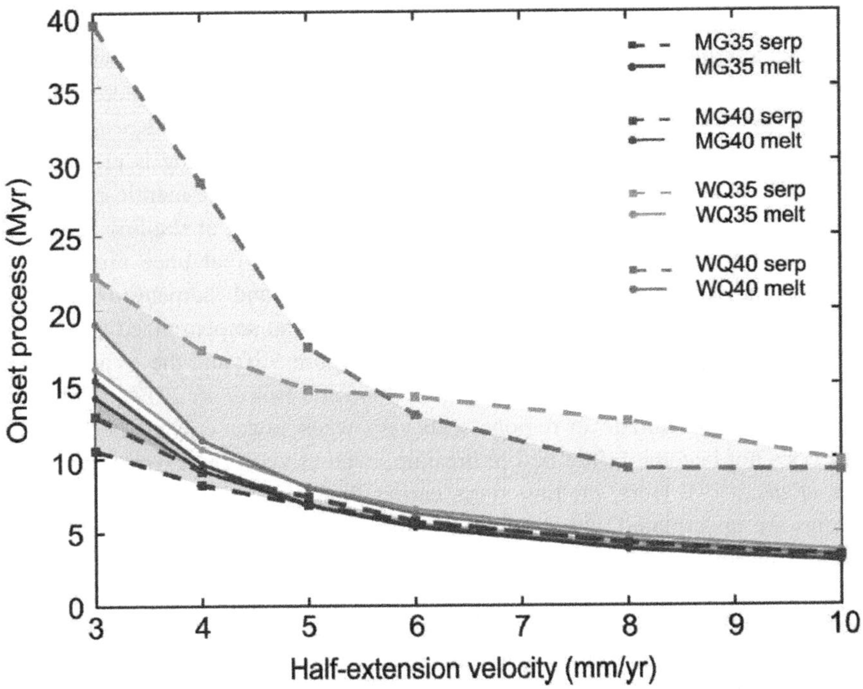

Figure 5.10. *Onset of melting and serpentinization. Results for the onset of melting and serpentinization for simulations ran at different half-extension velocities and with different initial rheologies and crustal thicknesses. The results for different models are depicted with their lower crustal rheologies: WQ: wet quartzite, MG: mafic granulite and crustal thicknesses of 35 and 40 km crust. For example, MG35 is a model with mafic granulite and 35 km thick crust. The solid and dashed lines depict the time at which the model starts to produce melting or serpentinization during the simulation, respectively. Serpentinization only starts before melting for very slow half-extension velocities and strong rheologies. In this case, it is hypothesized that the melt produced will pond beneath the impermeable serpentinite layer and a continent–ocean transition with serpentinized mantle may exist. In other cases, melting starts before serpentinization and the transition from thinned continental to oceanic crust is predicted to be abrupt and magmatic. The predictions based on this model are shown in Figure 5.8 (source: adapted from Ros* et al. *2017)*

In contrast to magma-poor margins, a large amount of magma is produced during the formation of magma-rich margins. The origin and formation mechanisms of these margins have also been discussed in several works using numerical models. Nielsen and Hopper (2002) discussed the contribution of small-scale convection versus high-temperature anomalies to the formation of magma-rich margins. They found that small-scale convection alone could reproduce the large amounts of magmatism observed, but did not result in stable oceanic spreading after breakup. They concluded that magma-rich margins could only be reproduced if there is a finite, exhaustible mantle reservoir of high temperature.

Increasing observational evidence shows that the formation of the magma-rich margin is also a function of the interaction between the incoming plume material and the topography of the lithosphere's base (Lizarralde *et al.* 2007; Minshull *et al.* 2008). Armitage *et al.* (2010) used a 2D model to show the timing of plume impact on the lithosphere's base in relation to the onset of lithospheric extension, with more spatially distributed magmatism predicted when the plume impact occurs on already thinned lithosphere. If a plume impacts beneath a stable continental lithosphere, hot and less dense plume material will tend to flow towards areas where the lithosphere is thin and surrounds thicker cratons (Koptev *et al.* 2015). Lateral flow of plume material towards areas that were already thinning may have led to the formation of magmatic margins south of the Santos-Benguela basins, in the South Atlantic, and more magma-poor margins where cratons inhibited plume flow to the north of those basins (Morgan *et al.* 2020).

5.4.5. *Surface processes*

Erosion and sedimentation occur as the rift evolves. Higher topographic areas are eroded, and mass is redistributed, modifying loads and influencing both the architecture and evolution of passive margins (Andrés-Martínez *et al.* 2019). Sediment loading changes the crustal strength and flexural stresses, and can trigger lower crustal flow that enhances subsidence and rifting (Burov and Cloetingh 1997; Burov and Poliakov 2001). Early models coupling rifting and surface processes (SP) lacked nonlinear and/or temperature-dependent viscosities of the lower crust, or lacked elasticity and/or a realistic landscape evolution model (e.g. lack of erosion and filling the basin up to the base level). Nevertheless, they showed important interactions between rifting and SP. For example, Bialas and Buck (2009) showed that narrow rifting may be promoted as a result of an input of sediments from far-field sources, because the loads reduce the buoyancy force between the shoulders and the center of the rift, which avoids rift migration. Also, sedimentation may control the transition from asymmetric to symmetric rifting due to the loading of the hanging wall (Buiter *et al.* 2008). Surface processes increase the life span of faults and the amount of horizontal strain they can accommodate. This occurs because surface processes reduce the topographic and flexural force working against the active fault (Olive *et al.* 2014). Recently, some authors have developed more realistic surface process functions and coupled them with mechanical and thermal solvers using nonlinear non-elasto-visco-plastic rheologies (Andrés-Martínez *et al.* 2019; Theunissen and Huismans 2019). The coupling is done by first solving the mechanical and thermal problems for a given time step and then solving for sediment transport and erosion within that time step (see Andrés-Martínez *et al.* (2019)). Sedimentation has two effects (see Figure 5.11): 1) it increases the pressure on the basement, leading to longer offset faults, and 2) it thermally insulates the basement, which weakens the crust and leads to more ductile deformation. Determining which of these effects dominates depends on the underlying rheology, which is controlled by the initial lithospheric rheology and the stage of rift development. Andrés-Martinez *et al.* (2019) showed that the first effect dominates in the initial phases of rifting; thus, the rift bounding faults and those in the proximal margins have larger offsets with increasing sedimentation and increasing underlying initial lower crustal strength. In the last extension phases, when the geothermal gradients are high, and especially for initially weak rheologies, the thermal blanketing effect dominates. This leads to the formation of unstructured wide distal domains and eventually to a switch from asymmetric to symmetric extension. Thus, passive margin architecture can be strongly shaped by sea level and climatic variations during extension (Andrés-Martínez *et al.* 2019).

Sedimentary architecture is typically used to constrain and map the tectonic evolution of basins and rifted margins, as sediments are more easily imaged seismically and are reached by drilling the underlying crust and mantle. Traditionally, sediments deposited during rifting had been assumed to show diverging bedding towards normal faults. Those deposited during the post-rift should show sub-horizontal geometries and overlap on the previous topography. Also, syn-rift and post-rift sediments had been assumed to be separated by an unconformity marking the lithospheric breakup event. This should correspond to a moment of large horizontal stress relief reflected as an uplift of the whole margin, which would show up as an apparent sea-level decrease at breakup (Braun and Beaumont 1989). However, in some South Atlantic margins, such as the Angola margin, post-rift-looking sag sediments with syn-rift age have been imaged. Also, other margins, such as the West Iberia-Newfoundland margins, do not show an unconformity at crustal breakup time (Tucholke *et al.* 2007). Recently, it has been shown that classical models have failed to properly identify the full geometry of syn-rift sequences because they ignore the importance of basinward migration of deformation during rifting, the potential for large lateral transport of sediments and the role of lithospheric strength in vertical movements during rifting (Figure 5.8, Pérez-Gussinyé *et al.* 2020). As the deformation moves oceanward, the age of the syn-tectonic sediment showing fan-wedge geometries becomes younger towards the center of the basin. Simultaneously, thermal subsidence starts in the abandoned proximal basins. This results in the deposition of post-rift looking sequences with syn-rift age in the proximal margin sections (Pérez-Gussinyé *et al.* 2020). We call these sequences post-tectonic, syn-rift sequences. When the rifting period is very long, these sequences can be very thick (as in the Angola margin, see Figure 5.8 and Pérez-Gussinyé *et al.* (2020)). Rift migration also produces rift migration unconformities that pre-date breakup and a break-up unconformity that only extends over the outermost margins, since breakup is not related to a sudden stress drop. Finally, sedimentary architecture, that is, the age, geometry and spatial distribution of sediments, closely mimics rift architecture, that is, when the rift is symmetric so is the sediment architecture and vice-versa (see Figure 5.8).

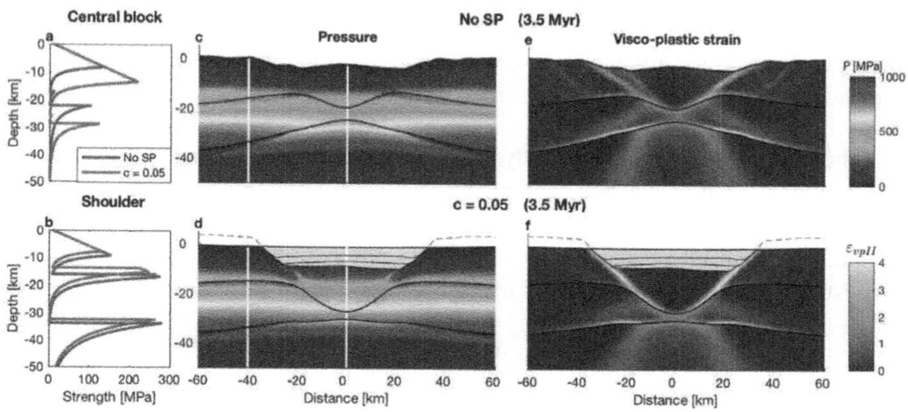

Figure 5.11. *Impacts of surface processes on pressure and strain localization. Strength profiles of a model with no surface processes (top) and with surface processes (bottom) are shown for the central block, a), and the rift shoulders, b). Panels c) and d) show pressures (P) for a model with no SP and a model with surface processes. Panels e) and f) show the second invariant of the visco-elasto-plastic strain for both models. Sediments are colored in light brown. The black lines inside the sediments represent depositional times lines at 1 Myr time intervals. The white lines in c) and d) indicate the position of the strength profiles in a) and b). The pressures in the central block are larger in the model with surface processes, c), than in the model with no surface processes, d), due to loading. This forcing results in larger displacement and faster weakening in the model with surface processes. Hence, surface processes produce larger offset faults (source: Andrés Martínez* et al. *2019)*

5.5. Summary and future perspectives

With the large increase in computational power in the last 40 years, quantitative predictions of geological processes have become increasingly accessible to geoscientists. Numerical models of rifting have been pivotal in defining how and which key factors control rift and sedimentary architecture and in unraveling the heat flow and subsidence history of margins. Models have also allowed us to better understand the interactions between sub-lithospheric processes, lithospheric deformation and sedimentation during extension. The main lessons learned are that the geological terrane in which rifting develops strongly influences margin and sedimentary architecture, faulting patterns and the nature of the transition from continental to oceanic crust. The influence of the initial geological terrain can be through the composition of its upper, lower crust (i.e. more mafic or felsic), mantle (i.e. wet or dry) or the initial thermal state of the lithosphere (i.e. hot or cold). Initial weaknesses in the models also influence the modeled rift evolution, with weaknesses located in weak layers distributing the deformation in time and space more than those weaknesses located in strong layers, which lead to more rapid localization and narrower rifts.

The near future will bring dynamic models that are able to incorporate observations, such as seismic sections and/or local petrological/thermal data from iODP drill holes, to guide the deformation in order to be able to compare models with data at fault block scale. A similar philosophy has already been adopted in the reconstruction of past mantle flow history, where mantle flow models are updated when data constraints on plate motion and seafloor ages are available (e.g. Bunge *et al.* 2003; Zhou and Liu 2017). In the study of rifting, this will bring an increased understanding of subsidence, uplift and paleo-water depths during margin formation.

Also, with the rapid development of computing power, the existing large-spatial-scale, long-time-scale geodynamic rift models have the potential to incorporate more realistic, complex geological processes at different time scales and spatial scales, for example, how magmatic processes interact with tectonic processes and influence the morphology of the eventual rift valley and its margins. In addition, a more accurate discussion of mantle serpentinization during rifting requires precise knowledge of the temperature field at the time of rifting. This requires combining the long-time-scale thermomechanical tectonic models with shorter-time-scale fluid flow models. This will shed light on the role of hydrothermal flow in tectonics, crust–ocean element exchange and the contribution of rifts in the global element cycle.

Finally, 3D models will provide a better understanding of how rifting processes are driven and influenced by their surrounding geodynamic setting (e.g. Glerum *et al.* 2020), including the interaction between distributed extension and ridge propagation (Le Pourhiet *et al.* 2018), the interaction between plumes and the shape of the lithosphere's base (Steinberger *et al.* 2019; Morgan *et al.* 2020), and how this affects subsidence, uplift and magmatism during extension.

Further reading.– The above descriptions are abbreviated and often simplified. If interested in reading and learning further, the reader is referred to the following list of publications and references.
– General: (McKenzie 1978; Buck 1991; Bassi *et al.* 1993; Lavier and Manatschal 2006; Rey *et al.* 2009; Armitage *et al.* 2010; Huismans and Beaumont 2011; Brune *et al.* 2014; Koptev *et al.* 2015; Le Pourhiet *et al.* 2018; Pérez-Gussinyé *et al.* 2020).

5.6. References

Andrés-Martínez, M., Morgan, J.P., Pérez-Gussinyé, M., Rüpke, L. (2015). A new free-surface stabilization algorithm for geodynamical modelling: Theory and numerical tests. *Physics of the Earth and Planetary Interiors*, 246, 41–51.

Andrés-Martínez, M., Pérez-Gussinyé, M., Armitage, J., Morgan, J.P. (2019). Thermomechanical implications of sediment transport for the architecture and evolution of continental rifts and margins. *Tectonics*, 38, 641–665.

Armitage, J.J., Collier, J.S., Minshull, T.A. (2010). The importance of rift history for volcanic margin formation. *Nature*, 465, 913–917.

Armitage, J.J., Whittaker, A.C., Zakari, M., Campforts, B. (2018). Numerical modelling of landscape and sediment flux response to precipitation rate change. *Earth Surface Dynamics*, 6, 77–99.

Bassi, G. (1991). Factors controlling the style of continental rifting: Insights from numerical modelling. *Earth and Planetary Science Letters*, 105, 430–452.

Bassi, G., Keen, C.E., Potter, P. (1993). Contrasting styles of rifting: Models and examples from the eastern Canadian margin. *Tectonics*, 12, 639–655.

Beaumont, C., Keen, C.E., Boutilier, R. (1982). On the evolution of rifted continental margins: Comparison of models and observations for the Nova Scotian margin. *Geophysical Journal International*, 70, 667–715.

Behn, M.D. and Ito, G. (2008). Magmatic and tectonic extension at mid-ocean ridges 1: Controls on fault characteristics. *Geochemistry, Geophysics, Geosystems*, 9, 1–22.

Bialas, R.W. and Buck, W.R. (2009). How sediment promotes narrow rifting: Application to the Gulf of California. *Tectonics*, 28.

Blaich, O.A., Faleide, J.I., Tsikalas, F. (2011). Crustal breakup and continent-ocean transition at South Atlantic conjugate margins. *Journal of Geophysical Research: Solid Earth*, 116, 1–38.

Block, L. and Royden, L.H. (1990). Core complex geometries and regional scale flow in the lower crust. *Tectonics*, 9(4), 557–567.

Boillot, G. and Winterer, E. (1988). Drilling on the Galicia margin: Retrospect and prospect. *Proceedings of the Ocean Drilling Program, Scientific Results*, 103, 809–828.

Boillot, G., Recq, M., Winterer, E.L., Meyer, A.W., Applegate, J., Baltuck, M., Baltuck, M., Bergen, J.A., Comas, M.C., Davies, T.A., Dunham, K., Evans, C.A., Girardeau, J., Goldberg, G., Haggerty, J., Jansa, L.F., Johnson, J.A., Kasahara, J., Loreau, J.P., Luna-Sierra, E., Moullade, M., Ogg, J., Sarti, M., Thurow, J., Williamson, M. (1987). Tectonic denudation of the upper mantle along passive margins: A model based on drilling results (ODP leg 103, western Galicia margin, Spain). *Tectonophysics*, 132(4), 335–342.

Bos, B. and Spiers, C.J. (2002). Frictional-viscous flow of phyllosilicate-bearing fault rock: Microphysical model and implications for crustal strength profiles. *Journal of Geophysical Research: Solid Earth*, 107, ECV 1-1–ECV 1-13.

Bott, M. (1971). Evolution of young continental margins and formation of shelf basins. *Tectonophysics*, 11, 319–327.

Bown, J.W. and White, R.S. (1995). Effect of finite extension rate on melt generation at rifted continental margins. *Journal of Geophysical Research: Solid Earth*, 100(B9), 18011–18029.

Braun, J. and Beaumont, C. (1989). A physical explanation of the relation between flank uplifts and the breakup unconformity at rifted continental margins. *Geology*, 17, 760–764.

Brune, S., Heine, C., Pérez-Gussinyé, M., Sobolev, S.V. (2014). Rift migration explains continental margin asymmetry and crustal hyper-extension. *Nature Communications*, 5, 4014.

Brune, S., Williams, S.E., Butterworth, N.P., Müller, R.D. (2016). Abrupt plate accelerations shape rifted continental margins. *Nature*, 536, 201–204.

Brune, S., Heine, C., Clift, P.D., Pérez-Gussinyé, M. (2017). Rifted margin architecture and crustal rheology: Reviewing Iberia-Newfoundland, Central South Atlantic, and South China Sea. *Marine and Petroleum Geology*, 79, 257–281.

Buck, W.R. (1985). When does small-scale convection begin beneath oceanic lithosphere? *Nature*, 313, 775–777.

Buck, W.R. (1991). Modes of continental lithospheric extension. *Journal of Geophysical Research: Solid Earth*, 96, 20161–20178.

Buck, W.R. (1993). Effect of lithospheric thickness on the formation of high-and low-angle normal faults. *Geology*, 21, 933–936.

Buck, W.R., Martinez, F., Steckler, M.S., Cochran, J.R. (1988). Thermal consequences of lithospheric extension: Pure and simple. *Tectonics*, 7, 213–234.

Buiter, S.J., Huismans, R.S., Beaumont, C. (2008). Dissipation analysis as a guide to mode selection during crustal extension and implications for the styles of sedimentary basins. *Journal of Geophysical Research: Solid Earth*, 113(B6).

Bunge, H.-P., Hagelberg, C., Travis, B. (2003). Mantle circulation models with variational data assimilation: Inferring past mantle flow and structure from plate motion histories and seismic tomography. *Geophysical Journal International*, 152, 280–301.

Burov, E. and Cloetingh, S. (1997). Erosion and rift dynamics: New thermomechanical aspects of post-rift evolution of extensional basins. *Earth and Planetary Science Letters*, 150, 7–26.

Burov, E. and Poliakov, A. (2001). Erosion and rheology controls on synrift and postrift evolution: Verifying old and new ideas using a fully coupled numerical model. *Journal of Geophysical Research: Solid Earth*, 106, 16461–16481.

Cameselle, A.L., Ranero, C.R., Barckhausen, U. (2020). Understanding the 3D formation of a wide rift: The Central South China Sea rift system. *Tectonics*, 39, e2019TC006040.

de Charpal, O., Guennoc, P., Montadert, L., Roberts, D. (1978). Rifting, crustal attenuation and subsidence in the Bay of Biscay. *Nature*, 275, 706–711.

Chenin, P., Jammes, S., Lavier, L.L., Manatschal, G., Picazo, S., Müntener, O., Karner, G.D., Figueredo, P.H., Johnson, C. (2019). Impact of mafic underplating and mantle depletion on subsequent rifting: A numerical modeling study. *Tectonics*, 38, 2185–2207.

Cochran, J.R. (1983). Effects of finite rifting times on the development of sedimentary basins. *Earth and Planetary Science Letters*, 66, 289–302.

Crameri, F., Schmeling, H., Golabek, G., Duretz, T., Orendt, R., Buiter, S., May, D., Kaus, B., Gerya, T., Tackley, P. (2012). A comparison of numerical surface topography calculations in geodynamic modelling: An evaluation of the "sticky air" method. *Geophysical Journal International*, 189, 38–54.

Davison, I. (1997). Wide and narrow margins of the Brazilian South Atlantic. *Journal of the Geological Society*, 154, 471–476.

Deng, H., Ren, J., Pang, X., Rey, P.F., McClay, K.R., Watkinson, I.M., Zheng, J., Luo, P. (2020). South China Sea documents the transition from wide continental rift to continental break up. *Nature Communications*, 11, 1–9.

Drucker, D.C. and Prager, W. (1952). Soil mechanics and plastic analysis or limit design. *Quarterly of Applied Mathematics*, 10, 157–165.

Dunbar, J.A. and Sawyer, D.S. (1989). How preexisting weaknesses control the style of continental breakup. *Journal of Geophysical Research: Solid Earth*, 94, 7278–7292.

Dyksterhuis, S., Rey, P., Müller, R., Moresi, L. (2007). Effects of initial weakness on rift architecture. *Geological Society, London, Special Publications*, 282, 443–455.

England, P. (1983). Constraints on extension of continental lithosphere. *Journal of Geophysical Research: Solid Earth*, 88, 1145–1152.

Forsyth, D.W. (1992). Finite extension and low-angle normal faulting. *Geology*, 20, 27–30.

Gerya, T. (2019). *Introduction to Numerical Geodynamic Modelling*. Cambridge University Press, New York.

Hansen, L., Zimmerman, M., Kohlstedt, D. (2012). The influence of microstructure on deformation of olivine in the grain-boundary sliding regime. *Journal of Geophysical Research: Solid Earth*, 117, 1–17.

Hirth, G. and Kohlstedt, D. (2003). Rheology of the upper mantle and the mantle wedge: A view from the experimentalists. In *Inside the Subduction Factory*, Hirth, G. and Kohlstedt, D. (eds). American Geophysical Union, Washington, DC [Online]. Available at: https://doi.org/10.1029/138GM06.

Huismans, R.S. and Beaumont, C. (2003). Symmetric and asymmetric lithospheric extension: Relative effects of frictional-plastic and viscous strain softening. *Journal of Geophysical Research: Solid Earth*, 108, 1–13.

Huismans, R. and Beaumont, C. (2007). Roles of lithospheric strain softening and heterogeneity in determining the geometry of rifts and continental margins. *Geological Society, London, Special Publications*, 282, 111–138.

Huismans, R. and Beaumont, C. (2011). Depth-dependent extension, two-stage breakup and cratonic underplating at rifted margins. *Nature*, 473, 74–78.

Huismans, R., Buiter, S.J., Beaumont, C. (2005). Effect of plastic-viscous layering and strain softening on mode selection during lithospheric extension. *Journal of Geophysical Research: Solid Earth*, 110(B2), 1–17.

Gleason, G.C. and Tullis, J. (1995). A flow law for dislocation creep of quartz aggregates determined with the molten salt cell. *Tectonophysics*, 247(1), 1–23 [Online]. Available at: https://doi.org/10.1016/0040-1951(95)00011-B.

Glerum, A., Brune, S., Stamps, S., Strecker, M. (2020). Victoria continental microplate dynamics controlled by the lithospheric strength distribution of the East African Rift. *Nature Communications*, 11, 2881 [Online]. Available at: https://doi.org/10.1038/s41467-020-16176-x.

Ismail-Zadeh, A. and Tackley, P. (2010). *Computational Methods for Geodynamics*. Cambridge University Press, New York.

Jarvis, G.T. and McKenzie, D.P. (1980). Sedimentary basin formation with finite extension rates. *Earth and Planetary Science Letters*, 48, 42–52.

Kaus, B.J.P. (2010). Factors that control the angle of shear bands in geodynamic numerical models of brittle deformation. *Tectonophysics*, 484, 36–47.

Kaus, B.J.P., Mühlhaus, H., May, D.A. (2010). A stablization algorithm for geodynamic numerical simulations with a free surface. *Physics of The Earth and Planetary Interiors*, 181(102), 12–20.

Keen, C. (1985). The dynamics of rifting: Deformation of the lithosphere by active and passive driving forces. *Geophysical Journal International*, 80, 95–120.

Keir, D., Pagli, C., Bastow, I.D., Ayele, A. (2011). The magma-assisted removal of Arabia in Afar: Evidence from dike injection in the Ethiopian rift captured using InSAR and seismicity. *Tectonics*, 30, 1–13.

Kendall, J.-M., Stuart, G., Ebinger, C., Bastow, I., Keir, D. (2005). Magma-assisted rifting in Ethiopia. *Nature*, 433, 146–148.

Kooi, H., Cloetingh, S., Burrus, J. (1992). Lithospheric necking and regional isostasy at extensional basins 1. Subsidence and gravity modeling with an application to the Gulf of Lions margin (SE France). *Journal of Geophysical Research: Solid Earth*, 97, 17553–17571.

Koptev, A., Calais, E., Burov, E., Leroy, S., Gerya, T. (2015). Dual continental rift systems generated by plume–lithosphere interaction. *Nature Geoscience*, 8, 388–392.

Krawczyk, C., Reston, T., Boillot, G., Beslier, M.-O., Riegel, R. (1996). Evidence for detachment tectonics on the Iberia Abyssal Plain rifted margin. *Proceedings of the Ocean Drilling Program: Scientific Results*, 603–615.

Kusznir, N. and Park, R. (1987). The extensional strength of the continental lithosphere: Its dependence on geothermal gradient, and crustal composition and thickness. *Geological Society, London, Special Publications*, 28, 35–52.

Kusznir, N. and Ziegler, P. (1992). The mechanics of continental extension and sedimentary basin formation: A simple-shear/pure-shear flexural cantilever model. *Tectonophysics*, 215, 117–131.

Larsen, H.C., Mohn, G., Nirrengarten, M., Sun, Z., Stock, J., Jian, Z., Klaus, A., Alvarez-Zarikian, C., Boaga, J., Bowden, S. (2018). Rapid transition from continental breakup to igneous oceanic crust in the South China Sea. *Nature Geoscience*, 11, 782–789.

Lavier, L.L. and Buck, W.R. (2002). Half graben versus large-offset low-angle normal fault: Importance of keeping cool during normal faulting. *Journal of Geophysical Research: Solid Earth*, 107, ETG 8-1–ETG 8-13.

Lavier, L.L. and Manatschal, G. (2006). A mechanism to thin the continental lithosphere at magma-poor margins. *Nature*, 440, 324.

Lavier, L.L., Buck, W.R., Poliakov, A.N. (2000). Factors controlling normal fault offset in an ideal brittle layer. *Journal of Geophysical Research*, 105, 431–423.

Le Pichon, X. and Sibuet, J.C. (1981). Passive margins: A model of formation. *Journal of Geophysical Research: Solid Earth*, 86, 3708–3720.

Le Pourhiet, L., Chamot-Rooke, N., Delescluse, M., May, D.A., Watremez, L., Pubellier, M. (2018). Continental break-up of the South China Sea stalled by far-field compression. *Nature Geoscience*, 11, 605–609.

Lister, G.S. and Davis, G.A. (1989). The origin of metamorphic core complexes and detachment faults formed during Tertiary continental extension in the northern Colorado River region, USA. *Journal of Structural Geology*, 11(1–2), 65–94.

Lister, G.S., Etheridge, M.A., Symonds, P.A. (1986). Detachment faulting and the evolution of passive continental margins. *Geology*, 14(3), 246–250.

Lister, G., Etheridge, M., Symonds, P. (1991). Detachment models for the formation of passive continental margins. *Tectonics*, 10, 1038–1064.

Lizarralde, D., Axen, G.J., Brown, H.E., Fletcher, J.M., González-Fernández, A., Harding, A.J., Holbrook, W.S., Kent, G.M., Paramo, P., Sutherland, F. (2007). Variation in styles of rifting in the Gulf of California. *Nature*, 448, 466–469.

Manatschal, G., Müntener, O., Lavier, L., Minshull, T., Peron-Pinvidic, G. (2007). Observations from the Alpine Tethys and Iberia–Newfoundland margins pertinent to the interpretation of continental breakup. *Geological Society, London, Special Publications*, 282, 291–324.

McKenzie, D. (1978). Some remarks on the development of sedimentary basins. *Earth and Planetary Science Letters*, 40, 25–32.

Minshull, T.A., Lane, C.I., Collier, J.S., Whitmarsh, R.B. (2008). The relationship between rifting and magmatism in the northeastern Arabian Sea. *Nature Geoscience*, 1, 463–467.

Montadert, L., Roberts, D., Auffret, G., Bock, W., DuPeuble, P., Hailwood, E., Harrison, W., Kagami, H., Lumsden, D., Muller, C. (1977). Rifting and subsidence on passive continental margins in the North East Atlantic. *Nature*, 268, 305–309.

Morgan, J.P., Reston, T., Ranero, C. (2004). Contemporaneous mass extinctions, continental flood basalts, and "impact signals": Are mantle plume-induced lithospheric gas explosions the causal link? *Earth and Planetary Science Letters*, 217, 263–284.

Morgan, J.P., Reston, T.J., Ranero, C.R. (2005). Contemporaneous mass extinctions, continental flood basalts, and "impact signals": Are mantle plume-induced lithospheric gas explosions the causal link? *Earth and Planetary Science Letters*, 3(236), 938–941.

Morgan, J.P., Taramón, J.M., Araujo, M., Hasenclever, J., Pérez-Gussinyé, M. (2020). Causes and consequences of asymmetric lateral plume flow during South Atlantic rifting. *Proceedings of the National Academy of Sciences*, 117, 27877–27883.

Naliboff, J.B. and Buiter, S.J. (2015). Rift reactivation and migration during multiphase extension. *Earth and Planetary Science Letters*, 421, 58–67.

Naliboff, J.B., Buiter, S.J., Péron-Pinvidic, G., Osmundsen, P.T., Tetreault, J. (2017). Complex fault interaction controls continental rifting. *Nature Communications*, 8(1), 1–9.

Nielsen, T.K. and Hopper, J.R. (2002). Formation of volcanic rifted margins: Are temperature anomalies required? *Geophysical Research Letters*, 29, 1811–1814.

Olive, J.A., Behn, M.D., Malatesta, L.C. (2014). Modes of extensional faulting controlled by surface processes. *Geophysical Research Letters*, 41, 6725–6733.

Osmundsen, P. and Ebbing, J. (2008). Styles of extension offshore mid-Norway and implications for mechanisms of crustal thinning at passive margins. *Tectonics*, 27.

Pelletier, D., Fortin, A., Camarero, R. (1989). Are fem solutions of incompressible flows really incompressible? (or how simple flows can casuse headaches!). *International Journal for Numerical Methods in Fluids*, 9, 99–112.

Pérez-Gussinyé, M. and Reston, T.J. (2001). Rheological evolution during extension at nonvolcanic rifted margins: Onset of serpentinization and development of detachments leading to continental breakup. *Journal of Geophysical Research: Solid Earth*, 106(B3), 3961–3975.

Pérez-Gussinyé, M., Reston, T., Morgan, J.P. (2001). Serpentinization and magmatism during extension at non-volcanic margins: The effect of initial lithospheric structure. *Geological Society, London, Special Publications*, 187, 551–576.

Pérez-Gussinyé, M., Morgan, J.P., Reston, T.J., Ranero, C.R. (2006). The rift to drift transition at non-volcanic margins: Insights from numerical modelling. *Earth & Planetary Science Letters*, 244, 458–473.

Pérez-Gussinyé, M., Andrés-Martínez, M., Araújo, M., Xin, Y., Armitage, J., Morgan, J. (2020). Lithospheric strength and rift migration controls on synrift stratigraphy and breakup unconformities at rifted margins: Examples from numerical models, the Atlantic and South China Sea margins. *Tectonics*, 39, e2020TC006255.

Peron-Pinvidic, G., Manatschal, G., Osmundsen, P.T. (2013). Structural comparison of archetypal Atlantic rifted margins: A review of observations and concepts. *Marine and Petroleum Geology*, 43, 21–47.

Pichot, T., Delescluse, M., Chamot-Rooke, N., Pubellier, M., Qiu, Y., Meresse, F., Sun, G., Savva, D., Wong, K., Watremez, L. (2014). Deep crustal structure of the conjugate margins of the SW South China Sea from wide-angle refraction seismic data. *Marine and Petroleum Geology*, 58, 627–643.

Poliakov, A.N. and Buck, W.R. (1998). Mechanics of stretching elastic-plastic-viscous layers: Applications to slow-spreading mid-ocean ridges. *Geophysical Monograph-American Geophysical Union*, 106, 305–324.

Ranalli, G. (1995). *Rheology of the Earth*. Springer, Netherlands.

Ranero, C.R. and Pérezgussinyé, M. (2010). Sequential faulting explains the asymmetry and extension discrepancy of conjugate margins. *Nature*, 468, 294–299.

Reston, T. (1996). The S reflector west of Galicia: The seismic signature of a detachment fault. *Geophysical Journal International*, 127, 230–244.

Rey, P., Teyssier, C., Whitney, D.L. (2009). Extension rates, crustal melting, and core complex dynamics. *Geology*, 37, 391–394.

Ros, E., Pérez-Gussinyé, M., Araújo, M., Romeiro, M.T., Andrés-Martínez, M., Morgan, J.P. (2017). Lower crustal strength controls on melting and serpentinization at magma-poor margins: Potential implications for the South Atlantic. *Geochemistry Geophysics Geosystems*, 18, 4538–4557.

Rowley, D.B. and Sahagian, D. (1986). Depth-dependent stretching: A different approach. *Geology*, 14, 32–35.

Royden, L. and Keen, C. (1980). Rifting process and thermal evolution of the continental margin of eastern Canada determined from subsidence curves. *Earth and Planetary Science Letters*, 51, 343–361.

Sawyer, D.S. (1985). Total tectonic subsidence: A parameter for distinguishing crust type at the US Atlantic continental margin. *Journal of Geophysical Research: Solid Earth*, 90, 7751–7769.

Schmeling, A., Grundmann, C., Fuhrmann, A., Kaatsch, H.-J., Knell, B., Ramsthaler, F., Reisinger, W., Riepert, T., Ritz-Timme, S., Rösing, F. (2008). Criteria for age estimation in living individuals. *International Journal of Legal Medicine*, 122, 457–460.

Sleep, N.H. (1971). Thermal effects of the formation of Atlantic continental margins by continental break up. *Geophysical Journal International*, 24, 325–350.

Steckler, M.S. (1985). Uplift and extension at the Gulf of Suez: Indications of induced mantle convection. *Nature*, 317, 135–139.

Steinberger, B., Bredow, E., Lebedev, S., Schaeffer, A., Torsvik, T.H. (2019). Widespread volcanism in the Greenland–North Atlantic region explained by the Iceland plume. *Nature Geoscience*, 12(1), 61–68.

Sun, Z., Jian, Z.M., Stock, J.M., the Expedition 367/368 Scientists (2018). South China Sea Rifted Margin. *Proc. IODP*, 367/368.

Svartman Dias, A.E., Lavier, L.L., Hayman, N.W. (2015). Conjugate rifted margins width and asymmetry: The interplay between lithospheric strength and thermomechanical processes. *Journal of Geophysical Research: Solid Earth*, 120, 8672–8700 [Online]. Available at: https://doi.org/10.1002/2015JB012074.

Tetreault, J.L. and Buiter, S.J.H. (2018). The influence of extension rate and crustal rheology on the evolution of passive margins from rifting to break-up. *Tectonophysics*, 746, 155–172.

Theunissen, T. and Huismans, R.S. (2019). Long-term coupling and feedback between tectonics and surface processes during non-volcanic rifted margin formation. *Journal of Geophysical Research: Solid Earth*, 124, 12323–12347.

Tucholke, B., Sawyer, D., Sibuet, J.-C. (2007). Breakup of the Newfoundland–Iberia rift. *Geological Society, London, Special Publications*, 282, 9–46.

Unternehr, P., Peron-Pinvidic, G., Manatschal, G., Sutra, E. (2010). Hyper-extended crust in the South Atlantic: In search of a model. *Petroleum Geoscience*, 16, 207–215.

Watts, A. and Ryan, W. (1976). Flexure of the lithosphere and continental margin basins In *Developments in Geotectonics*, 12, 25–44.

Weissel, J.K. and Karner, G.D. (1989). Flexural uplift of rift flanks due to mechanical unloading of the lithosphere during extension. *Journal of Geophysical Research: Solid Earth*, 94, 13919–13950.

Wernicke, B. (1985). Uniform-sense normal simple shear of the continental lithosphere. *Revue canadienne des sciences de la terre*, 22, 108–125.

Wernicke, B. and Burchfiel, B.C. (1982). Modes of extensional tectonics. *Journal of Structural Geology*, 4(2), 105–115.

Whitmarsh, R.B., Manatschal, G., Minshull, T.A. (2001). Evolution of magma-poor continental margins from rifting to seafloor spreading. *Nature*, 413(6852), 150–154.

Wilks, K. and Carter, N. (1990). Rheology of some continental lower crustal rocks. *Tectonophysics*, 182(1–2), 57–77.

van Zelst, I., Crameri, F., Pusok, A.E., Glerum, A., Dannberg, J., Thieulot, C. (2021). 101 geodynamic modelling: How to design, carry out, and interpret numerical studies. *Solid Earth Discussions*, 1–80.

Zhong, S. (1996). Analytic solutions for Stokes' flow with lateral variations in viscosity. *Geophysical Journal International*, 124, 18–28.

Zhong, S. and Gurnis, M. (1996). Interaction of weak faults and non-Newtonian rheology produces plate tectonics in a 3D model of mantle flow. *Nature*, 383, 245–247.

Zhou, Q. and Liu, L. (2017). A hybrid approach to data assimilation for reconstructing the evolution of mantle dynamics. *Geochemistry, Geophysics, Geosystems*, 18, 3854–3868.

6

Analogue Modeling of Continental Rifting: An Overview

Frank ZWAAN and Guido SCHREURS
University of Bern, Switzerland

6.1. Introduction

When studying the evolution of rifts and rifted margins, geologists have to face various challenges. Such tectonic systems extend over large parts of the globe, making it hard to chart them in detail. Furthermore, large portions of these systems are buried under thick layers of sediment or are covered by water, strongly reducing their accessibility. These problems have been mitigated to a certain degree using geophysical techniques, in particular seismic methods, and deep borehole drilling. Yet perhaps the greatest obstacle on the path to a thorough understanding of rift processes is posed by the timescale at which they operate; it is simply not possible to directly observe the development of a rift system over millions of years, which prevents us from fully understanding the dynamics involved.

Recognizing these challenges, geologists have long since turned to analogue models. By choosing the correct materials and experimental setup, it is possible to simulate tectonic processes stretching over vast spatial and temporal scales within a

---

For a color version of all the figures in this chapter, see www.iste.co.uk/peronpinvidic/margins1.zip.

Continental Rifted Margins 1,
coordinated by Gwenn PERON-PINVIDIC.

couple of hours or days in the laboratory. This approach provides an easy and relatively cheap method for testing various parameters that may affect tectonic systems, yielding unique insights in associated dynamics and kinematics that are not readily deduced from static field examples, which is especially relevant for petroleum geologists (e.g. Naylor *et al.* 1994; Brun and Fort 2004). The first such analogue model (or experiment), simulating tectonic folding, was published by Sir James Hall in 1815 (Hall 1815). In the 200 years since the first experiments, a wide variety of methods have been used to simulate a vast spectrum of tectonic processes (e.g. Koyi 1997; Gravelau *et al.* 2012). It must however be stressed that analogue modeling applications are not restricted to tectonics, but these techniques have also been applied to study, for instance, sedimentary processes, magmatic events, slope collapse and tsunami hazards (e.g. Donnadieu *et al.* 2003; Wang *et al.* 2015; McFall and Fritz 2016; Poppe *et al.* 2019). Meanwhile, a broad arsenal of methods has been developed to not only observe and describe but also quantify surface- and internal model deformation, so that the analogue modeling community remains at the forefront of geological innovation.

Although the first models were conducted to simulate compressional tectonics, numerous experimental studies have addressed extensional tectonics over the years. Ample references to these studies can be found in the reviews and overview papers by Vendeville *et al.* (1987), McClay (1990, 1996), Allemand and Brun (1991), Beslier (1991), Naylor *et al.* (1994), Koyi (1997), Brun (1999, 2002), Michon and Merle (2000, 2003), Corti *et al.* (2003), Bahroudi *et al.* (2003), Corti (2012), Brun *et al.* (2018) and Zwaan *et al.* (2019). The aim of this chapter is therefore not to present an exhaustive review of all preceding publications but to provide an overview of analogue modeling of rift tectonics, describing the general methodology (materials and scaling, setups and state-of-the-art analysis techniques), and to illustrate how these can be applied for studying a variety of aspects of rifts and rifted margins. We also describe the current challenges and opportunities in the field, which revolve around key topics such as rheology, structural inheritance and kinematics, and hope that this work may serve as a guide and inspiration for future analogue modeling studies.

6.2. Methodology

Before running an experiment, analogue modelers need to carefully consider factors such as scaling, model materials and setup in order to ensure that the model simulates the chosen natural tectonic setting as well as possible.

6.2.1. *Scaling principles*

When using analogue modeling techniques, proper scaling is necessary to guarantee (1) geometrical, (2) kinematic and (3) dynamic similarity between a model and its natural equivalent. These similarities can be expressed by means of simple scaling equations (Hubbert 1937; Ramberg 1981; Le Calvez 2002; Corti *et al.* 2003) and references therein). Geometrical similarity implies that all dimensions (length, width, height, layer thickness, fault angles) in the analogue model have the same proportions as in the natural prototype. Kinematic similarity means that the model and the natural example maintain geometric similarity during their deformation without developing any temporal distortions along the way (i.e. structures develop at the correct moment in time). Finally, dynamic similarity is established when all forces, stresses and the rheology of the materials are properly translated from the natural example to the model scale (i.e. all forces maintain the same relative proportions as in nature). Although it is practically impossible to incorporate all detailed complexities that characterize natural geological settings into a small laboratory experiment, a correct scaling of the dominant factors controlling deformation will allow the scaling criteria to be fulfilled. In order to achieve this, it is important to select the proper analogue materials that reproduce the behavior of the lithosphere, and to choose an experimental setup and model dimensions, as well as a deformation rate, appropriate for simulating a specific tectonic setting (e.g. Bahroudi *et al.* 2003; Zwaan *et al.* 2019).

6.2.2. *Materials*

For modeling the brittle parts of the lithosphere, granular materials such as fine quartz sand are commonly used, but other materials such as wet clay or wheat flour can also be applied (e.g. Krantz 1991; Schellart and Strak 2016; Reber *et al.* 2020). These materials, of which the rheological properties can be tested with a ring-shear tester (e.g. Panien *et al.* 2006b) (or with a rheometer in the case of wet clay, Eisenstadt and Sims 2005), have angles of internal friction similar to those of materials in the brittle upper crust or upper lithospheric mantle (Panien *et al.* 2006b; Klinkmüller *et al.* 2016; Ritter *et al.* 2016), meaning that they develop similar structures to their natural counterpart when subject to deformation. If necessary, it is possible to mix granular materials or to wet them to adjust their properties (e.g. cohesion) (Van Mechelen 2004; Abdelmalak *et al.* 2016; Montanari *et al.* 2017). Some granular materials (e.g. glass beads) have a lower angle of internal friction and can serve to represent structural weaknesses such as detachment layers. Since deformation of these materials is strain rate-independent, experimental deformation rates of models composed of only brittle materials can be selected at will.

For the modeling of ductile parts of the lithosphere such as the lower crust, the lower lithospheric mantle or crustal decollements (e.g. migmatites in the deeper parts of the crust and shales or salt layers in sedimentary sequences within the upper crust), a wide variety of viscous materials can be used. Silicones are a common choice, but an overview of alternatives can be found in, for example, Schellart and Strak (2016) and Reber *et al.* (2020). Often substances are mixed to obtain a viscous material with the correct density and properties. These viscous materials can have various rheologies, from Newtonian (linear) to power-law types, which can be tested using a coni-cylindrical viscometer (e.g. Mooney and Ewart 1934; Lee and Warren 1940) or a rheometer (e.g. Rudolf *et al.* 2016). Yet, one thing they have in common is that their behavior is strain rate-dependent (generally strengthening with increasing strain rates (Brun 1999, 2002)). Therefore, it is essential to properly scale deformation rates when applying viscous materials. When simulating the whole lithosphere, a low-viscosity material such as honey or glucose syrup is often used to incorporate the isostatic effects of the underlying (asthenospheric) mantle (Figure 6.1d). It is however important to note that analogue materials do not generally incorporate the effects of temperature variations in the lithosphere (e.g. melting and phase changes), which poses some limitations to their application. For more information on (viscous) materials, see Rudolf *et al.* (2016), Schellart and Strak (2016), Reber *et al.* (2020) and references therein.

6.2.3. *General experimental setups and boundary conditions*

Setups for experimental modeling concern the method of imposing deformation on the model (boundary conditions), and a first-order distinction can be made on the base of the gravity field that is applied. In the past, numerous researchers have used a centrifuge setup for enhanced-gravity experiments (e.g. Koyi 1997 and references therein). This method, in which an enhanced gravitational force (up to 200 g, Corti *et al.* 2003) is used to collapse the model layers to create extension, allows the application of relatively stiff viscous materials while respecting scaling laws, simplifying model construction. Drawbacks include the small size of the model and the challenges of observing the rotating model within the closed centrifuge, yet the centrifuge method is still being used today, yielding highly relevant results in the field of rift tectonics (e.g. Corti *et al.* 2003; Agostini *et al.* 2009; Corti 2012; Philippon *et al.* 2015; Zwaan *et al.* 2020a).

Running analogue experiments under normal gravity conditions is generally easier than the centrifuge approach. Nowadays, this is the most popular option, and therefore the focus of this chapter. As described by, for example, Vendeville *et al.* (1987), Allemand and Brun (1991) and Zwaan *et al.* (2019), there are various setups to model extensional tectonics under normal gravity conditions, depending on the model scale (upper crustal to lithospheric), tectonic setting and inferred lithospheric strength profile (Figure 6.1). An important difference between these and centrifuge models is that deformation in normal gravity models is generally driven by the mobile model base and/or sidewalls, that is, by an imposed extension velocity boundary condition, rather than by gravitational forces.

When studying deformation in the brittle upper parts of the lithosphere, from basin scale to upper crustal scale, modelers have often used a so-called plate base or conveyer base setup, on top of which the brittle model layers are applied (Figure 6.1a). By moving the base plate away with the use of precise (stepper) motors, its edge forms a so-called velocity discontinuity (VD). The VD is meant to simulate a fault in the basement that causes the brittle cover to deform locally (as both are directly "coupled", that is, the model base directly influences the brittle cover). A limitation of this "narrow rift" setup is that the "basement fault" does not allow vertical motion, which can, however, be solved with a basement block setup (see section 6.3.1.1). Another basal boundary condition for brittle-only models can be applied with a compressed foam base underlying a sand layer (Figure 6.1b, for example, (Schlagenhauf *et al.* 2008; Zwaan *et al.* 2019)). This setup may simulate a ductile lower crust directly coupled to the brittle crust. Here, distributed deformation is transmitted to the brittle layer as the model sidewalls move apart and the foam expands, leading to widespread faulting or "wide rifting". A rubber base can create a similar type of deformation (e.g. Bahroudi *et al.* 2003), but may also cause significant boundary effects due to the fact that rubber, when stretched, tends to contract perpendicularly to the stretching direction (see, for example, Zwaan *et al.* 2019). Note that modelers have also combined base plates and rubber sheets in the same model to create broader rift zones (e.g. Clifton *et al.* 2000; McClay *et al.* 2002).

A standard model setup for brittle-ductile settings involves a base plate system with a viscous layer representing the ductile lower crust and an overlying brittle layer simulating the upper crust (Figure 6.1c, for example, Allemand *et al.* 1989; Tron and Brun 1991; Michon and Merle 2000, 2003). Note that we could also use such a brittle-viscous layering for simulating a detachment (e.g. salt) within the brittle crust (see section 6.3.1.1). Here, the velocity discontinuity underlying the model materials represents a fault in the brittle lithospheric mantle, localizing deformation. Importantly, the viscous material, if sufficiently weak, can act as a detachment layer that decouples the brittle cover from the model base, so that the cover can deform independently. Depending on numerous factors, a single, double or no rift at all may develop (see section 6.3.1.2).

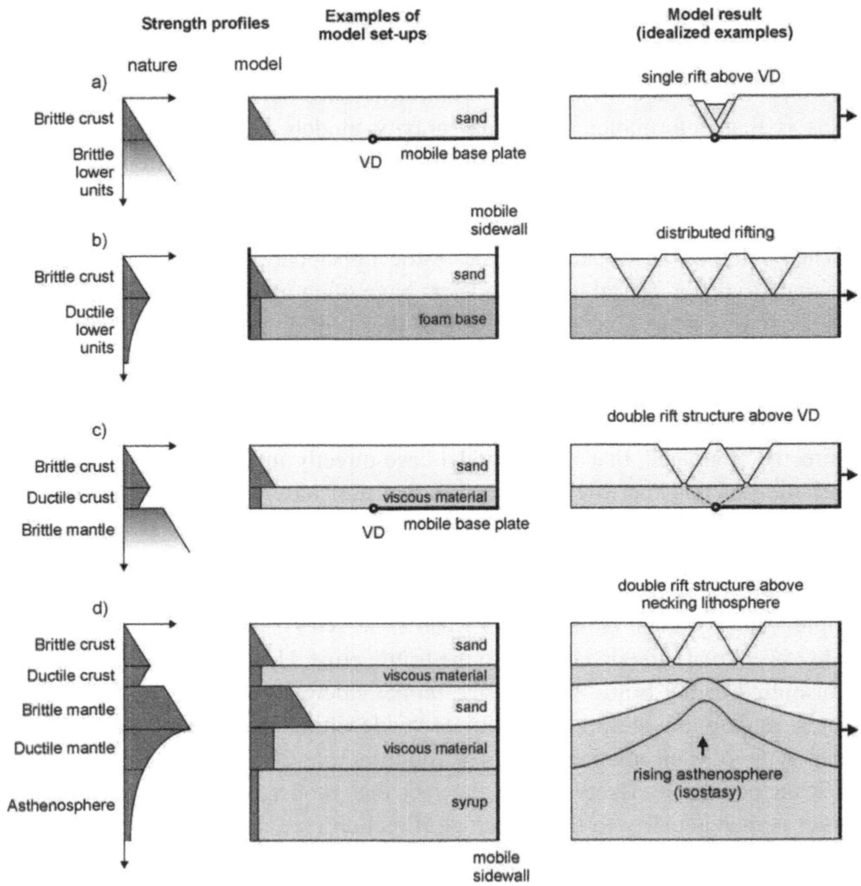

Figure 6.1. *Schematic section-view examples of normal-gravity experimental setups for simulating rifting at different scales, and examples of idealized results (without sedimentation). Note that slight variations in boundary conditions may have important effects on model evolution (see Figures 6.4 and 6.5, for example) (source: modified from Allemand and Brun 1991, Brun 1999, 2002, Michon and Merle 2003 and Zwaan* et al. *2019). a) Base plate setup with brittle cover, representing a brittle-only system, develops a graben at the edge of the basal plate, where a velocity discontinuity (VD) occurs; b) foam base setup, representing a brittle cover overlying a deeper (ductile) layer that evenly distributes faulting; c) brittle-viscous base plate model, representing a brittle-ductile crust above a very strong brittle mantle with a single fault (VD). Depending on various factors, a double rift may develop above the VD and d) four-layer lithospheric-scale experiment on top of syrup representing the asthenosphere, allowing isostatic compensation. Deformation is induced by moving the model sidewall. Compare with c)*

Finally, when simulating thinning and necking of the entire lithosphere, modelers need to include the mantle lithosphere and the underlying asthenosphere, which allows for isostatic compensation (Figure 6.1d). In this case, four-layer models representing the brittle and viscous parts of the lithosphere are mostly used, although modelers have also worked with three-layer systems (e.g. Allemand *et al.* 1989). A very weak viscous layer such as honey or glucose syrup acts as an analogue for the asthenosphere. By moving the sidewalls apart, the layers are stretched. In these models, numerous factors, especially the coupling between the various layers, which is a function of layer thickness, viscosity and model strain rate, may affect the style of rifting (see section 6.3.1.3).

When applying analogue models, their layering translates to a strength profile that should be comparable (i.e. properly scaled) in model and nature (e.g. Zwaan *et al.* 2019; Figure 6.1). Although the brittle part of the profile is mostly matched fairly well, the depth-dependent strength decrease in the ductile domain often remains an approximation, since the effects of increased heating and pressure are challenging to incorporate. Also, various other lithospheric strength profiles may occur in nature, which can be replicated with different materials and layer thicknesses.

Furthermore, it may be necessary to control where deformation occurs in a model to ensure reproducibility or to simulate pre-existing structures that reactivate. In the case of direct coupling, faulting will be directly affected by the model base (Figure 6.1a, b). Yet when viscous layers decouple the model components, modelers can, for instance, apply "seeds" (rods of viscous material) at the base of the brittle cover to weaken it locally (e.g. Le Calvez and Vendeville 2002; Zwaan *et al.* 2016; Molnar *et al.* 2019) or create pre-cut faults within the brittle layer (e.g. McClay *et al.* 2002; Bellahsen and Daniel 2005), to localize deformation. On a lithospheric scale, modelers have also used weak zones within the upper mantle analogue to focus deformation (Molnar *et al.* 2017).

The examples of model setups in Figure 6.1 are shown in section view, and analogue models have often been depicted and analyzed as such. This is permissible if the section is parallel to the direction of extension, and if no variations along the rift axis are included, so that deformation can be assumed to have occurred in plane. However, it is important to emphasize that any analogue model experiment is 3D and that numerous models incorporate processes that act in the third dimension. For these experiments, analysis cannot be limited to a single 2D section.

6.2.4. *Model analysis techniques*

Researchers have been using increasingly sophisticated techniques to monitor deformation in their analogue experiments. The most basic option, used since the early days of modeling, is photography (Figure 6.2a). Top view images are a great help for visualizing model evolution and are part of the standard toolkit of any analogue model laboratory. Automated time-lapse photography allows modelers to follow deformation in great detail. Moreover, if the model setup includes a transparent sidewall, side-view photographs provide valuable insights in how the model is deforming internally, although researchers must be aware of potential boundary effects due to sidewall friction. A very common alternative is to make cross-sections of the experiment. However, in order to do so, the model must be stabilized (e.g. wetted, frozen or impregnated with gelatine) and physically cut. Although several labs have elevated such techniques to an art form, cutting very fine sections, the photographs of which can be imported in structural interpretation software for detailed analysis (e.g. Wu *et al.* 2009; Withjack *et al.* 2017; Dooley and Hudec 2020), it means that the model must be destroyed and the resulting information only represents the final model state. Still, such sections provide instructive insights into model structures and form the basis of several figures presented in this chapter.

Researchers commonly add surface markers (e.g. a grid) or differently colored sand layers to visualize deformation in map-view and side-view/cross-section photographs, allowing a semi-quantitative assessment of deformation (Figure 6.2e). A more precise analysis of such data can be achieved by means of particle image velocimetry (PIV) or digital image correlation (DIC) of time-lapse series (e.g. Adam *et al.* 2005; Boutelier *et al.* 2019). These techniques compare photographs from different time steps, tracing and displaying 2D displacement (Figure 6.2f). The now fully quantified 2D displacement patterns furthermore allow detailed strain analysis (e.g. Boutelier and Oncken 2011).

Nevertheless, these results represent only 2D insights, whereas the processes in models and in nature are three-dimensional. By using laser/structured light scanners or stereoscopic camera configurations and photogrammetry software, it is possible to reconstruct detailed digital elevation models and capture vertical displacements and 3D surface deformation (e.g. Donnadieu *et al.* 2003; Michon and Sokoutis 2005; Schlagenhauf *et al.* 2008; Nestola *et al.* 2015; Maestrelli *et al.* 2020; Zwaan *et al.* 2020a). Yet more sophisticated is 3D surface analysis by means of PIV software. Similar to normal photogrammetry software, this 3D PIV technique reconstructs the surface of the model and goes a step further than 2D PIV methods by tracing vertical displacements as well (e.g. Adam *et al.* 2005; Molnar *et al.* 2017; Ge *et al.* 2019). As a result, this technique allows a unique and fully quantified 3D analysis of surface deformation.

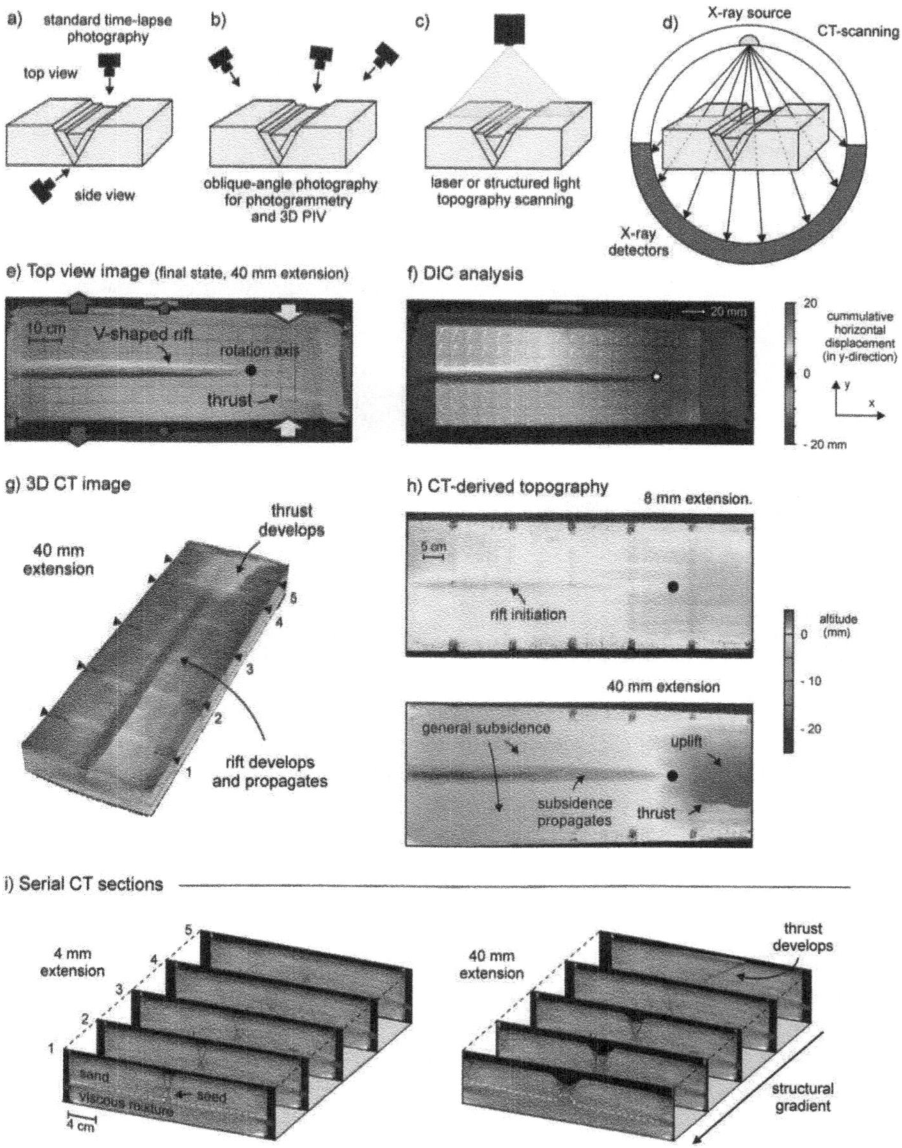

Figure 6.2. *Experimental monitoring techniques*

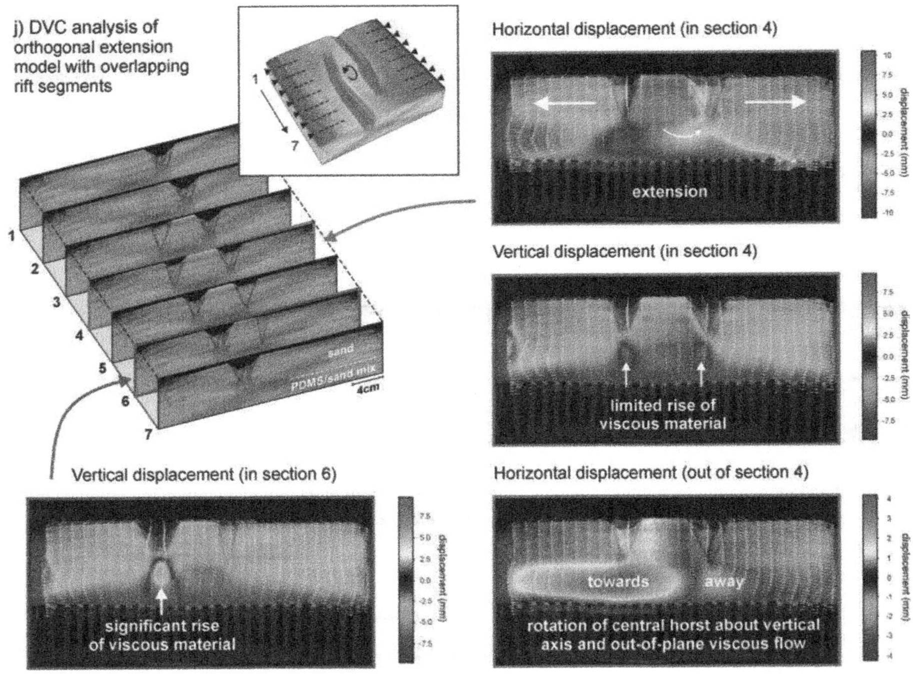

Figure 6.2. *(continued) Experimental monitoring techniques. a) Top- and side-view photographs, for 2D PIV/DIC analysis; b) oblique-angle photography for topographic and 3D PIV/DIC analysis; c) surface scanning for topographic evolution monitoring. d) x-ray CT-scanning allowing non-destructive inspection of internal model evolution, as well as internal PIV/DIC and DVC analysis. e–i) Application of various analysis techniques on a rotational extension experiment from Zwaan* et al. *(2020b): e) top view of final model state; f) quantification of cumulative horizontal displacement through DIC analysis; g) 3D CT imagery of the final deformation stage; h) CT-derived topography maps; i) serial CT sections of internal structures over time, showing the structural gradient in the early and final stages of the model run. Note the structural weakness (seed) that localizes deformation. Section locations are indicated in (g). j) Example of digital volume correlation (DVC) analysis on CT data from a model with interacting rift segments under orthogonal extension, illustrating internal displacement patterns. Note the differences in vertical displacement at different locations (compare vertical displacements in sections 4 and 6), as well as out-of-plane displacement of both brittle and viscous materials (indicating rotation of the central horst about a vertical axis), highlighting the 3D character of the system (source: modified after Zwaan* et al. *2018)*

However, these techniques do not provide a complete insight into internal model deformation. This can, so far, only be achieved by means of X-ray CT-scanning, during which the model is not physically disturbed (e.g. Colletta *et al.* 1991; Naylor *et al.* 1994; Schreurs *et al.* 2003). The method has some limitations in that both the setup and experimental materials need to be X-ray transparent, and the complete experiment needs to fit into a (medical) CT scanner. However, CT-scanning provides unrivaled potential for model analysis. The model can be visualized in 3D (Figure 6.2g) and it allows the extraction of detailed digital topography maps, much like photogrammetry or surface scanning (Figure 6.2h). Furthermore, the 3D CT volume allows modelers to make cross-sections in any direction they desire, for every time step at which the model was scanned. This provides detailed insights into internal model development (e.g. Zwaan *et al.* 2020b, Figure 6.2i) and allows 4D analysis when imported into structural interpretation software (e.g. Chauvin *et al.* 2018; Fedorik *et al.* 2019). However, these insights also remain semi-quantitative. The final step is then to apply PIV or DIC techniques on CT data, which can be done on 2D sections from different time intervals, yielding unique quantitative information (Zwaan *et al.* 2020b), and also on successive 3D volumetric CT datasets produced at different time steps during the evolution of one particular experiment. This so-called digital volume correlation (DVC) technique (Adam *et al.* 2013) uniquely allows the tracing of displacements and the quantification of strain throughout the complete model, clearly illustrating that rifting is a 3D process (e.g. Zwaan *et al.* 2018, Figure 6.2j).

6.3. Model application

As described in section 6.2, different setups and materials are used to study diverse aspects of rifting on different scales. In the following section, we shall provide an overview of various examples, ranging from quasi-2D models of crustal and lithospheric scale experiments, to models involving 3D rift processes, such as oblique extension, rift segment interaction and rotational rifting. As emphasized in section 6.2.3, all analogue models are by definition 3D objects. However, rift models are often analyzed in section parallel to the extension direction, which provides quasi-2D insights.

6.3.1. *A 2D perspective on rifting*

6.3.1.1. *Normal fault development in the upper crust*

Various authors have simulated the development of normal faulting in the upper (parts of the) crust (Figure 6.3). For instance, the CT-scanned experiments by Panien *et al.* (2006b) reveal that in a brittle base plate setup, normal faulting initiates at the basal velocity discontinuity (VD) and subsequently propagates towards the model surface to form a graben structure (Figure 6.3a). Furthermore, the authors show that, due to stress deflection, these initial faults may overturn towards the surface, becoming reverse faults in the upper few millimeters of the model (Figure 6.3a). Later on, the lower part of the normal fault is reactivated and continues upward in the footwall with the initial subvertical and reverse segments being abandoned. Such features are also found in nature (Trippanera *et al.* 2014), most spectacularly at the rims of collapsed calderas (Martí *et al.* 2008). When extension is asymmetric, the resulting fault pattern will also be asymmetric, as new faults develop above the edge of the moving base plate (e.g. McClay 1990; Allemand and Brun 1991; Beslier 1991 Figure 6.3b).

Other researchers have studied the effects of basement block subsidence on fault development in the sedimentary cover. Naylor *et al.* (1994) show how vertical basement faults can cause reverse faulting in the overburden (Figure 6.3c), in a process similar to the reverse faults described by Panien *et al.* (2006b) (Figure 6.3a). By contrast, models with low-angle basement faults create a graben structure above the fault (Naylor *et al.* 1994; Holland *et al.* 2006) (Figure 6.3d). Furthermore, if the basement fault is listric, a roll-over structure forms, and if syn-rift sedimentation is applied by stepwise filling up the generated accommodation space, a series of small grabens will develop in the accumulating hanging wall strata (e.g. McClay 1990, Figure 6.3e).

Moreover, adding a layer of relatively weak viscous material to simulate salt or shale detachments can decouple the brittle cover from the model basement (e.g. Vendeville *et al.* 1995; Dooley *et al.* 2003, Figure 6.3f). As a result, flexure can develop as the viscous material flows, and normal faulting in the brittle layer can be displaced sideways with respect to the underlying basement fault (Figure 6.3f).

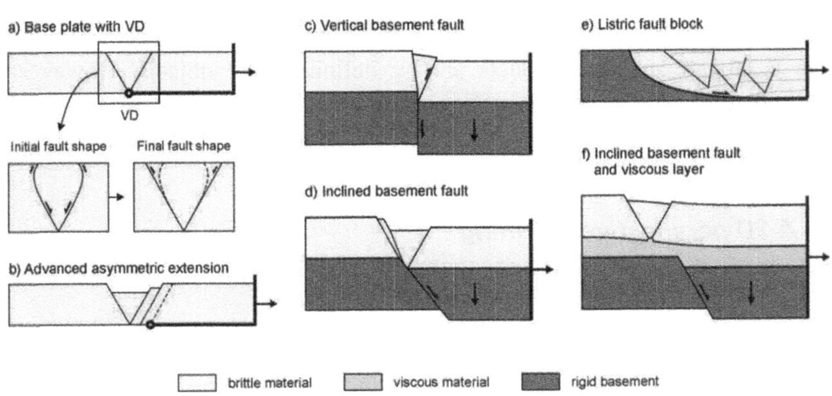

Figure 6.3. *Schematic examples of upper crustal scale models to study normal faulting*

CONTINUATION OF CAPTION FOR FIGURE 6.3.– *a) Base plate creating a graben in the brittle material above the velocity discontinuity (VD) at its plate edge (top). Initial fault may be curved, leading to local reverse kinematics (bottom left). Later on, full normal faulting is established (bottom right); b) when extension is asymmetric, an asymmetric graben will form after advanced deformation (source: modified after Allemand* et al. *1989 and Panien* et al. *2006b); c) vertical (high angle) basement fault causing reverse faulting in the brittle cover (source: modified after Naylor* et al. *1994); d) inclined (low angle) basement fault inducing normal faulting and graben formation in brittle cover (source: modified after Naylor* et al. *1994); e) listric fault model developing a roll-over anticline with internal grabens (source: modified after McClay 1990) and f) effect of a decoupling viscous layer on faulting in the brittle cover above a basement fault. Compare with the same setup without a viscous layer in d) (source: modified after Dooley* et al. *2003).*

6.3.1.2. *Brittle-viscous crustal scale models*

When zooming out and applying brittle-viscous layers representing the entire brittle-ductile crust, researchers have found various influences of lithospheric strength as well as model boundary conditions (i.e. layer thicknesses and extension rates) on the mode of rifting (Brun 1999; Corti *et al.* 2003; Zwaan *et al.* 2019 and references therein). A very thick viscous layer, representing, for instance, the effect of crustal thickening and radioactive heating in an orogen, creates a very weak crust (Figure 6.4a). As a result, the brittle cover is very much decoupled from the model base, which induces detachment faulting and the formation of exhumation structures analogue to metamorphic core complexes when subjected to standard rifting rates (Brun *et al.* 1994, Figure 6.4a).

By contrast, a thinner viscous layer (low brittle-to-viscous thickness ratio) representing a colder crust leads to localized rifting when extension rates are low (Figure 6.4b). However, when applying high extension rates, the viscous layer is stronger and coupling between the viscous and brittle layers is enhanced so that they start deforming together causing distributed faulting ("wide rift mode" Brun 1999) (Figure 6.4c).

When the viscous layer is even thinner, as may be the case in an older, cold crust, the basal boundary condition starts to affect the style of rifting (Figure 6.4d–f). This boundary condition was not particularly important in the previous cases due to a high degree of decoupling between model base and brittle layer (Figure 6.4a–c). However, increased coupling between the base and brittle cover causes the materials to deform in a similar way to that in the models without a viscous layer (Figures 6.1a, b, 6.3a–d and 6.4d–f). A foam base (simulating a ductile/weak mantle) causes distributed extension throughout the brittle cover (Figure 6.4d), whereas a plate base (simulating a strong/brittle mantle) causes more localized deformation (Zwaan *et al.* 2019, Figure 6.4e). Whether a single or double rift develops in the latter depends on the extension rate (Michon and Merle 2000).

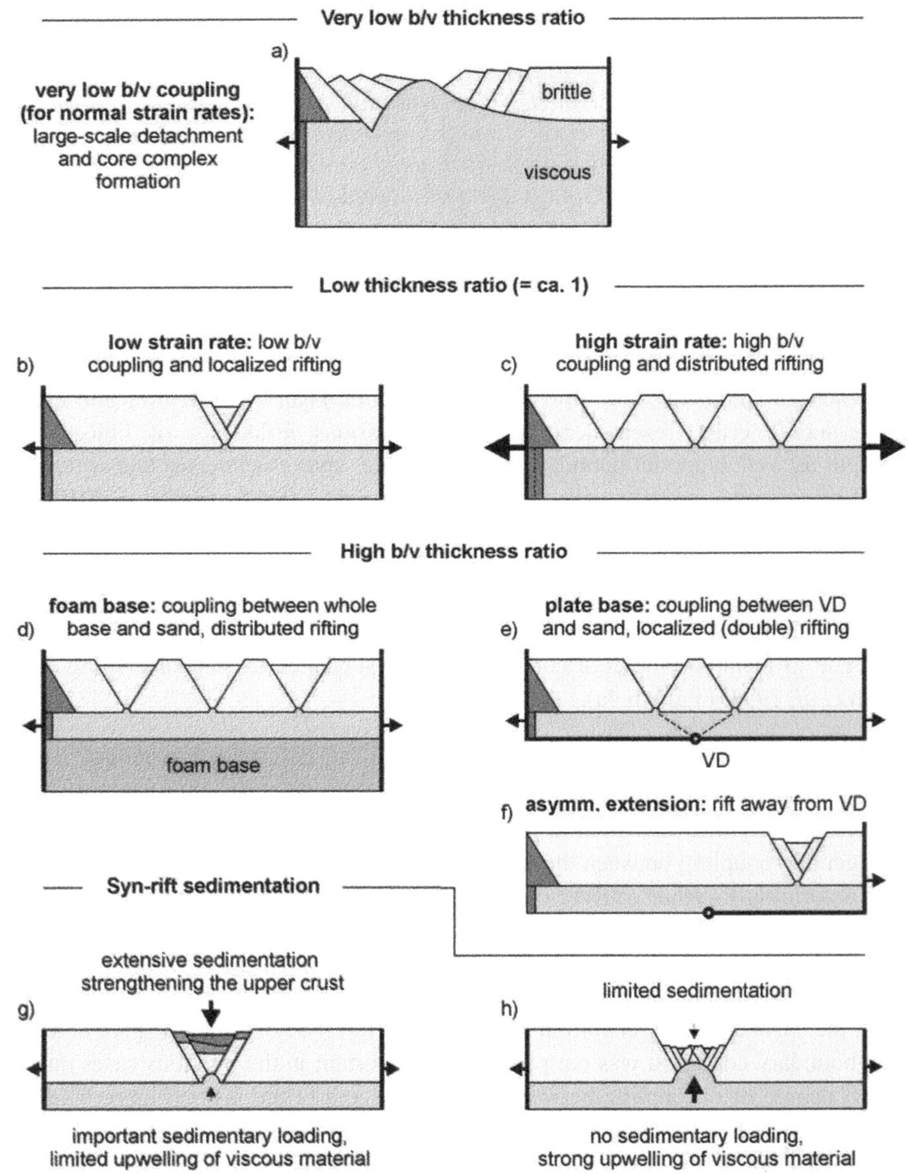

Figure 6.4. *Schematic examples of parameters affecting brittle-viscous experiments*

CONTINUATION OF CAPTION FOR FIGURE 6.4.– *a) Very low brittle-to-viscous (b/v) thickness ratios (i.e. very low strength crust) decouple the brittle layer from the base under normal extension velocities, leading to core complex formation (source: modified after Brun* et al. *1994); b and c) Low b/v thickness ratios still decouple the brittle layer from the base, yet high strain rates cause distributed faulting (wide rifting mode), whereas low strain rates localize deformation (narrow rifting mode) (source: modified after Brun 1999 and Zwaan* et al. *2019); d–f) A thin viscous layer leads to coupling between base and brittle cover. For foam base setups, the foam's distributed deformation transfers to the brittle cover, which develops widespread faulting. A base plate setup causes localized deformation above the velocity discontinuity (VD), yet depending on strain rate a single or double rift may develop (source: modified after Michon and Merle 2000, 2003). Furthermore, asymmetric extension may (in some cases) deflect deformation away from the VD (Allemand* et al. *1989); g and h) Effects of sedimentation on rift development. Sedimentary infill causes strengthening of the brittle layer, focusing fault activity along a few faults and suppressing upwelling of viscous material. Absence of sedimentation causes the rift wedge to split along numerous faults while the viscous layer rises (source: modified after Zwaan* et al. *2018). Note that the dark gray part on the left of the images represents the strength profile.*

Further influences on the evolution of brittle-viscous rift systems are caused by the application of asymmetric extension and sedimentation. Simply put, a symmetric extension setup will lead to symmetric rift structures, whereas an asymmetric extension setup, if coupling between the base and brittle cover is significant enough, may cause the rift to focus on the moving plate, away from the basal velocity discontinuity (Allemand and Brun 1991, Figure 6.4f). Moreover, synrift sedimentation can, in addition to preventing the brittle layer from breaking up as deformation remains focused along a few large normal faults (Figure 6.4h), prevent viscous material from rising below an otherwise thinned rift wedge due to the weight of the sediment infill (Zwaan *et al.* 2018, Figure 6.4g, h). Such flow of viscous material below the rift basin can be clearly visualized by means of displacement analysis on CT data (Zwaan *et al.* 2018, 2020b, Figure 6.2j).

6.3.1.3. *Lithospheric scale models*

For crustal-scale models, it is not always necessary to incorporate the (isostatic) effects of the deeper mantle layers. However, as plate tectonics involves the entire lithosphere, which is in isostatic equilibrium with the asthenosphere, both the lithosphere and asthenosphere are included in various large-scale modeling studies of rifting (e.g. Allemand *et al.* 1989; Brun and Beslier 1996; Nestola *et al.* 2015; Molnar *et al.* 2017; Beniest *et al.* 2018). These experiments show that the presence of a competent upper mantle layer is of great importance (Corti *et al.* 2003). When such a layer is absent, the strength of the lithosphere is dominated by the upper crust, leading to localized rifting (Figure 6.5a, similar to Figure 6.4b). When present, this competent upper mantle controls the strength of the lithosphere and its rupture creates either a single or double rift in the brittle upper crustal layer (Figure 6.5b, c, compare these structures with those in Figures 6.1c and 6.4e).

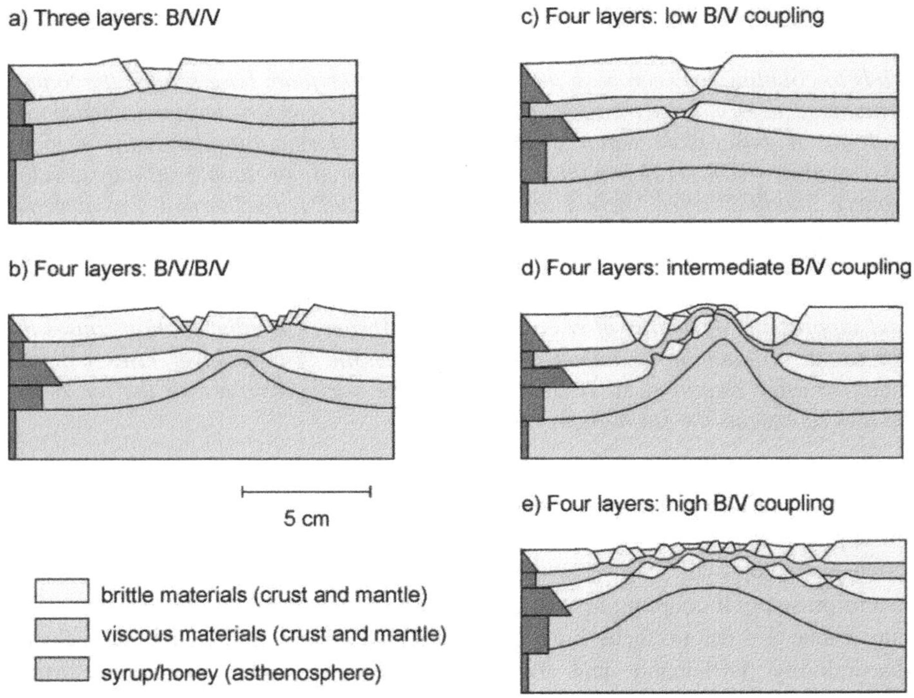

Figure 6.5. *Schematic examples of lithospheric-scale rift model results as a function of (a, b) lithospheric layering (three vs. four layers) and (c–e) degree of coupling between the brittle and viscous materials (B/V coupling) in four-layer models. B: brittle, V: viscous (ductile) (source: modified after Allemand* et al. *1989, Beslier 1991, Brun and Beslier 1996 and Brun 2002)*

In such four-layer lithospheric models, (de)coupling between brittle and viscous layers is highly important. When coupling is low, either due to low strain rates or low viscosities in the ductile domain (Brun 1999, 2002), deformation is localized in both the brittle mantle and crust, which may represent natural rift settings like the Upper Rhine Graben (Beslier 1991; Brun 1999, Figure 6.5c). By contrast, enhanced (intermediate) brittle-viscous coupling causes a more distributed style of rifting and may lead to exhumation of the model mantle as observed in nature along the Atlantic margin of Iberia (Figure 6.5d, Brun and Beslier 1996). On the other hand, extension of models with high coupling between their viscous and brittle components (e.g. due to a higher extension rate) may cause widespread deformation (Beslier 1991, Figure 6.5e, compare with high brittle-viscous coupling models in Figure 6.4c and d).

The models in Figure 6.5, however, do not consider structural inheritance. Other researchers have included crustal and mantle weaknesses to localize deformation (e.g. Agostini *et al.* 2009). Beniest *et al.* (2018) have studied lateral strength variations and show how rifting predominantly localizes in the weaker part of the lithosphere (i.e. away from competent areas such as cratons). Corti *et al.* (2003) include low-viscosity patches to represent melts and provide a schematic overview of the relative relations between various parameters and rift styles (their Figure 6.36). However, although these models yield valuable insights into rift processes, including mantle exhumation (Brun and Beslier 1996, Figure 6.5d), analogue models are limited to the continental rifting phase, given that commonly applied materials do not allow modelers to simulate the creation of new oceanic lithosphere.

6.3.2. *Exploring 3D rift processes*

Although many analogue modeling studies have approached lithospheric extension as a 2D phenomenon and analyzed it accordingly, various researchers have explored the third dimension as well. Their results show the importance of understanding the 3D aspects of rift processes, resulting from, for example, oblique extension, the interaction between individual rift segments, rotational rifting or rift propagation.

6.3.2.1. *Oblique extension*

Most extensional systems experience some form of oblique extension during their lifetime (Brune *et al.* 2018), and researchers have extensively used analogue models in order to understand the significance of this parameter (e.g. Withjack and Jamison 1986; Tron and Brun 1991; McClay and White 1995; Clifton *et al.* 2000; Agostini *et al.* 2009; Autin *et al.* 2010, 2013). These studies show that in orthogonal extension settings, large normal faults strike perpendicularly to the extension direction and thus parallel to the rift axis (Figure 6.6a). However, when extension is oblique, the faults strike at an angle to the rift axis (not exactly perpendicular to the extension direction, see Withjack and Jamison 1986), and the resulting rift zone is bordered by en-échelon boundary faults (Figure 6.6b). Although the boundary faults are oriented obliquely to the rift axis, their kinematics remain dominantly normal (Philippon *et al.* 2015). With increasing obliquity towards the strike-slip domain, however, normal faults will give way to strike-slip faults. Typical for such models is the development of initial Riedel shears above the rift axis, which subsequently link up to form a continuous strike-slip fault (e.g. Naylor *et al.* 1986; Dooley and Schreurs 2012 and references therein, Figure 6.6c).

Furthermore, Bonini *et al.* (1997), Keep and McClay (1997) and others explored the effects of multiphase oblique extension and showed how structures formed during the initial phases are often reactivated to dominate subsequent extension phases. However, the reactivation of pre-existing structural weaknesses in the crust or mantle under (oblique) extension normally only localizes deformation when these weaknesses are oriented favorably to the regional (oblique) extension direction (Zwaan and Schreurs 2017; Molnar *et al.* 2019). On a lithospheric scale, Autin *et al.* (2010) describe how their models suggest that oblique extension may protract continental breakup.

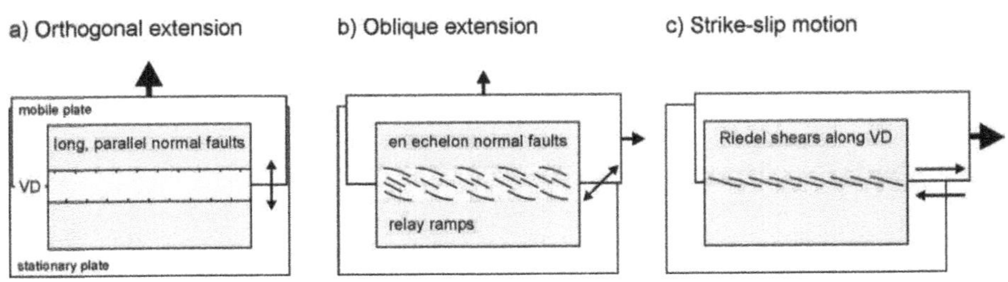

Figure 6.6. *Schematic map view examples of fault patterns related to different extension directions as observed in experiments. a) Orthogonal extension leads to long, velocity discontinuity (VD)-parallel normal faults flanking a graben (bright yellow); b) when extension is oblique, en-échelon normal faults linked via relay ramps develop along the VD, but their strike is not perfectly perpendicular to the extension direction (source: Withjack and Jamison 1986) and c) under strike-slip conditions, a series of Riedel shears appear above the VD, which in later stages link up to form a continuous strike-slip fault (source: modified after Tron and Brun 1991). Note that in the setups shown in a) and b), the diverging base plates are generally connected by a rubber sheet or partly overlain by a patch of viscous material*

6.3.2.2. *Rift segment interaction*

When the lithosphere is stretched, deformation often localizes along pre-existing structural weaknesses to form individual rift segments. In order to develop into a full-scale rift system, these segments need to interact, propagate and connect. Analogue modeling studies show that such rift interaction structures are affected by the horizontal distance (offset), the amount of underlap or overlap between rift segments and the presence of secondary pre-existing weaknesses linking the segments and the degree of extension obliquity (e.g. Acocella *et al.* 1999; Le Calvez and Vendeville 2002; Tentler 2003a, 2003b; Zwaan *et al.* 2016; Molnar *et al.* 2019). The effect of the latter is well visible if the offset between rift segments is sufficiently large (Figure 6.7a–e). Since normal faults tend to develop at high angles to the regional extension direction (see section 6.3.2.1), rift segments propagate away from, parallel to or towards each other. In the latter case, the central block may become isolated (rift pass structure) and can start rotating due to the interaction between the overlapping rift axes (Zwaan *et al.* 2018, Figures 6.7c and 6.2j), a process currently observed at the Victoria Plate in the East African Rift (Glerum *et al.* 2020; Zwaan and Schreurs 2020). In situations with small offsets, these effects will not occur as the rift segments readily grow into each other (Figure 6.7d). Furthermore, when the rift segments underlap, a series of minor en-échelon grabens may develop (e.g. Tentler 2003a, 2003b, Figure 6.7e). Some models show that secondary pre-existing weaknesses may help to connect rift segments, but only if these are oriented favorably to the extension direction; if they are aligned (sub-)parallel to the extension direction, they will not activate (Zwaan and Schreurs 2017; Molnar *et al.* 2019). Yet when applying strongly-controlling basal boundary conditions with a plate base setup, it is possible to force the development of transform fault-like structures along the edge of the plate, even if this velocity discontinuity is aligned with the extension direction (e.g. Acocella *et al.* 1999; Dauteuil *et al.* 2002, Figure 6.7f).

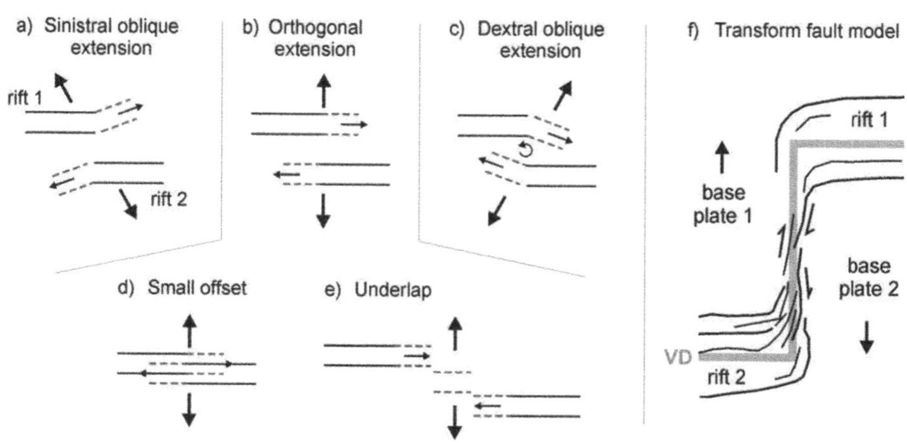

Figure 6.7. *Schematic examples of modeled rift interaction between right-stepping rift segments, depending on extension direction*

CONTINUATION OF CAPTION FOR FIGURE 6.7.– *Since the strike of new normal faulting tends to be at a high angle to the extension direction (Withjack and Jamison 1986), a) sinistral oblique extension causes the rifts to diverge; b) orthogonal extension has the rifts grow in a parallel fashion; and c) dextral oblique extension lets the rifts grow together. The block in between the rifts in c) can start rotating and even form a continental sliver or microcontinent. Note that the relations between rift development and extension direction as shown in this figure are opposite when the rift segments are arranged in a left-stepping fashion. d) Rifts directly grow into each other when their offset is small; e) rifts may create en-échelon interaction zone when underlap and offset are sufficiently large and f) transform fault modeled by a plate base setup. VD: velocity discontinuity (source: modified after Dauteuil* et al. *2002, Tentler 2003a, 2003b, Zwaan* et al. *2016, 2018 and Zwaan and Schreurs 2017).*

6.3.2.3. *Rotational rifting and rift propagation*

When running rifting experiments, tectonic modelers generally apply constant extension rates along the length of their models (Figures 6.6 and 6.8a). However, plate tectonics on a spherical earth demands rotational motion about an Euler pole, whereas various cases of smaller-scale plate rotations are recorded around the world as well (e.g. Zwaan and Schreurs 2020; Zwaan *et al.* 2020b, Figure 6.8a). A few analogue modelers have explored the implications of rotational boundary conditions (e.g. Souriot and Brun 1992; Benes and Scott 1996; Sun *et al.* 2009; Molnar *et al.* 2017). These models invariably show that rift development in rotational settings causes structural gradients, rift propagation and the formation of V-shaped basins, whereas constant along-strike extension rates lead to cylindrical structures (Figures 6.2e–j, 6.6a and 6.8a, b). In addition, pressure gradients in such models cause not only across-strike displacements (Figure 6.2j) but also along-strike flow of viscous material, highlighting the importance of a 3D mindset when studying rifting processes (e.g. Mondy *et al.* 2018; Zwaan *et al.* 2018, 2020b; Khalil *et al.* 2020; Maestrelli *et al.* 2020). DVC analysis of CT data clearly shows the interaction between deep-seated viscous deformation and surface deformation (Figure 6.2j).

The models depicted in Figure 6.8a and b represent a homogeneous layer cake. Molnar *et al.* (2017, 2018, 2019) describe how structural weaknesses of various types and orientations can (partially) reorient propagating rift systems. Benes and Scott (1996) test how such propagating rifts interact when encountering a competent domain, and describe how the rift struggles to penetrate this (Figure 6.8c, d). When the rheological contrast is perpendicular to the extension direction, the well-defined propagating rift spreads out over various faults (Figure 6.8d), yet if the rheological contrast is obliquely oriented, part of the deformation is deflected along the contact (Figure 6.8d). A similar result is obtained by Brune *et al.* (2017), who use both analogue and numerical models to study how an oblique weak zone affects rift interaction, although the authors use constant along-strike strain rates (Figure 6.8e).

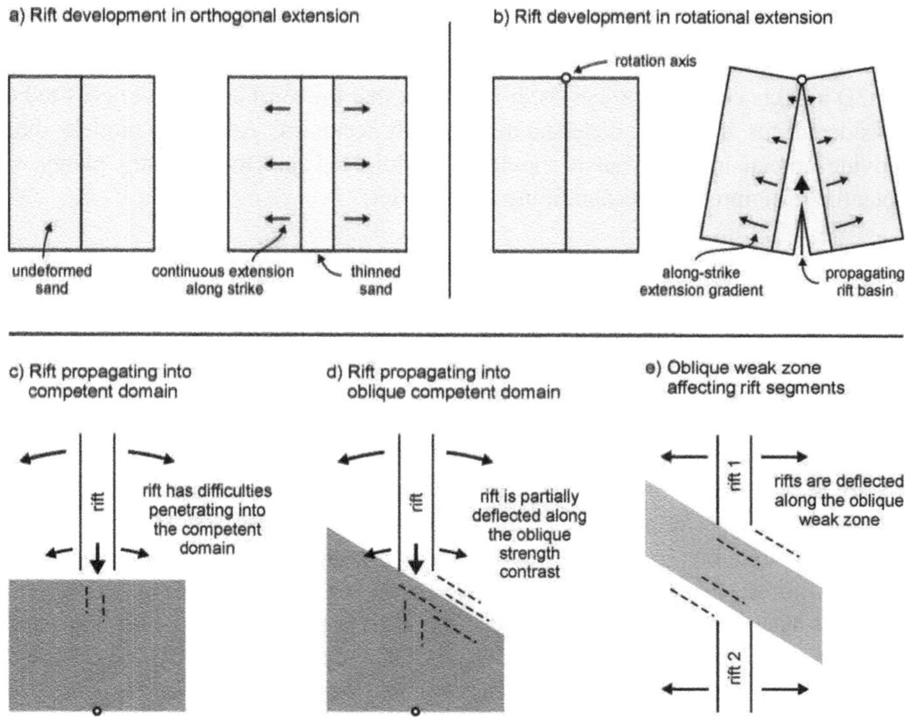

Figure 6.8. *Schematic examples of rift development and propagation under different boundary conditions. a) Orthogonal extension causes synchronous rift development along its axis. Oblique extension boundary conditions produce similar results (Figure 6.6); b) rotational rifting, inducing a strain rate and structural gradient, as well as rift propagation towards the rotation pole. Compare with the rotational rifting model analyses shown in Figure 6.2e–i; c) rift propagating into a competent domain orthogonal to the rift axes; d) rift propagating into a competent domain oblique to the rift axes and e) rift deflection along a zone of different rheological competence (source: modified from Vink 1982, Martin 1984, Benes and Scott 1996 and Brune* et al. *2017)*

6.4. Summary, challenges and future opportunities

In this chapter, we have described how analogue modeling techniques have been used to study a wide variety of aspects associated with rifting processes, from normal fault development to lithospheric necking towards continental breakup. The variety of analogue modeling methods and advanced analysis techniques provides researchers with a handy and highly versatile toolkit to explore factors that may affect rifting, yielding invaluable insights into the dynamic evolution and associated kinematics. A key message should be that although various models have focused on the 2D aspects of rifting processes, it is evident that the third dimension needs to be considered for a proper understanding of rift tectonics. Analogue models thus provide crucial insights, but it should be highlighted that there is still plenty of potential for improvements and future applications.

Recently, Peron-Pinvidic, Manatschal *et al.* (2019) listed six main focus points for rift and passive margin research efforts: rheology, structural inheritance, faulting, stratigraphy, kinematics and the influence of the mantle, amongst which various links exist. Analogue modeling work could contribute significantly to a better understanding of the role of these parameters during rifting.

1. The influence of rheology can be addressed through analogue modeling materials as researchers are constantly looking for new, adequate materials to improve model performance (e.g. Schellart and Strak 2016; Reber *et al.* 2020). A common limitation in analogue models is, for instance, that model material cannot undergo phase changes due to thermal effects or hydration of minerals. Yet modelers have, for instance, applied materials such as paraffin wax, which melts and solidifies when changing temperature (e.g. Oldenburg and Brune 1972; Brune and Ellis 1997; Katz *et al.* 2005). Other researchers have used layers of hydrocarbon-based materials with temperature-dependent rheology floating on water to represent the asthenosphere (Chemenda *et al.* 2002; Boutelier and Oncken 2011; Boutelier *et al.* 2012) and gelatine to simulate earthquakes (Corbi *et al.* 2013), although these materials often require very specialized setups. A major breakthrough would be the development of convenient analogue materials that would allow the simulation of both continental rifting until breakup and the subsequent creation of the oceanic lithosphere (see section 6.3.1.3).

2. The role of structural inheritance (compositional, structural or thermal) can be addressed in a variety of ways. Differences in lithospheric composition can be simulated by using different model materials, and this approach is strongly related to the topic of rheology. Structural inheritance includes discrete weaknesses (linear seeds, pre-cut faults), but it may also be possible to induce pervasive features such as a regional foliation (Chattopadhyay and Chakra 2012). Furthermore, the application of multiphase deformation can be of great value. Various experimental studies focus on multiphase rifting (e.g. Bonini *et al.* 1997; Keep and McClay 1997) or inversion of rift basins (e.g. Brun and Nalpas 1996; Panien *et al.* 2005; Mattioni *et al.* 2007;

Cerca *et al.* 2010) and show how pre-existing faults may, or may not, play a role during subsequent tectonic phases. Vice versa, we could model another part of the Wilson cycle by simulating a compressional phase that is followed by rifting.

3. Analogue models are excellent for studying faulting, but how faults develop in 3D space remains poorly constrained (Jackson *et al.* 2017). In order to get a more detailed grasp of fault development during rifting, analysis techniques should be optimized. Key is fully capturing model surface deformation by means of (3D) PIV/DIC methods. Detailed 3D analysis of fault populations over time by importing 3D CT data in structural analysis software (e.g. Fedorik *et al.* 2019) would also provide crucial insights. Such data may also be used to create (3D) synthetic seismic data that can be compared with seismic surveys from natural rift zones (Lindanger *et al.* 2004). Another intriguing application in this line is the use of CT-scanned models to test the validity of methods commonly used for tectonic restorations (Chauvin *et al.* 2018). This would allow the application of numerical analysis techniques and machine learning algorithms on analogue models (e.g. Corbi *et al.* 2019). Combining analogue and numerical methods to tackle a specific topic can especially help to calibrate modeling results and increase their reliability (e.g. Bellahsen *et al.* 2003; Buiter *et al.* 2006; Panien *et al.* 2006a; Corti *et al.* 2007; Quirk *et al.* 2012; Zwaan *et al.* 2016; Brune *et al.* 2017; Brun *et al.* 2018; Hughes *et al.* 2020). Since collaboration between analogue modeling laboratories improves research quality and output, the European Union-supported EPOS project has created a network between labs (Multi-scale Laboratories or MSL) to promote knowledge exchange and resource sharing (available at: https://www.epos-ip.org/tcs/multi-scale-laboratories).

4. The development of stratigraphy in rift and passive margin settings is strongly affected by faulting, but sedimentary infill, if thick enough, can itself also affect the structural architecture of a rift basin (Zwaan *et al.* 2018). The application of sedimentation processes in analogue models is generally rather coarse: often the basin is simply filled up to the brim. Hughes *et al.* (2020) recently developed a device that allows localized deposition of sedimentary units, opening the way to including detailed sedimentation patterns in analogue models. Another way in which sedimentation can affect a rift system is by the formation of layers with different properties, creating rheological variations and compositional inheritances. The most extreme case is the formation of thick salt deposits, which can decouple the sedimentary cover from the underlying basement, as modeled by many researchers (e.g. McClay *et al.* 1998; Brun and Fort 2004; Fort *et al.* 2004; Vendeville 2005; Adam and Krezsek 2012; Ge *et al.* 2019, Figure 6.3e). It may be noted that such work is of great interest for hydrocarbon exploration.

5. The kinematics of rifting concern factors such as extension directions, deformation rates and also changes in general deformation style on the path from incipient rifting to continental breakup. Analogue models are well suited to studying these topics, especially when addressing the 3D aspects of rift kinematics (Figures 6.2 and 6.6–6.8). An interesting challenge may concern extension boundary conditions, since extension rates are also known to change over time, especially prior to breakup, as shown by Brune *et al.* (2016). The authors attribute this effect to the fact that the forces that act on a rifting plate are fairly constant, but the dwindling strength of the lithosphere towards breakup translates in a rapid acceleration of extension. Analogue modelers should attempt to incorporate such changes in strain rate, at least when modeling large-scale rift systems. This could be done by either implementing different extension velocities over time, or perhaps more elegantly by applying some sort of force boundary condition by means of a weight-and-pulley system. In addition, coupling between analogue and numerical models may help to explore kinematic links with the deep earth and large-scale driving forces.

6. Finally, the influence of the mantle, which is considered a dominant factor during rifting and continental breakup, should be further explored. This topic is strongly associated with rheology and kinematics, and various analogue modelers have included the mantle in their lithospheric-scale experiments (e.g. Figure 6.5, Corti *et al.* 2003 and references therein). For these models, the inclusion of convenient analogue materials that can mimic temperature-dependent rheological changes during rifting should be a major objective. Furthermore, most lithospheric-scale models have focused on the 2D aspects of rifting. Fully 3D efforts including oblique or rotational rifting, such as those applied in the works by Agostini *et al.* (2009), Philippon *et al.* (2015) and Molnar *et al.* (2017) will yield important insights into mantle influence within rift systems. In particular, the application of advanced methods like CT scanning and DVC analysis would greatly help us to unravel the complex internal deformation of the lithosphere and the interactions between (different components of) the crust, the lithospheric mantle and the asthenosphere below.

In addition to the research focus points listed above, we must stress that considerable opportunities lie in rerunning models from previous studies. As a first step, this would allow an evaluation of experimental reproducibility (e.g. Schreurs *et al.* 2006, 2016), although this might be challenging as older publications often lack the necessary background information on rheology, setup and the practicalities surrounding model construction. It is therefore of great importance to provide all relevant information when publishing modeling work (Zwaan *et al.* 2019) and to make extensive supplementary material publicly available via online repositories (e.g. at GFZ Data Services, which is a partner in the EPOS MSL network, available at: http://dataservices.gfz-potsdam.de). As a second step, rerunning previously published models would help us to obtain more detailed and quantified insights, especially on the evolution of internal model deformation. In fact, models analyzed with state-of-the-art PIV, DIC and DVC methods provide a wealth of data, revealing processes and details, as well as boundary effects that may have gone previously unnoticed (e.g. Adam and Kreszek 2012; Molnar *et al.* 2017; Zwaan *et al.* 2018, 2020b). It is especially this kind of detailed observations made possible by the latest technical developments that constantly help us to revise and improve our interpretations of the natural world.

6.5. Acknowledgments

We thank the members of the Tectonics research group of the University of Bern's Institute of Geological Sciences (in particular Marco Herwegh, Veronica Peverelli, Timothy Schmid, Renata Schmitt, James Gilgannon, Vénice Akker and Ferdinado Musso Piantelli), as well as reviewers Mohammed Gouiza and Marco Bonini and editor Gwenn Peron-Pinvidic for providing valuable and constructive feedback that greatly helped us to improve the text.

Further Reading.– The above descriptions are abbreviated and often simplified. If interested in reading and learning further, the reader is referred to the following list of publications and references:

– General: (Vendeville *et al.* 1987; McClay 1990; Koyi 1997; Brun 1999; Corti *et al.* 2003; Schreurs *et al.* 2006; Corti 2012; Brun *et al.* 2018; Zwaan *et al.* 2019).

6.6. References

Abdelmalak, M.M., Bulois, C., Mourges, R., Galland, O., Legland, J.-B., Gruber, C. (2016). Description of new dry granular materials of variable cohesion and friction coefficient: Implications for laboratory modeling of the brittle crust. *Tectonophysics*, 684, 39–51. Available at: https://doi.org/10.1016/j.tecto.2016.03.003.

Acocella, V., Faccenna, C., Funiciello, R., Rossetti, F. (1999). Sand-box modelling of basement-controlled transfer zones in extensional domains. *Terra Nova*, 11, 149–156. Available at: https://doi.org/10.1046/j.1365-3121.1999.00238.x

Adam, J. and Kreszek, C. (2012). Basin-scale salt tectonic processes of the Laurentian Basin, Eastern Canada: Insights from integrated regional 2D seismic interpretation and 4D physical experiments. In *Salt Tectonics, Sediments and Prospectivity*, Alsop, G.I., Archer, S.G., Hartley, A.J., Grant, N.T., Hodgkinson, R. (eds). *Geological Society, London, Special Publications*, 363, 331–360. Available at: http://dx.doi.org/10.1144/SP363.15.

Adam, J., Urai, J.L., Wieneke, B., Oncken, O., Pfeiffer, K., Kukowski, N., Lohrmann, J., Hoth, S., van der Zee, W., Schmatz, J. (2005). Shear localisation and strain distribution during tectonic faulting – New insights from granular-flow experiments and high-resolution optical image correlation techniques. *Journal of Structural Geology*, 27, 283–301. Available at: https://doi.org/10.1016/j.jsg.2004.08.008.

Adam, J., Klinkmüller, M., Schreurs, G., Wieneke, B. (2013). Quantitative 3D strain analysis in analogue experiments simulating tectonic deformation: Integration of X-ray computed tomography and digital volume correlation techniques. *Journal of Structural Geology*, 55, 127–149. Available at: https://doi.org/10.1016/j.jsg.2013.07.011.

Agostini, A., Corti, G., Zeoli, A., Mulugeta, G. (2009). Evolution, pattern, and partitioning of deformation during oblique continental rifting: Inferences from lithospheric-scale centrifuge models. *Geochemistry, Geophysics, Geosystems*, 10, Q11015. Available at: https://doi.org/10.1029/2009GC002676.

Allemand, P. and Brun, J-.P. (1991). Width of continental rifts and rheological layering of the lithosphere. *Tectonophysics*, 188, 63–69. Available at: https://doi.org/10.1016/0040–1951(91)90314-I.

Allemand, P., Brun, J.-P., Davy, P., Van der Driessche, J. (1989). Symétrie et asymétrie des rifts et mécanismes d'amincissement de la lithopshère. *Bulletin de la Société Géologique de France*, 8, 445–451. Available at: https://doi.org/10.2113/gssgfbull.V.3.445.

Autin, J., Bellahsen, N., Husson, L., Beslier, M.-O., Leroy, S., d'Acremont, E. (2010). Analog models of oblique rifting in a cold lithosphere. *Tectonics*, 29, TC6016. Available at: https://doi.org/10.1029/2010TC002671.

Autin, J., Bellahsen. N., Leroy, S., Husson, L., Beslier, M.-O., d'Acremont, E. (2013). The role of structural inheritance in oblique rifting: Insights from analogue models and application to the Gulf of Aden. *Tectonophysics*, 607, 51–64. Available at: http://dx.doi.org/10.1016/j.tecto.2013.05.041.

Bahroudi, A., Koyi, H.A., Talbot, C.J. (2003). Effect of ductile and frictional décollements on style of extension. *Journal of Structural Geology*, 25, 1401–1423. Available at: https://doi.org/10.1016/S0191-8141(02)00201-8.

Bellahsen, N. and Daniel, J.M. (2005). Fault reactivation control on normal fault growth: An experimental study. *Journal of Structural Geology*, 27, 769–780. Available at: https://doi.org/10.1016/j.jsg.2004.12.003.

Bellahsen, N., Daniel, J.-M., Bollinger, L., Burov, E. (2003). Influence of viscous layers on the growth of normal faults: Insights from experimental and numerical models. *Journal of Structural Geology*, 25, 1471–1485. Available at: https://doi.org/10.1016/S0191-8141(02)00185-2.

Benes, V. and Scott, S.D. (1996). Oblique rifting in the Havre trough and its propagation into the continental margin of New Zealand: Comparison with analogue experiments. *Marine Geophysical Researches*, 18, 189–201. Available at: https://doi.org/10.1007/BF00286077.

Beniest, A., Willingshofer, E., Sokoutis, D., Sassi, W. (2018). Extending continental lithosphere with lateral strength variations: Effects on deformation localization and margin geometries. *Frontiers in Earth Science*, 6, 148. Available at: https://doi.org/10.3389/feart.2018.00148.

Beslier, M.-O. (1991). Formation des marges passives et remontée du manteau : modélisation expérimentale et exemple de la marge de la Galice. *Mémoires de Géosciences Rennes*, 45 [Online]. Available at: https://tel.archives-ouvertes.fr/tel-00594598.

Bonini, M., Souriot, T., Boccaletti, M., Brun, J.P. (1997). Successive orthogonal and oblique extension episodes in a rift zone: Laboratory experiments with application to the Ethiopian Rift. *Tectonics*, 16, 347– 362. Available at: https://doi.org/10.1029/96TC03935.

Boutelier, D. and Oncken, O. (2011). 3-D thermo-mechanical laboratory modeling of plate-tectonics: Modeling scheme, technique and first experiments. *Solid Earth*, 2, 35–51. Available at: https://doi.org/10.5194/se-2-35-2011.

Boutelier, D., Oncken, O., Cruden, S. (2012). Fore-arc deformation at the transition between collision and subduction: Insights from 3-D thermomechanical laboratory experiments. *Tectonics*, 31, TC2015. Available at: https://doi.org/10.1029/2011TC003060.

Boutelier, D., Schrank, C., Regenauer-Lieb, K. (2019). 2-D finite displacements and strain from particle imaging velocimetry (PIV) analysis of tectonic analogue models with TecPIV. *Solid Earth*, 10, 1123–1139. Available at: https://doi.org/10.5194/se-10-1123-2019.

Brun, J.-P. (1999). Narrow rifts versus wide rifts: Inferences for the mechanics of rifting from laboratory experiments. *Philosophical Transactions of the Royal Society of London A*, 357, 695–710. Available at: https://doi.org/10.1098/rsta.1999.0349.

Brun, J.-P. (2002). Deformation of the continental lithosphere: Insights from brittle-ductile models. In *Deformation Mechanisms, Rheology and Tectonics. Current Status and Future Perspectives*, De Meer, S., Drury, M.R., De Bresser, J.H.P., Pennoch, G.M. (eds). *Geological Society, London, Special Publications*, 200, 355–370. Available at: https://doi.org/10.1144/GSL.SP.2001.200.01.20.

Brun, J.-P. and Beslier, M.O. (1996). Mantle exhumation at passive margins. *Earth and Planetary Science Letters*, 142, 161–173. Available at: https://doi.org/10.1016/0012-821X(96)00080-5.

Brun, J.-P. and Fort, X. (2004). Compressional salt tectonics (Angolan margin). *Tectonophysics*, 382, 129–150. Available at: https://doi.org/10.1016/j.tecto.2003.11.014.

Brun, J.-P. and Nalpas, T. (1996). Graben inversion in nature and experiments. *Tectonics*, 15(2), 677–687. Available at: https://doi.org/10.1029/95TC03853.

Brun, J.-P., Sokoutis, D., Van Den Driessche, J. (1994). Analogue modeling of detachment fault systems and core complexes. *Geology*, 22, 319–322. Available at: https://doi.org/10.1130/0091-7613(1994)022<0319:AMODFS>2.3.CO;2.

Brun, J.-P., Sokoutis, D., Tirel, C., Gueydan, F., Van Den Driessche, J., Beslier, M.-O. (2018). Crustal versus mantle core complexes. *Tectonophysics*, 746, 22–45. Available at: http://dx.doi.org/10.1016/j.tecto.2017.09.017.

Brune, J.N. and Ellis, M.A. (1997). Structural features in a brittle–ductile wax model of continental extension. *Nature*, 387, 67–70. Available at: https://doi.org/10.1038/387067a0.

Brune, S., Williams, S., Butterworth, N., Müller, D. (2016). Abrupt plate accelerations shape rifted continental margins. *Nature*, 536, 201–204. Available at: https://doi.org/10.1038/nature18319.

Brune, S., Corti, G., Ranalli, G. (2017). Controls of inherited lithospheric heterogeneity on rift linkage: Numerical and analogue models of interaction between the Kenyan and Ethiopian rifts across the Turkana depression. *Tectonics*, 36, 1767–1786. Available at: https://doi.org/10.1002/2017TC004739.

Brune, S., Williams, S.E., Müller, R.D. (2018). Oblique rifting: The rule, not the exception. *Solid Earth*, 9, 1187–1206. Available at: https://doi.org/10.5194/se-9-1187-2018.

Buiter, S.J.H., Babeyko, A.Y., Ellis, S., Gerya, T.V., Kaus, B.J.P., Kellner, A., Schreurs, G., Yamada, Y. (2006). The numerical sandbox: Comparison of model results for a shortening and an extension experiment. In *Analogue and Numerical Modelling of Crustal-Scale Processes*, Buiter, S.J.H. and Schreurs, G. (eds). *Geological Society, London, Special Publications*, 253, 29–64. Available at: https://doi.org/10.1144/GSL.SP.2006.253.01.02.

Cerca, M., Ferrari, L., Corti, G., Bonini, M., Manetti, P. (2010). Analogue model of inversion tectonics explaining the structural diversity of Late Cretaceous shortening in southwestern Mexico. *Lithosphere*, 2, 172–187. Available at: https://doi.org/10.1130/L48.1.

Chattopadhyay, A. and Chakra, M. (2012). Influence of pre-existing pervasive fabrics on fault patterns during orthogonal and oblique rifting: An experimental approach. *Marine and Petroleum Geology*, 39, 74–91. Available at: http://dx.doi.org/10.1016/j.marpetgeo.2012.09.009.

Chauvin, B., Lovely, P., Stockmeyer, J., Plesch, A., Caumon, G., Shaw, J.H. (2018). Validating novel boundary conditions for three-dimensional mechanics-based restoration: An extensional sandbox model example. *AAPG Bulletin*, 102, 245–266. Available at: https://doi.org/10.1306/0504171620817154.

Chemenda, A., Déverchère, J., Calais, E. (2002). Three-dimensional laboratory modelling of rifting: Application to the Baikal Rift, Russia. *Tectonophysics*, 356, 253–273. Available at: https://doi.org/10.1016/S0040-1951(02)00389-X.

Clifton, A.E., Schlische, R.W., Withjack, M.O., Ackermann, R.V. (2000). Influence of rift obliquity on fault-population systematics: Results of experimental clay models. *Journal of Structural Geology*, 22, 1491–1509. Available at: https://doi.org/10.1016/S0191-8141(00)00043-2.

Colletta, B., Letouzey, J., Pinedo, R., Ballard, J.F., Balé, P. (1991). Computerized X-ray tomography analysis of sandbox models: Examples of thin-skinned thrust systems. *Geology*, 19, 1063–1067. Available at: https://doi.org/10.1130/0091-7613(1991)019<1063:CXRTAO>2.3.CO;2.

Corbi, F., Funiciello, F., Moroni, M., Van Dinther, Y., Mai, P.M., Dalguer, L.A., Facenna, C. (2013). The seismic cycle at subduction thrusts 1: Insights from laboratory models. *Journal of Geophysical Research: Solid Earth*, 118, 1483–1501. Available at: https://doi.org/10.1029/2012JB009481.

Corbi, F., Sandri, L., Bedford, J., Funiciello, F., Brizzi, S., Rosenau, M., Lallemand, S. (2019). Machine learning can predict the timing and size of analog earthquakes. *Geophysical Research Letters*, 46, 1303–1311. Available at: https://doi.org/10.1029/2018GL081251.

Corti, G. (2012). Evolution and characteristics of continental rifting: Analog modeling-inspired view and comparison with examples from the East African rift system. *Tectonophysics*, 522–523, 1–33. Available at: https://doi.org/10.1016/j.tecto.2011.06.010.

Corti, G., Bonini, M., Conticelli, S., Innocenti, F., Manetti, P., Sokoutis, D. (2003). Analogue modelling of continental extension: A review focused on the relations between the patterns of deformation and the presence of magma. *Earth-Science Reviews*, 63, 169–247. Available at: https://doi.org/10.1016/S0012-8252(03)00035-7

Corti, G., Van Wijk, J., Cloetingh, S., Morley, C.K. (2007). Tectonic inheritance and continental rift architecture: Numerical and analogue models of the East African Rift system. *Tectonics*, 26, TC6006. Available at: https://doi.org/10.1029/2006TC002086.

Dauteuil, O., Bourgeois, O., Mauduit, T. (2002). Lithosphere strength controls oceanic transform zone structure: Insights from analogue models. *Geophysical Journal International*, 150, 706–714. Available at: https://doi.org/10.1046/j.1365-246X.2002.01736.x.

Donnadieu, F., Kelfoun, K., Van Wijk de Vries, B., Cecchi, E., Merle, O. (2003). Digital photogrammetry as a tool in analogue modelling: Applications to volcano instability. *Journal of Volcanology and Geothermal Research*, 123, 161–180. Available at: https://doi.org/10.1016/S0377-0273(03)00034-9.

Dooley, T. and Hudec, M.R. (2020). Extension and inversion of salt-bearing rift systems. *Solid Earth*, 11, 1187–1204. Available at: https://doi.org/10.5194/se-11-1187-2020.

Dooley, T. and Schreurs, G. (2012). Analogue modelling of intraplate strike-slip tectonics: A review and new experimental results. *Tectonophysics*, 574–575, 1–71. Available at: http://dx.doi.org/10.1016/j.tecto.2012.05.030.

Dooley, T., McClay, K.R., Pascoe, R. (2003). 3D analogue models of variable displacement extensional faults: Applications to the Revfallet Fault system, offshore mid-Norway. In *New Insights into Structural Interpretation and Modelling*, Nieuwland, D.A. (ed.). *Geological Society, London, Special Publications*, 212, 151–167. Available at: https://doi.org/10.1144/GSL.SP.2003.212.01.10.

Eisenstadt, G. and Sims, D. (2005). Evaluating sand and clay models: Do rheological differences matter? *Journal of Structural Geology*, 27, 1399–1412. Available at: https://doi.org/10.1016/j.jsg.2005.04.010.

Fedorik, J., Zwaan, F., Schreurs, G., Toscani, G., Bonini, L., Seno, S. (2019). The interaction between strike-slip dominated fault zones and thrust belt structures: Insights from 4D analogue models. *Journal of Structural Geology*, 122, 89–105. Available at: https://doi.org/10.1016/j.jsg.2019.02.010.

Fort, X., Brun, J.-P., Chauvel, F. (2004). Salt tectonics on the Angolan margin, synsedimentary deformation processes. *AAPG Bulletin*, 88, 1523–1544. Available at: https://doi.org/10.1306/06010403012.

Ge, Z., Rosenau, M., Warsitzka, M., Gawthorpe, R.L. (2019). Overprinting translational domains in passive margin salt basins: Insights from analogue modelling. *Solid Earth*, 10, 1283–1300. Available at: https://doi.org/10.5194/se-10-1283-2019.

Glerum, A., Brune, S., Stamps, D.S., Strecker, M.R. (2020). Victoria continental microplate dynamics controlled by the lithospheric strength distribution of the East African Rift. *Nature Communications*, 11, 2881. Available at: https://doi.org/10.1038/s41467-020-16176-x.

Graveleau, F., Malavieille, J., Dominguez, S. (2012). Experimental modelling of orogenic wedges: A review. *Tectonophysics*, 538–540, 1–66. Available at: https://doi.org/10.1016/j.tecto.2012.01.027.

Hall, Sir J. (1815). On the vertical position and convolutions of certain strata and their relation with granite. *Royal Society of Edinburgh Transactions*, 7, 79–108. Available at: https://doi.org/10.1017/S0080456800019268.

Holland, M., Urai, J.L., Martel, S. (2006). The internal structure of fault zones in basaltic sequences. *Earth and Planetary Science Letters*, 248, 286–300. Available at: https://doi.org/10.1016/j.epsl.2006.05.035.

Hubbert, M.K. (1937). Theory of scaled models as applied to the study of geological structures. *Geological Society of America Bulletin*, 48, 1459–1520. Available at: https://doi.org/10.1130/GSAB-48-1459.

Hughes, A., Adam, J., Burgess, P. (2020). Integrating analogue and numerical modelling techniques for improved simulation of coupled regional tectonic processes and syn-depositional systems. *EGU General Assembly*, EGU2020-5059. Available at: https://doi.org/10.5194/egusphere-egu2020-5059.

Jackson, C.A.-L., Bell, R.E., Torevatn, A., Tvedt, A.B.M. (2017). Techniques to determine the kinematics of synsedimentary normal faults and implications for fault growth models. In *The Geometry and Growth of Normal Faults*, Childs, C., Holdsworth, R.E., Jackson, C.A.-L., Manzocchi, T., Walsch, J.J., Yielding, G. (eds). *Geological Society, London, Special Publications*, 439.

Katz, R.F., Ragnarsson, R., Bodenschatz, E. (2005). Tectonic microplates in a wax model of sea-floor spreading. *New Journal of Physics*, 7, 37.

Keep, M. and McClay, K.R. (1997). Analogue modelling of multiphase rift systems. *Tectonophysics*, 273, 239–270.

Khalil, H.M., Capitano, F.A., Betts, P.G., Cruden, A.R. (2020). 3-D analog modeling constraints on rifting in the Afar region. *Tectonics*, 39, e2020TC006339.

Klinkmüller, M., Schreurs, G., Rosenau, M., Kemnitz, H. (2016). Properties of granular analogue model materials: A community wide survey. *Tectonophysics*, 684, 23–38. Available at: https://doi.org/10.1016/j.tecto.2016.01.017.

Koyi, H. (1997). Analogue modelling: From a qualitative to a quantitative technique – A historical outline. *Journal of Petroleum Geology*, 20, 223–238.

Krantz, R.W. (1991). Measurements of friction coefficients and cohesion for faulting and fault reactivation in laboratory models using sand and sand mixtures. *Tectonophysics*, 188, 203–207.

Le Calvez, J.H. (2002). Physcial modelling of normal faults and graben relays above salt. PhD Thesis, University of Texas, Austin.

Le Calvez, J.H. and Vendeville, B.C. (2002). Experimental designs to model along-strike fault interaction. In *Analogue Modelling of Large-Scale Tectonic Processes*, Schellart, W.P. and Passchier, C. (eds). *Journal of the Virtual Explorer*, 7, 1–17.

Lee, A.R. and Warren, J.B. (1940). A coni-cylindrical viscometer for measuring the visco-elastic characteristics of highly viscous liquids. *Journal of Scientific Instruments*, 17, 63–67.

Lindanger, R., Øygaren, M., Gabrielsen, R.H., Mjelde, R., Randen, T., Tjøstheim, B.A. (2004). Analogue (plaster) modelling and synthetic seismic representation of hangingwall fault blocks above ramp-flat ramp extensional faults. *First Break*, 22, 33–41.

Maestrelli, D., Montanari, D., Corti, G., Del Ventisette, C., Moratti, G., Bonini, M. (2020). Exploring the interactions between rift propagation and inherited crustal fabrics through experimental modeling. *Tectonics*, 39, e2020TC006211.

Martí, J., Geyer, A., Folch, A., Gottsmann, J. (2008). A review on collapse caldera modelling. In *Caldera Volcanism: Analysis, Modelling and Response*, Gottsmann, J. and Martíacute, J. (eds). *Developments in Volcanology*, 10, 233–283.

Martin, A.K. (1984). Propagating rifts: Crustal extension during continental rifting. *Tectonics*, 3, 611–617.

Mattioni, L., Sassi, W., Callot, J.-P. (2007). Analogue models of basin inversion by transpression: Role of structural heterogeneity. In *Deformation of the Continental Crust. The Legacy of Mike Coward*, Ries, A.C., Butler, R.W.H., Graham, R.H. (eds). *Geological Society, London, Special Publications*, 272, 397417.

McClay, K.R. (1990). Extensional fault systems in sedimentary basins: A review of analogue model studies. *Marine and Petroleum Geology*, 7, 206–233.

McClay, K.R. (1996). Recent advances in analogue modelling: Uses in section interpretation and validation. In *Modern Developments in Structural Interpretation, Validation and Modelling*, Buchanan, E.G. and Nieuwland, D.A. (eds). *Geological Society Special Publications*, 99, 201–225.

McClay, K.R. and White, M.J. (1995). Analogue modelling of orthogonal and oblique rifting. *Marine and Petroleum Geology*, 12, 137–151.

McClay, K.R., Dooley, T., Lewis, G. (1998). Analog modeling of progradational delta systems. *Geology*, 26, 771–774.

McClay, K.R., Dooley, T., Whitehouse, P., Mills, M. (2002). 4-D evolution of rift systems: Insights from scaled physical models. *AAPG Bulletin*, 86, 935–959.

McFall, B.C. and Fritz, H.M. (2016). Physical modelling of tsunamis generated by three-dimensional deformable granular landslides on planar and conical island slopes. *Proceedings of the Royal Society A*, 472, 20160052.

Michon, L. and Merle, O. (2000). Crustal structures of the Rhinegraben and the Massif Central grabens: An experimental approach. *Tectonics*, 19, 896–904.

Michon, L. and Merle, O. (2003). Mode of lithospheric extension: Conceptual models from analogue modeling. *Tectonics*, 22, 1028.

Michon, L. and Sokoutis, D. (2005). Interaction between structural inheritance and extension direction during graben and depocentre formation: An experimental approach. *Tectonophysics*, 409, 125–146.

Molnar, N.E., Cruden, A.R., Betts, P.G. (2017). Interactions between propagating rotational rifts and linear rheological heterogeneities: Insights from three-dimensional laboratory experiments. *Tectonics*, 36, 420–443.

Molnar, N.E., Cruden, A.R., Betts, P.G. (2018). Unzipping continents and the birth of microcontinents. *Geology*, 46, 451–454.

Molnar, N.E., Cruden, A.R., Betts, P.G. (2019). Interactions between propagating rifts and linear weaknesses in the lower crust. *Geosphere*, 15, 1617–1640.

Mondy, L.S., Patrice, F.R., Duclaux, G., Mores, L., (2018). The role of asthenospheric flow during rift propagation and breakup. *Geology*, 46, 103–106.

Montanari, D., Agostini, A., Bonini, M., Corti, G., Del Ventisette, C. (2017). The use of empirical methods for testing granular materials in analogue modelling. *Materials*, 10, 635. Available at: https://doi.org/10.3390/ma10060635.

Mooney, M. and Ewart, R.H. (1934). The conicylindrical viscometer. *Physics*, 5, 350–354.

Naylor, M.A., Mandl, G., Sijpesteijn, C.H.K. (1986). Fault geometries in basement-induced wrench faulting under different initial stress states. *Journal of Structural Geology*, 8, 737–752.

Naylor, M.A., Laroque, J.M., Gauthier, B.D.M. (1994). Understanding extensional tectonics: Insights from sandbox models. In *Geodynamic Evolution of Sedimentary Basins*, Roure, F., Ellouz, N., Shein, V.S., Skvortsov, I.I. (eds). *International Symposium on Geodynamic Evolution of Sedimentary Basins*, Technip, 92, 69–83.

Nestola, Y., Storti, F., Cavozzi, C. (2015). Strain rate-dependent lithosphere rifting and necking architectures in analog experiments. *Journal of Geophysical Research, Solid Earth*, 120, 584–594.

Oldenburg, D.W. and Brune, J.N. (1972). Ridge transform fault spreading pattern in freezing wax. *Science*, 178, 301–304.

Panien, M., Schreurs, G., Pfiffner, A. (2005). Sandbox experiments on basin inversion: Testing the influence of basin orientation and basin fill. *Journal of Structural Geology*, 27, 433–445.

Panien, M., Buiter, S.J.H., Schreurs, G., Pfiffner, O.A. (2006a). Inversion of a symmetric basin: Insights from a comparison between analogue and numerical experiments. In *Analogue and Numerical Modelling of Crustal-Scale Processes*, Buiter, S.J.H. and Schreurs, G. (eds). *Geological Society, London, Special Publications*, 253, 253–270.

Panien, M., Schreurs, G., Pfiffner, A. (2006b). Mechanical behaviour of granular materials used in analogue modelling: Insights from grain characterisation, ring-shear tests and analogue experiments. *Journal of Structural Geology*, 28, 1710–1724.

Peron-Pinvidic, G., Manatschal, G. and IMAGinING RIFTING Workshop Participants (2019). Rifted margins: State of the art and future challenges. *Frontiers in Earth Sciences*, 7, 208. Available at: https://doi.org/10.3389/feart.2019.00218.

Philippon, M., Willingshofer, E., Sokoutis, D., Corti, G., Sani, F., Bonini, M., Cloetingh, S. (2015). Slip re-orientation in oblique rifts. *Geology*, 43, 147–150.

Poppe, S., Holohan, E.P., Galland, O., Buls, N., Van Gompel, G., Keelson, B., Tournigand, P.-Y., Brancart, J., Hollis, D., Nila, A., Kervyn, M. (2019). An inside perspective on magma intrusion: Quantifying 3D displacement and strain in laboratory experiments by dynamic X-ray computed tomography. *Frontiers in Earth Science*, 7, 62.

Quirk, D.G., Schødt, N., Lassen, B., Ings, S.J., Hsu, D., Hirsch, K.K., Von Nicolai, C. (2012). Salt tectonics on passive margins: Examples from Santos, Campos and Kwanza basins. In *Salt Tectonics, Sediments and Prospectivity*, Alsop, G.I., Archer, S.G., Hartley, A.J., Grant, N.T., Hodgkinson, R. (eds). *Geological Society, London, Special Publications*, 363, 207–244.

Ramberg, H. (1981). *Gravity, Deformation and the Earth's Crust*. Academic Press, London.

Reber, J., Cooke, M.L., Dooley, T.P. (2020). What model material to use? A review on rock analogs for structural geology and tectonics. *Earth-Science Reviews*, 202, 103107. https://doi.org/10.1016/j.earscirev.2020.103107.

Ritter, M.C., Leever, K., Rosenau, M., Oncken, O. (2016). Scaling the sandbox – Mechanical (dis)similarities of granular materials and brittle rock. *Journal of Geophysical Research: Solid Earth*, 121, 6863–6879.

Rudolf, M., Boutelier, D., Rosenau, M., Schreurs, G., Oncken, O. (2016). Rheological benchmark of silicone oils used for analog modeling of short- and long-term lithospheric deformation. *Tectonophysics*, 684, 12–22.

Schellart, W.P. and Strak, V. (2016). A review of analogue modelling of geodynamic processes: Approaches, scaling, materials and quantification, with an application to subduction experiments. *Journal of Geodynamics*, 100, 7–32.

Schlagenhauf, A., Dominguez, S., Malavieille, J., Manighetti, I. (2008). Incremental growth of normal faults: Insights from a laser-equipped analog experiment. *Earth and Planetary Science Letters*, 273, 299–311.

Schreurs, G., Hänni, R., Panien, M., Vock, P. (2003). Analysis of analogue models by helical X-ray computed tomography. In *Applications of X-ray Computed Tomography in the Geosciences*, Mees, F., Swennen, R., Van Geet, M., Jacobs, P. (eds). *Geological Society, London, Special Publications*, 215, 213–223.

Schreurs, G., Buiter, S.J.H., Boutelier, D., Corti, G., Costa, E., Cruden, A., Daniel, J.-M., Hoth, S., Koyi, H., Kukowski, N., Lohrmann, J., Ravaglia, A., Schlische, R.W., Withjack, M.O., Yamada, Y., Cavozzi, C., Del Ventisette, C., Brady, J.A.E., Hoffmann-Rothe, A., Mengus, J.-M., Montanari, D., Nilforoushan, F. (2006). Analogue benchmarks of shortening and extension experiments. In *Analogue and Numerical Modelling of Crustal-Scale Processes*, Buiter, S.J.H. and Schreurs, G. (eds). *Geological Society, London, Special Publications*, 253, 1–27.

Schreurs, G., Buiter, S.J.H., Boutelier, J., Burberry, C., Callot, J.-P., Cavozzi, C., Cerca, M., Chen, J.-J., Cristallini, E., Cruden, A.R. *et al.* (2016). Benchmarking analogue models of brittle thrust wedges. *Journal of Structural Geology*, 92, 116–139.

Souriot, T. and Brun, J.-P. (1992). Faulting and block rotation in the Afar triangle, East Africa: The Danakil "crank-arm" model. *Geology*, 20, 911–914.

Sun, Z., Zhong, Z., Keep, M., Zhou, D., Cai, D., Li, X., Wu, S., Jiang, J. (2009). 3D analogue modelling of the South China Sea: A discussion on breakup pattern. *Journal of Asian Earth Sciences*, 34, 544–556.

Tentler, T. (2003a). Analogue modeling of overlapping spreading centers: Insights into their propagation and coalescence. *Tectonophysics*, 376, 99–115.

Tentler, T. (2003b). Analogue modeling of tension fracture pattern in relation to mid-ocean ridge propagation. *Geophysical Research Letters*, 30, 1268.

Trippanera, D., Acocella, V., Ruch, J. (2014). Dike-induced contraction along oceanic and continental divergent plate boundaries. *Geophysical Research Letters*, 41, 7098–7104.

Tron, V. and Brun, J.-P. (1991). Experiments on oblique rifting in brittle-ductile systems. *Tectonophysics*, 188, 71–88.

Van Mechelen, J.L.M. (2004). Strength of moist sand controlled by surface tension for tectonic analogue modelling. *Tectonophysics*, 384, 275–284.

Vendeville, B.C. (2005). Salt tectonics driven by sediment progradation: Part I – Mechanics and kinematics. *AAPG Bulletin* 89, 1071–1079.

Vendeville, B.C., Cobbold, P.R., Davy, P., Brun, J.-P., Choukroune, P. (1987). Physical models of extensional tectonics at various scales. In *Continental Extensional Tectonics*, Coward, M.P., Dewey, J.F., Hancock, P.L. (eds). *Geological Society, London, Special Publications*, 28, 95–107.

Vendeville, B.C., Ge, H., Jackson, M.P.A. (1995). Scale models of salt tectonics during basement-involved extension. *Petroleum Geoscience*, 1, 179–183. Available at: https://doi.org/10.1144/petgeo.1.2.179.

Vink, G.E. (1982). Continental rifting and the implications for plate tectonic reconstructions. *Journal of Geophysical Research*, 87, 10667–10688.

Wang, J., Jiang, Z., Zhang, Y., Gao, L., Wei, X., Zhang, W., Liang, Y., Zhang, H. (2015). Flume tank study of surface morphology and stratigraphy of a fan delta. *Terra Nova*, 27, 42–53.

Withjack, M.O. and Jamison, W.R. (1986). Deformation produced by oblique rifting. *Tectonophysics*, 126, 99–124.

Withjack, M.O., Henza, A.A., Schlische, R.W. (2017). Three-dimensional fault geometries and interactions within experimental models of multiphase extension. *AAPG Bulletin*, 101, 1767–1789.

Wu, J.E., McClay, K., Whitehouse, P., Dooley, T. (2009). 4D analogue modelling of transtensional pull-apart basins. *Marine and Petroleum Geology*, 26, 1608–1623.

Zwaan, F. and Schreurs, G. (2017). How oblique extension and structural inheritance influence rift segment interaction: Insights from 4D analog models. *Interpretation*, 5, SD119–SD138.

Zwaan, F. and Schreurs, G. (2020). Rift segment interaction in orthogonal and rotational extension experiments: Implications for the large-scale development of rift systems. *Journal of Structural Geology*, 140, 104119.

Zwaan, F., Schreurs, G., Naliboff, J., Buiter, S.J.H. (2016). Insights into the effects of oblique extension on continental rift interaction from 3-D analogue and numerical models. *Tectonophysics*, 693, 239–260.

Zwaan, F., Schreurs, G., Adam, J. (2018). Effects of sedimentation on rift segment evolution and rift interaction in orthogonal and oblique extensional settings: Insights from analogue models analysed with 4D X-ray computed tomography and digital volume correlation techniques. *Global and Planetary Change*, 171, 110–133.

Zwaan, F., Scheurs, G., Buiter, S.J.H. (2019). A systematic comparison of experimental set-ups for modelling extensional tectonics. *Solid Earth*, 10, 1063–1097.

Zwaan, F., Corti, G., Keir, D., Sani, F. (2020a). Analogue modelling of marginal flexure in Afar, East Africa: Implications for passive margin formation. *Tectonophysics*, 796, 228595.

Zwaan, F., Schreurs, G., Rosenau, M. (2020b). Rift propagation in rotational versus orthogonal extension: Insights from 4D analogue models. *Journal of Structural Geology*, 135, 103946. Available at: https://doi.org/10.1016/j.jsg.2019.103946.

List of Authors

Julia AUTIN
Université de Strasbourg
CNRS
Institut Terre et Environnement de Strasbourg
UMR 7063
France

Dominique FRIZON DE LAMOTTE
GEC
CY Cergy Paris Université
Cergy-Pontoise
France

Zhonglan LIU
MARUM – Center for Marine Environmental Sciences
University of Bremen
Germany

Anders MCCARTHY
GEC
CY Cergy Paris Université
Cergy-Pontoise
France
and
Department of Earth Sciences
ETH-Zürich
Switzerland

Geoffroy MOHN
GEC
CY Cergy Paris Université
Cergy-Pontoise
France

Marta PÉREZ-GUSSINYÉ
MARUM – Center for Marine Environmental Sciences
University of Bremen
Germany

Gwenn PERON-PINVIDIC
Geological Survey of Norway (NGU)
and
Department of Geoscience and Petroleum
Norwegian University of Science and Technology
Trondheim
Norway

Benoît PETRI
EOST – Institut Terre et Environnement de Strasbourg
CNRS UMR 7063
Université de Strasbourg
France

Guido SCHREURS
University of Bern
Switzerland

Julie TUGEND
GEC
CY Cergy Paris Université
Cergy-Pontoise
and
Sorbonne Université
CNRS-INSU
Institut des Sciences de la Terre Paris
ISTeP UMR 7193
Paris
France

Mark ROWAN
Rowan Consulting, Inc.
Boulder
USA

Louise WATREMEZ
Université de Lille
CNRS
Université Littoral Côte d'Opale
IRD, UMR 8187 – LOG
Laboratoire d'Océanologie
et de Géosciences
Lille
France

Frank ZWAAN
University of Bern
Switzerland

Index

T, U, W

Summary of Volume 2

Chapter 6. Arctic Continental Margins

Carmen GAINA

Chapter 7. Alpine Tethys Margins

Gianreto MANATSCHAL and Pauline CHENIN

Chapter 8. A Fossil Magma-rich Rifted Margin in the Scandinavian Caledonides

Johannes JAKOB, Hans Jørgen KJØLL and Torgeir Bjørge ANDERSEN

Chapter 9. Cretaceous Rifting at the Pyrenean Iberia–Eurasia Border: Crustal Thinning under a Pre-rift Cover – Application to Passive Margins

Yves LAGABRIELLE

Printed and bound by CPI Group (UK) Ltd, Croydon, CR0 4YY

13/09/2026

14960994-0007